The Ecology of Pastoralism

The Ecology of Pastoralism

EDITED BY
P. Nick Kardulias

UNIVERSITY PRESS OF COLORADO
Boulder

Published by University Press of Colorado
5589 Arapahoe Avenue, Suite 206C
Boulder, Colorado 80303

Printed in the United States of America

The University Press of Colorado is a proud member of
The Association of American University Presses.

The University Press of Colorado is a cooperative publishing enterprise supported, in part, by Adams State University, Colorado State University, Fort Lewis College, Metropolitan State University of Denver, Regis University, University of Colorado, University of Northern Colorado, Utah State University, and Western State Colorado University.

∞ This paper meets the requirements of the ANSI/NISO Z39.48-1992 (Permanence of Paper).

ISBN: 978-1-60732-342-6 (cloth)
ISBN: 978-1-60732-343-3 (ebook)

Chapter 7, "FulBe Pastoralists and the Neo-Patrimonial State in the Chad Basin" by Mark Moritz, was originally published in *Geography Research Forum*, vol. 25 (2005): 83–104, and is reprinted with permission.

Library of Congress Cataloging-in-Publication Data
The ecology of pastoralism / edited by P. Nick Kardulias (College of Wooster).
pages cm
Includes bibliographical references and index.
ISBN 978-1-60732-342-6 (cloth : alkaline paper) — ISBN 978-1-60732-343-3 (ebook)
1. Pastoral systems—Environmental aspects. 2. Pastoral systems—History. 3. Adaptation (Biology) 4. Adaptability (Psychology) 5. Human ecology. 6. Ethnology. 7. Ethnoarchaeology. 8. Social archaeology. 9. Landscape archaeology. I. Kardulias, P. Nick.
SF140.P38E36 2015
636.08'45—dc23

2014029020

24 23 22 21 20 19 18 17 16 15 10 9 8 7 6 5 4 3 2 1

Cover photograph © Tamara Kulikova / Shutterstock.

To the memory of Mark T. Shutes (1947–2001),
a superb colleague and friend

Contents

List of Figures *ix*

List of Tables *xi*

Preface *xiii*

1. Introduction: Pastoralism as an Adaptive Strategy
   P. Nick Kardulias *1*

2. The Study of Nomads in the Republic of Kazakhstan
   Claudia Chang *17*

3. The Ecology of Inner Asian Pastoral Nomadism
   Nikolay N. Kradin *41*

4. Agropastoralism and Transhumance in Hunza
   Homayun Sidky *71*

5. Animals, Identity, and Mortuary Behavior in Late Bronze Age–Early Iron Age Mongolia: A Reassessment of Faunal Remains in Mortuary Monuments of Nomadic Pastoralists
   Erik G. Johannesson *97*

6. Kalas and Kurgans: Some Considerations on Late Iron Age Pastoralism within the Central Asian Oasis of Chorasmia
Michelle Negus Cleary 117

7. FulBe Pastoralists and the Neo-Patrimonial State in the Chad Basin
Mark Moritz 171

8. Flexibility in Navajo Pastoral Land Use: A Historical Perspective
Lawrence A. Kuznar 195

9. Accidental Dairy Farmers: Social Transformations in a Rural Irish Parish
Mark T. Shutes 211

10. Real Milk from Mechanical Cows: Adaptations among Irish Dairy Cattle Farmers
Mark T. Shutes 225

11. Island Pastoralism, Isolation, and Connection: An Ethnoarchaeological Study of Herding on Dokos, Greece
P. Nick Kardulias 243

12. The Ecology of Herding: Conclusions, Questions, Speculations
Thomas D. Hall 267

About the Contributors 281

Index 285

Figures

3.1. Map of Central Asia, with environmental zones and cultural groups indicated 43
4.1. Hunza's location between Central and South Asia 73
5.1. Locations of sites in Mongolia and Siberia mentioned in the text 99
5.2. Khirigsuurs, illustrating differences in size and shape of central mound and perimeter fence 105
5.3. Slab burial at Baga Gazaryn Chuluu 107
5.4. Xiongnu ring tomb, with the broad band of stones in the superstructure exposed 109
5.5. Crania of cattle, sheep, and goats buried in a niche in northern section of Xiongnu tomb at Baga Gazaryn Chuluu 110
6.1. Chorasmia and western Central Asia regional map 118
6.2. GIS map of ancient Chorasmia (Amu Darya delta oasis) showing all sites dating between seventh century BC and fourth century AD 120
6.3. GIS map of archaeological sites in the Sarykamysh delta area, seventh century BC–fourth century AD 121
6.4. GIS map of the Kalaly-gyr-area oasis 140
6.5. Kalaly-gyr 2 site plan; Kalaly-gyr 1 site plan 141

6.6. Tarym-kaya 1 fortified settlement and kurgan cemetery 144
6.7. Tarym-kaya to Kanga-gyr GIS landscape map 145
6.8. Kanga-kala plan 147
6.9. Digital Globe satellite image showing Kanga-kala 1 fortified site and kurgans as pale circles to the immediate west and northwest of the fortress 149
6.10. Mangyr-kala and kurgans site plan, with information from satellite imagery 150
7.1. Map of FulBe area 175
8.1. Map showing the location of Navajo land in the Four Corners region of the US Southwest 200
9.1. Map of Ireland showing the location of County Kerry 212
11.1. Map showing the location of Dokos in relation to other islands and the southern Argolid in Greece 244
11.2. View of the narrow channel between Dokos on left and the mainland (southern Argolid) on right 245
11.3. Chapel of Saint John the Theologian near the Douskos family complex 247
11.4. Nikos Douskos demonstrating the use of a glass sherd as a wood scraper 253
11.5. The saddle at the north end of Dokos, where the chapel and Douskos house are located 254
11.6. Mules and donkeys grazing on an abandoned agricultural terrace 255
11.7. Douskos complex on Dokos 257
11.8. Late Roman to Early Byzantine period cistern built around a natural cave on the island of Evraionisos in the Saronic Gulf 260

Tables

6.1. List of sites from the Sarykamysh delta area arranged in groups that appear to be spatially and/or temporally related 128

6.2. List of sites from the Sarykamysh delta area including ceramic types 132

8.1. Historic population figures, livestock numbers, and intrinsic population growth rates for the Navajo people 199

9.1. Number of cattle and price of milk in Kilcastle Parish, Ireland, 1925–70 216

Preface

This volume has had a long period of gestation as the result of a number of circumstances. The original impetus was a session at the 1999 Annual Meeting of the Central States Anthropological Society in Chicago. My colleague Mark Shutes and I organized the session and delivered papers, along with Lawrence Kuznar and Daniel Ayana. Shutes discussed dairy farming in western Ireland, Kuznar examined alterations in Navajo land use and herd size, Ayana detailed the transformation from pastoralism to plow farming by the Oromo of Ethiopia during the seventeenth century, and Kardulias described the use of small islands for pasture in the Aegean. We followed this initial effort with a session at the 1999 Annual Meeting of the American Anthropological Association, also in Chicago, at which Homayun Sidky added a paper on the nature of pastoral activities in the Karakoram Mountains of Pakistan and Thomas Hall contributed a commentary on the presentations.

The general idea from the outset had been to examine the various ways pastoralism serves as a highly flexible system. The flexibility derives from two basic sources: (1) the ability of various domesticated animals to adapt to differing environments and still provide a subsistence base for people and (2) the webs of intricate relationships pastoralists develop to accommodate their political, social, and economic needs. Shutes provided the inspiration for the sessions and argued vigorously both there and in other venues for the value of situating people in their environment as most broadly construed. As a student of both cultural ecology and cross-cultural comparative work, he focused on the networks that allow people to gain sustenance. His was not a formulaic approach, however. Shutes repeatedly noted the ways people innovated on the base provided by their dependence on particular animals, shifting approaches as necessary as conditions changed. This strategic approach laid bare the practical matters that impinged on human action. Our goal in putting together an edited volume from the two sessions was to explore the ways pastoralism is embedded in an environment whose framework is set by the terrain, the distribution of vital resources, and the needs of the animals herders tend. Equally important in understanding this dynamic relationship is the cultural landscape, as reflected in economic transactions for acquiring sufficient pasture and access to other resources, kinship ties, political structure (e.g., how the relationship between local and national administrations affects the movement of herders within and between political units), and the worldview that includes the values and beliefs of the pastoral mind-set. To enhance the comparative component of the collection, we decided to include contributions from scholars working in a number of world areas, including Northern Europe, Central Asia, the Mediterranean, South Asia, Africa, and the Americas. In this way, we hoped to explore the crosscutting similarities in pastoral adaptations while also pointing out significant differences.

After our initial solicitation of additional manuscripts, Claudia Chang submitted a contribution, but Daniel Ayana had to withdraw his paper. As we were seeking additional contributors, Shutes fell ill and died suddenly in February 2001. His untimely passing and a number of other commitments on my part brought progress on the volume to a halt for three years. On several occasions over the next seven years, I renewed work on the volume sporadically as time permitted. During that period, I contacted and received contributions from Nikolay Kradin, Mark Moritz, and, most recently, Michelle Negus Cleary and Erik Johannesson. A research sabbatical during the 2010–11 academic year provided the opportunity to finalize work on the volume. While not as diverse geographically as Shutes and I had originally hoped, the contributions do represent a wide range of areas and, it is hoped, will provide readers with interesting case studies demonstrating both the basic structure and the individual variation that characterize the herding lifestyle. A brief summary of the individual studies demonstrates the range of the volume.

The Introduction (chapter 1) discusses some key elements of a pastoral ecology to lay a foundation for the studies that follow. The stress is on the mutual dependence between herders and the animals they keep; the physiological needs of domesticated animals dictate certain types of actions by herders, most notably high levels of mobility, and these people in turn develop a series of cultural accommodations to meet those needs. The chapter also deals with the ways anthropologists have studied pastoralists and their reasons for doing so.

In chapter 2, Chang provides historical background on pastoral studies to demonstrate how schools of thought can influence what scholars say about herding societies. She traces the intellectual history of studies of Eurasian steppe nomads conducted in the twentieth century by Soviet-trained archaeologists, ethnographers, and historians. By way of examining the Soviet tradition of pastoral nomadic studies, Chang also charts the political and ideological divides between Western and Soviet-based scholarship on Eurasian steppe nomads.

Nikolay Kradin provides a historical overview of pastoral activities in Eurasia in chapter 3. He explores the significance of climate, the types of livestock, the diet of nomads, mobility, and political structure. The key point is the need to employ an array of evidence in the effort to reconstruct nomadic lifestyles. Since nomads rarely write their own historical accounts, at least initially, scholars must use ethnohistoric sources and archaeological data in a judicious manner.

Sidky (chapter 4) discusses the Hunzakutz people who live in the western Karakoram Mountains of northern Pakistan, in what has been called one of the most formidable upland regions on earth. Relatively isolated in their remote, inhospitable, and resource-scarce high-mountain environment, the Hunzakutz have had to find practical solutions to basic problems, such as a shortage of arable land, lack of sufficient water for irrigation, and ever-increasing population pressure on resources. Their solution has been to operate a subsistence economy that combines the cultivation of cereal crops with animal husbandry and transhumant pastoralism. Traditionally, this production system—based on a range of complementary plant and animal species and the exploitation of multiple resource clusters spread over different altitudinal zones—has been adjusted to ecological circumstances. Sidky focuses specifically on the ecological dimensions of Hunzakutz pastoralism and how pastoral production is regulated through the careful management of herd size and composition, as well as the ability of various species (cattle, sheep, goats, and yaks) to utilize different ecological niches. Attention is also given to the overall articulation of pastoralism and farming as complementary food production systems.

Examining issues of specific importance to archaeology, Johannesson (chapter 5) argues that faunal remains in mortuary contexts are often a significant challenge to archaeologists, at least in part because of the often symbolic nature of funerary assemblages and ritual behavior. This has a particular impact on the interpretation of the evidence for the adoption of nomadic pastoralism in Mongolia and the subsequent emergence of the first nomadic state in the region, the Xiongnu, in the third century BCE. To identify the full range of animal exploitation during this time, which included both economic and symbolic utilization of faunal resources, Johannesson suggests it is necessary to adopt an inclusive interpretive framework that considers animals as constituting one line of evidence of mortuary practice to be contextualized in reference to other aspects of funerary behavior, such as mortuary monumental types and placement, as well as associated funerary assemblages. He employs such an approach to demonstrate that faunal resources were used strategically by the Xiongnu in mortuary ritual to convey cultural unity and political legitimacy.

Negus Cleary moves the discussion to the western end of Eurasia in chapter 6. The Late Iron Age fortified enclosures of ancient Central Asia, specifically those of the oasis of Chorasmia, present an architecture of interaction between "steppe" and "sown." This interaction has been assumed to have involved conflict, constraint, and probably also commerce. A closer examination of these fortified sites reveals that they may not have been solely the constructions of sedentary agriculturalists and that they most likely played a different role than that of urban centers. Analysis of the ancient settlement pattern in Chorasmia reveals a low-density, non-nucleated system of buildings, enclosures, and facilities organized around water-supply canals. The fortresses and canals were integral to both the domination of the oasis territory and control of the local environment. This manipulated landscape appeared relatively suddenly and represents an interesting transitional period from the small preceding mobile, or semi-mobile, steppic communities toward a more sedentary, urbanized existence in the medieval period. While agriculture is clearly present in association with the fortresses, the lack of evidence for significant permanent domestic habitation and other factors suggest a greater role for pastoralists in the oasis during this period.

Moritz (chapter 7) argues that studies of African pastoral societies should consider the informal politics of the neo-patrimonial state in their analyses of pastoralists' relations with the state rather than focus on the official laws and policies of an ideal bureaucratic state. To illustrate his argument, Moritz examines the role of the neo-patrimonial state in the lives of nomadic FulBe Mare'en pastoralists. He discusses pastoral development, access to grazing land, and insecurity, situating the analysis within the historical and geographical contexts of the Chad Basin.

Kuznar (chapter 8) moves us to the New World. He examines the Navajo (Dineh) of the American Southwest who are best known as traditional sheepherders. However, their subsistence economy has ranged from foraging to horticulture to small stock herding to cattle ranching. Today, while herding ceases to be an important subsistence activity for most Navajo, many still herd livestock; and modern pastoralism, like the herding that preceded it, is still characterized by flexibility. Kuznar examines the causes of historic shifts in Navajo pastoralism, as well as the mechanisms that enable flexibility in land use.

In chapter 9, Shutes presents a general argument and uses data from Ireland to support his view. He argues that despite wide geographic diversity, pastoralist societies exhibit a relatively narrow range of internal social organizational forms based principally on localized agnatic kinship bonds. This pattern of social unity within geographic diversity strongly suggests that the strategic elements inherent in any form of animal husbandry demand certain specific kinds of social formation regardless of the particular ecological circumstances within which it is being carried out. Shutes argues that if this suggestion is correct, then, logically, it should follow that an agricultural community that moves from mixed-crop farming to specialized animal husbandry production should experience significant changes in its existing patterns of social organization. Drawing upon ethnographic data from a rural parish in southwestern Ireland (chapter 10), Shutes examines the relationships between animal husbandry and local social organization and offers evidence in support of the idea that the adoption of such strategies by agricultural communities results in a transformation of existing social relationships.

Shifting to the other end of Europe, in chapter 11 Kardulias examines a specific pattern of land use in Greece. Substantial settlements from the Early Bronze Age to the Early Modern period on the arid island of Dokos off the eastern coast of the Peloponnesos attest to the human ability to cope successfully with the lack of freshwater sources. The ethnoarchaeological study of a resident herding family on Dokos provides important insights into water management and other subsistence activities in an austere environment. The study examines strategic planning in terms of herd management, use of local resources, and contacts with the mainland and provides analogues for understanding past human adaptation to island settings. Since the use of such islands is a common feature of Greek antiquity, the study can contribute to a broader understanding of human adaptation to island settings.

In his commentary, Hall (chapter 12) places the individual studies into the broad framework of macro-analysis, using the work of Gerhard Lenski as a starting point. Lenski's scheme is evolutionary in nature, and he lays out a developmental pattern in which pastoralism has an important role because of its ability to produce a surplus of resources. Hall argues that among the reasons to study pastoral societies is the often unique position they hold as peripheral to sedentary groups, with whom they have complex relationships. In addition, herding societies have frequently played major roles in social change, rising and falling in complexity and size in concert with neighboring sedentary groups. Hall argues that world-systems analysis is one approach that is well-suited to exploring the interactions that define the relationships of pastoralists both internally and externally. In short, he suggests that studying pastoral societies is fundamentally important to understanding the structure of past and present civilizations.

While all books are collaborative efforts at some level, edited volumes are especially so. In putting together this volume, I have received able assistance from a number of individuals. The contributors to the original symposia have been patient as this project worked through various phases, including a long hiatus following Shutes's death. They and the other authors have provided valuable original data that they could have presented in other venues. All of the contributors were responsive to the various requests for changes to their chapters; their collegial good humor made it possible to bring this book to fruition. I thank the two anonymous reviewers who made valuable suggestions that have enhanced the quality of the volume. It was not possible to follow all of their recommendations, but the individual authors made every reasonable effort to do so. Jessica d'Arbonne, acquisitions editor at the University Press of Colorado, has been a great help in guiding the book through the various steps of preparation leading to publication. Her responses to questions were always upbeat and timely. I thank Avinoam Meir for permission to reprint Mark Moritz's article that initially appeared in *Geography Research Forum* (volume 25, 2005).

The College of Wooster has provided assistance that facilitated the completion of this project. Course release time made available through a grant from the Luce Fund for Distinguished Scholarship in 2007–8 and a sabbatical leave in 2010–11 made it possible to collect, edit, and revise the chapter manuscripts and to rewrite my own contributions. Stephanie Bosch, Brittany Rancour, Chelsea Fisher, and James Torpy aided in formatting the manuscript, hunting down references, and other tasks; these students were funded by the College's Sophomore Research Assistant program between 2007 and 2013. Steve Flynn, emerging technologies librarian at the College of Wooster, provided valuable assistance in preparing the final figures, along with Stephanie Bosch. Bosch continued her work on the book in the capacity of an archaeology laboratory research assistant; she was most helpful in the final stages of editing and in compiling the index. Brett Arnold, Emily Butcher, Jacob Dinkelaker, Catie Gullett, Renee Hennemann, and Sarah Tate read drafts of the chapters carefully.

Finally, I extend deep gratitude to my late colleague and collaborator on other projects, Mark Shutes. His passing was a significant loss for his family, friends, and the discipline of anthropology that he loved and respected. Mark had an unquenchable curiosity about all aspects of the human condition, but he reserved most of his immense energy for the investigation of the intricate relationship between humans and domesticated animals. He was a comparative anthropologist with a wide-ranging intellect and remarkable curiosity about how people think and act in social settings. With his engaging teaching style, infectious enthusiasm for fieldwork, and astute observations laid out in clear prose, Mark was the consummate teacher-scholar. It is with an enduring sense of admiration for his many skills and of loss at being deprived of his invigorating company that this book is dedicated to his memory.

P. NICK KARDULIAS
Wooster, Ohio

The Ecology of Pastoralism

1

Introduction

Pastoralism as an Adaptive Strategy

P. Nick Kardulias

A PASTORAL ECOLOGY

Animal husbandry has been one of the main subsistence patterns for many cultures around the world since the Neolithic period. In the past, pastoral peoples have proved to be central players in major historical transformations, including the emergence of major empires such as those of the Mongols and Arabs. In the study of pastoralists, anthropologists and archaeologists pay specific attention to the ecological factors that govern pastoral activities, unlike analysis performed by economists or specialists in development. The studies in this volume demonstrate the careful way pastoral peoples past and present have organized their relationship with certain animals to maximize their ability to survive and adapt to a wide range of environmental conditions over time. In addition, the contributors demonstrate that pastoralism has a significant impact not only on basic subsistence but also on the network of social, political, and religious institutions of the respective societies. The book builds on the work of others who have studied herding cultures from an anthropological perspective (e.g., Campbell 1964; Dyson-Hudson and Dyson-Hudson 1980; Galaty and Johnson 1990; see also Spooner 1973; Khazanov 1984; Barfield 1989, 1993; Chang and Koster 1994; Kradin, Bondarenko, and Barfield 2003; Salzman 2004; Parman 2005).

The contributors take a broad view of ecology. The biological approach to ecology considers the relationships between organisms and their surroundings, with a focus on the utilization of the various resources that sustain life (hydrology, soils, geology,

DOI: 10.5876/9781607323433.c001

fauna, and flora) within particular climatic regimes. The balances among these various elements determine the ability of a certain zone to sustain a given number of organisms (i.e., carrying capacity). Human ecology is specifically concerned with how people fit into local environments and as a result ties the physical features of a region to the cultural mechanisms people deploy in their efforts to adapt to the environment (Bates 2001:28). For several reasons, the study of pastoralism is an ideal way to explore ecological relationships because it involves symbiotic connections between humans and domesticated animals, with a series of cascading effects in political, social, economic, and religious organization. Conversely, an ecological framework offers perhaps the best way to explain the inherent flexibility of herding/animal husbandry systems. Such groups adapt well to a wide range of environmental and social conditions. On the one hand, they can never really be completely isolated or self-sufficient. They need, or at least seek out, links to other groups and areas through trade, migration, and raiding. On the other hand, pastoral folk often demonstrate a remarkable ability to thrive in marginal zones and exhibit political autonomy in doing so. While they are often participants in market systems, they can manipulate such economic systems through the pliable network of social and political relationships they possess. This ability to mediate their involvement with outside groups is the key to the endurance of such cultures. Dyson-Hudson and Dyson-Hudson (1980:27) opined that "to advance our understanding of the complex relationships between ecology and human social organization, we need detailed studies both of animal and of human behavior." A key goal of the various contributors to this volume is to present just such analyses, providing information about environmental conditions while also heeding Fratkin's (1997:236) call to consider the political setting.

While herding/animal husbandry systems retain their cultural integrity despite links with outside groups, they are also syncretic cultures, that is, they transform rather easily from less structured to more formal organizations. Such groups have undergone more adjustments than many other cultures, yet they have been able to retain their sense of place and purpose. Pastoral strategies are easily transferred from place to place and are not linked to territory in the same way as sedentary farmers. The products of herding and animal husbandry systems (such as leather, meat, milk) are universally desired by other groups and constitute a form of fluid wealth that can be converted into multiple forms of capital, savings, and credit, thus enhancing the economic flexibility of such groups. In addition, the animals themselves possess an inherent adaptability. They are subject to easy mutations, can live in cold or intense heat, and can adjust to resource depletion through migration. In addition, sheep and goats are non-selective grazers, so they can live in many different climates, some of which are completely unsuitable for agriculture. In this manner, herding/

animal husbandry creates a viable ecological niche for humans where none would have existed otherwise. This is not to say there are no dangers associated with a pastoral system. At times, pastoralism can be as sensitive or more sensitive to seasonal variations in temperature. For example, an early or late frost can lead to the deaths of many lambs in sheepherding societies, creating a problem of volatility in animal population and structure.

Politically, such groups can exhibit strong central organization, but without a massive concentration of resources, because of the need to disperse resources to build alliances, form factions, and so on. As a result, pastoral groups are usually able to organize hierarchically between groups yet retain egalitarian structures within groups. While their contrasting subsistence patterns can bring pastoralists and sedentary agriculturalists into conflict, there are also significant instances of an almost symbiotic cooperation. Despite the difference in mobility, herders can develop social and political organization that often mirrors that of farmers in complexity if not in exact form. Just as their herds can expand and contract in number under various environmental conditions, pastoral societies can fluctuate in size and structure depending on a number of factors. The segmentary lineages of East African cattle pastoralists are perhaps the prime example of this process (Evans-Pritchard 1940). Lineal groups that normally compete with one another may unify against an external threat but then dissolve back to the looser association once the outside danger has passed. In eastern and Central Asia there have been times when the process of consolidation crossed a threshold and chiefdoms and states arose out of a pastoral context. In short, pastoral societies have been and continue to be dynamic and critical to our understanding of cultural evolution. The chapters in this book explore some of these dimensions of herding/animal husbandry systems in various areas of the world, with a focus on Eurasia, Africa, and North America.

STUDYING PASTORALISM

While studies of pastoral societies have occupied the anthropological imagination almost since the inception of the discipline, research on the topic has increased dramatically since the mid-1970s. Several factors account for this expansion. One is that, in some regions, pastoral people represent a traditional form of life that is rapidly disappearing under pressure from the modern world, which restricts their movement across political boundaries or entices members away from animal husbandry with promises of a better life in settled communities. At the same time, many countries present the image of the traditional herder as a unifying cultural theme, part of a national ethnic identity. As a result of this interest, some have come to study pastoralists as representatives of the customary life that had existed for millennia but is rapidly vanishing. A second reason is the opening of regions to investigation by Western scholars. The collapse of the Soviet regime in particular has made it possible for scholars to study regions largely closed to the West for much of the twentieth century. This is particularly true for the vast swathe of the Eurasian steppe that is home to many pastoral groups. There has been a veritable explosion of research by Western scholars in the region from Mongolia to the Ukraine. Among the excellent studies of this area are those by Anthony (2007), Barfield (1993), Chang (2006), Chang et al. (2003), and Frachetti (2008); and long-term projects across this vast area have become the norm. Symposia on current research in Eurasia, often with a focus on pastoralism, have become annual occurrences at major conferences, such as the Annual Meeting of the Society for American Archaeology. This growth in studies of Eurasian pastoralism is reflected in the present volume by the presence of four chapters dealing with this area. I should note that excellent work had also been done by non-Western scholars, especially Russians, prior to 1991 (e.g., Khazanov 1984; Kradin 1987), but only a portion of this scholarship reached a broader audience.

Some definitions are in order to lay the groundwork for this discussion. The basis of pastoralism is animal husbandry, "the breeding, care, and use of herd animals such as sheep, goats, camels, cattle, horses, llamas, reindeer, and yaks" (Bates 2001:104). Salzman (2004:1) adds that the animals are raised "on 'natural' pasture unimproved by human intervention." However, he notes that direct and indirect human action can create pasturelands; these activities can include deforestation and burning to suppress tree growth. The reliance on domesticated animals for subsistence is not monolithic, since many pastoral groups also grow some crops, for both human and animal consumption. This basic fact means that pastoralists have to view their landscape in a composite fashion, considering the needs of both the herds they tend and the plants that can form an important part of the food base. This expansive vision of their subsistence base leads pastoral people to think at an extra-local level. Because the substantial herds on which they depend require sufficient pasturage that is often at a premium in any particular locale, they must be aware of available land in several regions. This necessity requires broader geographic knowledge than many farmers possess. Whereas a farmer may be most concerned with one area and invests substantial energy in plowing, building terrace walls to reduce erosion, and erecting permanent structures such as houses, barns, and threshing floors and thus places great emphasis on the location where these features exist, a pastoralist sees the value of land in a somewhat more transient fashion—as a place to be used and valued, to be certain, but in a system with multiple foci instead of one. In addition, while the farmer needs immobile structures for everything from residence to storage, the pastoralist places a premium value on portability, a fact that complicates the study of pastoral archaeological sites (see Cribb 1991). As with most things in anthropology, however, there is overlap between the farming and pastoral lifestyles. As Barfield (1993:4) suggests:

> One of the most enduring stereotypes is the myth of the "pure nomad," one who subsists entirely on meat, milk, or blood, abhors farmers, farming, and grain, despises sedentary life in general, and never has contact with villages or cities except when he loots and burns them. Nothing could be further from the truth. The historical and ethnographic record is full of nomads who also farm, trade, serve as soldiers, smuggle, or drive trucks, just to mention a few occupations . . . even those nomads who did appear purely pastoral, such as the Bedouin of the Empty Quarter or horse riders of Mongolia, maintained an ideal of "purity" largely by means of the subsidies they received from neighboring sedentary societies with whom they had important political and economic relationships. While nomadic pastoralists have always viewed animal husbandry as the culturally ideal way of making a living, and the movement of all or part of the society as a normal and natural part of life, they never rejected other opportunities. However, these activities were always viewed as adjuncts to pastoralism which remained the key element of their cultural and social identity.

One of the key features of pastoralism reflected in Barfield's description is thus an inherent and necessary flexibility. While the keeping of animals may be paramount, in thought if not completely so in action, pastoral peoples combine strategic thinking and pragmatism in deciding what particular activities are appropriate under given conditions. When we consider the often fragile or marginal environments pastoralists occupy, keeping various options open is to be expected. While pastoralists are often listed among the traditional societies of the world, it is important to think of them in dynamic rather than static terms.

As Salzman (2004:2–3) notes, pastoralists must be well attuned to their environment to raise their animals successfully. Among the key factors they must understand are elements of climate, topography, vegetation, human and animal population profiles, and sources of disease. In an ecological calculus, "The pastoralists try to identify for their particular environment the optimal combination of location and timing to maximize benefit for the animals—high quality and quantity of pasture, good water, and favorable temperatures—and minimize detrimental influences—extreme temperatures, lack of water or pasture, exposure to disease, and vulnerability to human or animal predators" (ibid.:3). These considerations lead to decisions concerning how and when to move herds. It is in the context of such strategic thinking that we see enacted the distinctions between horizontal and vertical forms of pastoralism. The available resources will dictate the conditions, and pastoralists decide where to move based on their assessment of the situation on the ground. In areas with little topographic relief and relatively uniform growing conditions for plant cover, people tend to adopt an extensive strategy in which they roam across substantial distances. The areas covered may become larger if the necessary resources are especially patchy, as in desert regions. High-relief zones offer the attraction of using highland pasture zones that otherwise provide few resources of use to humans. Transhumant pastoralism thus makes it possible to convert an otherwise nonproductive zone into an ecological niche useful to humans.

A key to a successful pastoral adaptation is to maintain flexibility. As Bates (2001:106) observes, people determine the degree of mobility based on an assessment of prevailing conditions. These decisions involve not only the herds' immediate needs for water and pasture but also determinations of appropriate group size and structure. To make appropriate choices, pastoral people must have an intimate knowledge not only of the environment where their herds graze at any particular time but also of the prevailing conditions in those other areas to which they may move in search of additional pastures. This fact makes the pastoral environmental view more expansive than that of others who are locked into one location.

Barfield (1993:7–9) divides the Old World region into five pastoral zones based on the types of animals each region can support. He identifies a cattle pastoral zone lying east-west across the sub-Saharan Sahel and in the savanna region of East Africa. These people also raise sheep and goats. Pastoralists of the desert zone of the Sahara and Arabian Deserts focus on the dromedary camel, which provides them with the ability to utilize desert zones others cannot reach. While the products extracted from camels are important, the Bedouin also consume and transport dates and provide animals to caravans. Third is the sheep and goat zone that covers a large expanse from areas bordering the Mediterranean to rough terrain in Central Asia. Barfield describes the pastoralists of this region as economic specialists who practice horizontal movements with horses, camels, and donkeys in addition to caprids. Equestrian nomads dominated the fourth zone, the Eurasian steppe zone from Mongolia to the Black Sea. In addition to horses, these people raised sheep, goats, cattle, and Bactrian camels and lived in yurts. As mounted warriors, these people established powerful empires. The fifth zone is the Tibetan Plateau, where the dominant domesticated animal is the yak, but people also raise sheep, goats, horses, and, in lower elevations, cattle. While a forbidding environment because of severe weather and high winds, the lack of agricultural activity makes for abundant pasture. These pastoralists largely consume milk and meat of yaks and dzos (a yak-cattle hybrid) and also exchange animal products for barley with farmers in the low-lying valleys.

For the purposes of this volume, I also define several pastoral zones in the New World. Pastoralism was extremely limited in the pre-Columbian Americas by comparison to the Old World largely because of the dearth of domesticable herd species. As a result, the only true pastoral region in prehistoric times was the Andean zone, where people tended camelids—especially llamas and alpacas—in the altiplano, which ecologically is similar to the Tibetan Plateau; while potatoes can grow in the Andean highlands, the zone is in many ways better suited to raising herds that move between pastures.

The second zone is in the Great Plains of North America, where indigenous groups became equestrian nomads after the Spanish brought horses to Mexico and eventually to the Southwest. One important similarity between the equestrian nomads of the Great Plains and those of the Eurasian steppe is that both became formidable military forces, albeit the former never reached the level of political complexity of the latter. This difference is in part at least a result of the fact that the Native Americans used horses to enhance their foraging lifestyle, with a significant emphasis on large game—especially buffalo—rather than depending directly on horse products for subsistence. This might be seen as an example of cultural divergence or variation in which the way people exploited similar environments (steppe grasslands) in two locations varied and produced different styles of political and social organization. I would not, however, overemphasize the differences, since numerous parallels can also be drawn between the areas, such as the high level of mobility and attendant aggressive military nature of the societies at the social level and the use of highly portable structures for shelter at the level of material culture. One could add a sheep zone in the Southwest after the Spanish introduced these animals (see Kuznar, chapter 8, this volume). This situation raises an interesting question concerning the nature of what one might call primary and secondary pastoral societies, that is, groups that domesticate animals and then formulate a pastoral lifestyle around those creatures versus people who receive the animals from elsewhere and develop pastoralism as the result of contact. In essence, the North American cases, whether in the Great Plains or the Southwest, can be seen as instances of ethnogenesis, since the cultures in the respective regions were revamped as the result of the introduction of foreign animal species. Furthermore, the adoption of horses demonstrates the flexibility cultures can exhibit in that some groups moved into new hunting territories within a relatively short span of time. DeMallie (2001:727) states that the acquisition of horses by the Sioux in the late eighteenth century "can be characterized as an intensifier of earlier cultural patterns. Horses were the major factor shaping changes in Sioux culture."

While pastoralists exhibit significant flexibility in their adaptations, they do confront certain constraints. Central among these limitations are the conditions required by the animals they tend. Barfield (1993:9–11) argues that while pastoralists rarely, if ever, depend on only one species, typically one particular type of animal is the primary focus of their activities. As a result, that animal's needs figure greatly in decisions about where they reside, how often they move, and how they structure their economic system (with effects on political and social organization). For a particular type of animal to take on this central role, Barfield says it must possess four key traits. First, to sustain large populations, it must be physically well adjusted to the prevailing environment in an area. Second, the animal must be ubiquitous, distributed broadly among all the pastoralists in a region. Third, the needs of the animal dictate general herd conditions at the expense of other species. Fourth, pastoralists must use the animal to structure their "social, political, or economic relation to the world" (ibid.:11).

As an economic system, pastoralism exhibits certain key traits. Rarely do pastoral people depend exclusively on domesticated animals for their subsistence. If they do not engage in cultivation of crops themselves, they can trade for such foodstuffs. The extensive use of animals, however, does dictate certain activities to which pastoralists must attend with regularity and provide an investment of time and labor in a fashion different from the raising of plants. First, of course, is the need to move on at least a seasonal basis to access sufficient pasture for their animals, in contrast to agriculture that requires a sedentary lifestyle. This mobility often restricts the size and variety of material culture, since everything from food-processing tools to the shelters that provide protection from the elements must be portable. In some instances of transhumance, the herders may maintain two separate sets of houses and facilities, one in the lowlands and the other in the highlands, and travel between them seasonally. Such arrangements necessitate a range of social relationships to retain control of two residential zones simultaneously, in some cases by separating households into distinct units.

A second major feature is the ability of the animals that form the primary capital resource to increase through sexual reproduction (Salzman 2004:10), unlike arable land, which is essentially a static commodity (although terracing and irrigation can expand the amount of land that can be brought under cultivation, there is not really a net increase in the extent of terrain). Third, pastureland and water are the other vital resources, and access to them is the focus of much social and political maneuvering. In some areas, public land is available to all members of a community, and people have developed appropriate means to share it without overtaxing the plant growth. Contrary to Hardin's (1968) "Tragedy of the Commons" thesis, Koster (1997; see also Koster and Forbes 2000) has found that Greek pastoralists safeguard the communal zones in various regions through a creative balance with privately owned pastures. A final point Salzman (2004:10–11) makes about pastoral systems is that economic goals can vary by group. In some societies pastoral production is geared to meet the subsistence needs of the herders, while in others animal products are traded in a market system. Salzman notes that in East Africa pastoralism satisfied the former goal, while in West Africa meat and leather were exchanged for various commodities. At a more general level, Kuznar (1995:119) argues that the central tenet for pastoralists in the Andes is to avoid risk, so they concentrate on animals "that not only produce income, but that are particularly resilient in the face of hazards such as drought, disease, and predation." This type of strategy seems applicable to pastoralists throughout the world.

The political and economic conditions to which Barfield and Salzman refer impose another set of limitations. The establishment of modern national borders has disrupted some migratory patterns as pastoralists are restricted from crossing between countries where in the past there was either a unified political system that lacked such boundaries or lax enforcement of such zones. The economic linkages that constitute globalization also play a role, in that pastoralists may gain access to key resources without the need to move. Some opt to settle in towns that offer various jobs rather than continue a rigorous lifestyle on the road.

One way to think of this process is through the world-systems paradigm, in which pastoralists form a peripheral group that is increasingly incorporated into the modern world economy. In this case the incorporation is generally accomplished through a process of enticement, as pastoralists see opportunities to gain more material goods and property by taking jobs that remove them from the system of herd movements. Connection to the larger regional, national, and even international economic system may provide supplemental income that permits people to retain a pastoral lifestyle on at least a part-time basis (see Kardulias, chapter 11, this volume). In certain instances, connecting to the modern world-system facilitates some aspects of pastoral life. For example, Chang (1997:127) describes the situation in Greece: "Since 1982 the European Community has provided Greek herders with increased benefits in the form of social insurance, subsidies, price supports for the marketing of produce, credit and loans, improved infrastructure, subsidized truck transport of herds between summer and winter pastures, and incentives for improved veterinary care." These telling observations indicate pastoralists' ability to take advantage of circumstances to continue and in some cases enhance their economic standing. An issue that bears investigation is what response herders make when such supports are removed or diminished. If incorporation has been thorough and pastoralists become dependent on the outside assistance, herders may lose a significant level of control. However, pastoralists are often at the mercy of outside forces and have tended to cope with them reasonably well. For example, transhumant pastoralists historically have had to deal with the issue of renting pasture and worked out viable plans that involved exchanging animal products for land-use rights.

WHY STUDY PASTORALISTS?

Scholars conduct substantial research on pastoral societies. Beyond the general interest people have in such groups, there are other reasons for this level of work. One major concern is to explain where pastoralists fit in the continuum of subsistence practices humans have developed. For a variety of reasons, different groups opt to depend largely on the products of their domesticated animals for food and various by-products, while others use a substantial portion of those products in an exchange system to obtain the primary food they consume as well as other commodities. Ethnographic work indicates that where pastoralists fall along this continuum may oscillate, as they adjust the level of self-sufficiency according to a number of factors, including environmental conditions that affect herd size and the desire to retain autonomy. As a result of the need to be open to opportunities and conversely to handle adverse events, pastoralists demonstrate flexibility through various practices such as herd management.

These actions lead directly to a second major interest centered on what we can learn about social evolution. Bondarenko and colleagues (2003) argue for a model of societal development that is not only multilinear but also capable of encompassing fluctuations that can move groups from hierarchical to egalitarian systems as well as from less to more complex political and economic structures over time. Examination of this complex process benefits substantially from the information archaeologists can add to what we know from ethnographic and historical sources, and in this we see the third key reason to study pastoralists. The confluence of ethnography and archaeology in pastoral studies demonstrates the viability of, perhaps one could even say the necessity for, the four-field approach in anthropology. The mutually reinforcing lines of investigation provided by the examination of contemporary and past groups enhance the quality of interpretations by lending time depth to the analyses. In particular, pastoralism is ideally suited for ethnoarchaeological study because the ethnographic data collected both about the often ephemeral sites and features herders utilize and the social and political rules that inform their behavior are vital for identifying and explaining the archaeological remains associated with animal husbandry in the past. Work by Chang (1981, 1992, 1993; Chang and Koster 1994), Kuznar (1995), Cribb (1991), and others has provided important insights to guide research about how pastoralism functions, thus giving clues for identifying certain elements in the material record. Archaeologists in turn use ethnographic and ethnoarchaeological information in the effort to understand how the static data they collect from material remains reflect the dynamic social systems of the past. Archaeological interest in pastoralism is reflected in numerous publications over the past several decades, including work by Cribb (1991), Bar-Yosef and Khazanov (1992), Kuznar (1995), Harris (1996), Marshall and Hildebrand (2002),

Anthony (2007), Barnard and Wendrich (2008), and Frachetti (2008). There have also been thematic issues of journals dedicated to pastoralism: *Transhumance and Archaeology* (*World Archaeology* 15, no. 1 [1983]) and *The Dawn of African Pastoralisms* (*Journal of Anthropological Archaeology* 17, no. 2 [1998]).

The study of pastoral groups is also useful in reinforcing or refining theories and methods. For example, Frachetti (2008:15–30) makes good use of the landscape archaeology concept, demonstrating that people construct mental maps to guide their activities in an environment. He adds nuances by exploring the relationship between social interaction and landscape and relating that finding to the "extension and renegotiation of the landscape over time," with a focus on the Bronze Age. Miller (2008) argues that the transition from hunting and gathering to agropastoralism in North Africa was a major economic shift in which gender roles would have been much more fluid than the situation revealed in ethnographic work on contemporary pastoralists. McCorriston and colleagues (2012:47) link cattle sacrifice during the Neolithic in the southern Arabian peninsula to "social boundary defense behavior among independent pastoralists" who ritually demarcated territory in a region where sedentary agriculture was not possible.

Studies of pastoralism also contribute to the ongoing development of world-systems analysis. Several scholars explore the expansion, contraction, and variation in steppe pastoralists in ways that further our understanding of the linkages that engender significant cultural change. Sherratt (2003) argues that the differentiation that developed between pastoralists and agricultural societies in and near the Eurasian steppe during the Bronze and Iron Age was the result of regular interaction and not isolation. Turchin and Hall (2003) develop a model of spatial synchrony, arguing that events such as political collapse occur at about the same time in widely separated areas such as the eastern and western ends of Eurasia, with the major similarities between pastoral peoples of the region playing a major role in this process. In an effort to accommodate the role of individuals in interactions that occur on the edges of world-systems, Kardulias (2007) uses the concept of negotiated peripherality in which local populations select those elements of foreign origin that best fit their perceived needs. This concept has direct applicability to the way modern pastoralists engage with the process of globalization—they are at least to some extent active players in deciding what elements to accept or reject while acknowledging that other factors affect these societies in major ways today (Galvin 2009). As this previous statement implies, the study of pastoralism provides an important way to understand the process of modernization. Salzman (2004:125–36) suggests that we can draw larger conclusions from studying herding societies. On the issue of whether equality and freedom are compatible, he notes that pastoralists, unlike most other groups, have both elements as structural components in their societies, but with the accompanying drawbacks of significant internal strife and lack of economic development. The second issue turns on the question of whether pastoralism is responsible for the "widespread deficit of democracy in contemporary Middle Eastern and African states" (ibid.:125). He notes that political organization in pastoral societies is strongly democratic, with consensual decision-making, and so the absence of democratic institutions in the modern states of the regions is not the result of some native tyrannical strain. One implication is that the present authoritarian regimes are the product of Western colonialism, which would draw us back to examination by way of world-systems analysis and its emphasis on the complex process of incorporation.

REFERENCES

Anthony, David W. 2007. *The Horse, the Wheel, and Language: How Bronze-Age Riders from the Eurasian Steppes Shaped the Modern World.* Princeton: Princeton University Press.

Bar-Yosef, Ofer, and Anatoly Khazanov, eds. 1992. *Pastoralism in the Levant: Archaeological Materials in Anthropological Perspectives.* Madison, WI: Prehistory Press.

Barfield, Thomas J. 1989. *The Perilous Frontier: Nomadic Empires and China.* Cambridge: Basil Blackwell.

Barfield, Thomas J. 1993. *The Nomadic Alternative.* Englewood Cliffs, NJ: Prentice-Hall.

Barnard, Hans, and Willeke Wendrich, eds. 2008. *The Archaeology of Mobility: Nomads in the Old and in the New World.* Los Angeles: Cotsen Institute of Archaeology, UCLA.

Bates, Daniel G. 2001. *Human Adaptive Strategies: Ecology, Culture, and Politics.* 2nd ed. Needham Heights, MA: Allyn and Bacon.

Bondarenko, Dmitri M., Andrey V. Korotayev, and Nikolay N. Kradin. 2003. "Introduction: Social Evolution, Alternatives, and Nomadism." In *Nomadic Pathways in Social Evolution*, ed. Nikolay N. Kradin, Dmitri M. Bondarenko, and Thomas J. Barfield, 1–24. Moscow: Russian Academy of Sciences.

Campbell, John K. 1964. *Honour, Family, and Patronage: A Study of Institutions and Moral Values in a Greek Mountain Community.* Oxford: Clarendon.

Chang, Claudia. 1981. "The Archaeology of Contemporary Herding Sites in Didyma, Greece." PhD diss., Department of Anthropology, State University of New York, Binghamton.

Chang, Claudia. 1992. "Archaeological Landscapes: The Ethnoarchaeology of Pastoral Land Use in the Grevena Province of Greece." In *Space, Time, and Archaeological Landscapes*, ed. Jacqueline Rossignol and LuAnn Wandsnider, 65–89. New York: Plenum.

Chang, Claudia. 1993. "Pastoral Transhumance in the Southern Balkans as a Social Ideology: Ethnoarchaeological Research in Northern Greece." *American Anthropologist* 95 (3): 687–703. http://dx.doi.org/10.1525/aa.1993.95.3.02a00080.

Chang, Claudia. 1997. "Greek Sheep, Albanian Shepherds: Hidden Economies in the European Community." In *Aegean Strategies: Studies of Culture and Environment on the European Fringe*, ed. P. Nick Kardulias and Mark T. Shutes, 123–39. Lanham, MD: Rowman and Littlefield.

Chang, Claudia. 2006. "The Grass Is Greener on the Other Side: A Study of Pastoral Mobility on the Eurasian Steppe of Southeastern Kazakhstan." In *Archaeology and Ethnoarchaeology of Mobility*, ed. Frederic Sellet, Russell Greaves, and Pei-Lin Yu, 184–200. Gainesville: University Press of Florida.

Chang, Claudia, Perry A. Tourtellotte, Karl Baipakov, and Feydor P. Grigoriev. 2003. *The Social Evolution of Eurasian Steppe Communities in Southeastern Kazakhstan: The Kazakh-American Talgar Project 1994–2001*. Almaty: Institute of Archaeology, Kazakh National Academy of Sciences.

Chang, Claudia, and Harold A. Koster, eds. 1994. *Pastoralists at the Periphery: Herders in a Capitalist World*. Tucson: University of Arizona Press.

Cribb, Roger. 1991. *Nomads in Archaeology*. Cambridge: Cambridge University Press. http://dx.doi.org/10.1017/CBO9780511552205.

DeMallie, Raymond J. 2001. "Sioux until 1850." In *Handbook of North American Indians*, vol. 13, part 2, ed. Waldo Wedel and George C. Frison, 718–60. Washington, DC: Smithsonian Institution Press.

Dyson-Hudson, Rada, and Neville Dyson-Hudson. 1980. "Nomadic Pastoralism." *Annual Review of Anthropology* 9 (1): 15–61. http://dx.doi.org/10.1146/annurev.an.09.100180.000311.

Evans-Pritchard, Edward E. 1940. *The Nuer: A Description of the Modes of Livelihood and Political Institutions of a Nilotic People*. Oxford: Clarendon.

Frachetti, Michael D. 2008. *Pastoralist Landscapes and Social Interaction in Bronze Age Eurasia*. Berkeley: University of California Press.

Fratkin, Elliot. 1997. "Pastoralism: Governance and Development Issues." *Annual Review of Anthropology* 26 (1): 235–61. http://dx.doi.org/10.1146/annurev.anthro.26.1.235.

Galaty, John G., and Douglas L. Johnson, eds. 1990. *The World of Pastoralism*. New York: Guilford.

Galvin, Kathleen. 2009. "Transitions: Pastoralists Living with Change." *Annual Review of Anthropology* 38 (1): 185–98. http://dx.doi.org/10.1146/annurev-anthro-091908-164442.

Hardin, Garrett. 1968. "The Tragedy of the Commons." *Science* 162 (3859): 1243–48. http://dx.doi.org/10.1126/science.162.3859.1243.

Harris, David R., ed. 1996. *The Origins and Spread of Agriculture and Pastoralism in Eurasia*. Washington, DC: Smithsonian Institution Press.

Kardulias, P. Nick. 2007. "Negotiation and Incorporation on the Margins of World-Systems: Examples from Cyprus and North America." *Journal of World-Systems Research* 13 (1): 55–82.

Khazanov, Anatoly M. 1984. *Nomads and the Outside World*. Trans. Julia Crookenden. Cambridge: Cambridge University Press.

Koster, Harold A. 1997. "Yours, Mine, and Ours: Private and Public Pasture in Greece." In *Aegean Strategies: Studies of Culture and Environment on the European Fringe*, ed. P. Nick Kardulias and Mark T. Shutes, 123–39. Lanham, MD: Rowman and Littlefield.

Koster, Harold A., and Hamish A. Forbes. 2000. "The 'Commons' and the Market: Ecological Effects of Communal Land Tenure and Market Integration on Local Resources in the Mediterranean." In *Contingent Countryside: Settlement, Economy, and Land Use in the Southern Argolid since 1700*, ed. Susan B. Sutton, 262–74. Stanford: Stanford University Press.

Kradin, Nikolay N. 1987. *Socio-Economic Relations among the Nomads in Soviet Anthropology*. Institute of Scientific Information of the Social Sciences 29892. Moscow: Russian Academy of Sciences.

Kradin, Nikolay N., Dmitri M. Bondarenko, and Thomas J. Barfield, eds. 2003. *Nomadic Pathways in Social Evolution*. Moscow: Russian Academy of Sciences.

Kuznar, Lawrence A. 1995. *Awatimarka: The Ethnoarchaeology of an Andean Herding Community*. Orlando, FL: Harcourt Brace.

Marshall, Fiona, and Elisabeth Hildebrand. 2002. "Cattle before Crops: The Beginnings of Food Production in Africa." *Journal of World Prehistory* 16 (2): 99–143. http://dx.doi.org/10.1023/A:1019954903395.

McCorriston, Joy, Michael Harrower, Louise Martin, and Eric Oches. 2012. "Cattle Cults of the Arabian Neolithic and Early Territorial Societies." *American Anthropologist* 114 (1): 45–63. http://dx.doi.org/10.1111/j.1548-1433.2011.01396.x.

Miller, Alexandra. 2008. "Changing Responsibilities and Collective Action: Examining Early North African Pastoralism." *Archeological Papers of the American Anthropological Association* 18 (1): 76–86. http://dx.doi.org/10.1111/j.1551-8248.2008.00006.x.

Parman, Susan. 2005. *Scottish Crofters: A Historical Ethnography of a Celtic Village*. 2nd ed. Belmont, CA: Thomson Wadsworth.

Salzman, Philip C. 2004. *Pastoralists: Equality, Hierarchy, and the State*. Boulder: Westview.

Sherratt, Andrew. 2003. "The Horse and the Wheel: The Dialectics of Change in the Circum Pontic Region and Adjacent Areas, 4500–1500 BC." In *Prehistoric Steppe Adaptation and the Horse*, ed. Marsha Levine, Colin Renfrew, and Katie Boyle, 233–52. Cambridge: McDonald Institute for Archaeological Research.

Spooner, Brian. 1973. *The Cultural Ecology of Pastoral Nomads*. Addison-Wesley Modules in Anthropology 45. Reading, MA: Addison-Wesley.

Turchin, Peter, and Thomas D. Hall. 2003. "Spatial Synchrony among and within World-Systems: Insights from Theoretical Ecology." *Journal of World-Systems Research* 9: 37–64.

2

The Study of Nomads in the Republic of Kazakhstan

Claudia Chang

In the main square of the former capital city, Almaty, in the Republic of Kazakhstan stands a new bronze statue of Golden Warrior, the seventeen-year-old Saka (Indo-Iranian) youth. Golden Warrior is dressed in a cloak of gold-plated armor and a tall pointed hat adorned with four gold bird figurines and spears, and he wields a golden dagger as he sits astride a prancing steppe horse. The remains of Golden Warrior were discovered in 1969 by Kazakh[1] archaeologists during salvage excavations of a large burial tomb in Issyk, a small town about 40 km east of Almaty (Akishev 1978). This splendid archaeological find dates to the Saka-Scythian period of the mid–first millennium BC and has become the national emblem of Kazakhstan. What few Kazakhstani citizens comprehend is that Golden Warrior, attributed to the Saka culture of Indo-Iranian origins, is separated in time by at least 1,000 years from the origins of the Turkic-speaking Kazakh nomads, whose origins date back to the 1500s. This statue serves as an icon for extolling the virtues of a pastoral nomadic past and representing a nation-state's break from its Russian Tsarist and Soviet past. Such simple details about Golden Warrior's true identity (Indo-Iranian) or attention to the inextricable relationship between the warrior and his horse or even how the nomadic past is seen as the guiding force behind a nation's independence and the identity of the Kazakh people inform the models, methods, and theories used to explain the history of nomadic pastoralism in the Republic of Kazakhstan today. In this chapter I trace the intellectual history of nomadic studies of the Eurasian steppe nomads conducted in the twentieth century by Soviet-trained archaeologists,

DOI: 10.5876/9781607323433.c002

ethnographers, and historians. By way of examining the Soviet tradition of pastoral nomadic studies, I also chart the political and ideological divides between Western and Soviet-based scholarship on Eurasian steppe nomads.

I start, however, with a less grandiose and more modest theme—my own intellectual development as a student of nomadic pastoralism. That familiar narrative of "professional training" starts with my doctoral research in a Mediterranean village on an occupational class of shepherds and goatherds in the late 1970s and ends with my most recent archaeological investigations of Iron Age pastoralists (Saka-Wusun) along the northern Tian Shan Mountains, about 25 km from the city of Almaty in the Republic of Kazakhstan. This exercise in autobiography underscores a central theme in this chapter—how Western scholars conceptualize pastoral nomadism. After all, it is impossible to understand the development of an intellectual history of Soviet historical science in the twentieth century without a clear understanding of the similarities and differences between Soviet historical science and Western anthropology.

In 1978 and 1979 I spent thirteen months studying the ethnography and ethnoarchaeology of Greek village shepherds. Most of my doctoral research consisted of collecting data on the material culture of herding households (their animal enclosures, campsites, corrals, and huts) and interviewing members of each herding household (Chang 2000). After I wrote my dissertation and published a series of essays on this research, which was labeled "the ethnoarchaeology of village pastoralism," I discovered that my research was not only narrowly defined but had no clear categories within the field of nomadic pastoral studies. First, the Didymiote herding households in southern Greece were not considered nomadic pastoralists, since these shepherds lived in a year-round sedentary village. Not only that, they were barely considered "pastoralists," since the term *pastoralism* itself was defined as those groups of people who maintained livestock on a regular basis and moved periodically in search of pasture and water. The fact that many of these Greek shepherds also owned wheat fields and olive groves only pushed "my people" further into the category of "peasants" and away from the category of "pastoralists." Moreover, if I tried to interest prehistoric archaeologists in my ethnoarchaeological research on contemporary Greek village herders by telling them that mixed herding and farming communities were no doubt the mainstay of Neolithic and Bronze Age communities in Europe and the Near East, there was further silence. What could a study of contemporary Greek village herders tell prehistoric archaeologists about the "origins of animal herding" and the emergence of specialized nomadic pastoralism? After all, the evolution of such specialized forms of animal husbandry as long-distance pastoral transhumance or nomadic pastoralism required the use of ethnographic analogies of contemporary nomads such as the

Basseri, Kermani, or Turkmen. The study of contemporary Greek village herders was seen as hardly a good analogy for understanding the evolution and development of specialized pastoralism.

Ten years later, to remedy the inappropriateness of my "ethnographic analogy," my husband, Perry Tourtellotte, and I conducted an ethnoarchaeological survey in northern Greece among a well-known group of long-distance transhumant herders, the Koutsovlach and the Kutpatshari, a Hellenized group of the same ethnic origins (Chang and Tourtellotte 1993). We spent six seasons examining the material culture of transhumant herders who used the Grevena region of the eastern Pindos Mountains as their summer grazing lands. We were able to document the pastoral use of three major elevation and vegetational zones of the eastern Pindos Mountains: (1) the high Pindos Mountains (ca. 1,300–1,500 m), used in the summer by transhumant herders known ethnically as the Koutsovlach; (2) the lower Pindos Mountains (ca. 1,000–1,300 m), used in the summer by transhumant herders known as Hellenized Koutsovlach or Kutpatshari and by year-round agropastoralists; and (3) the foothills of the Pindos Mountains (below 1,000 m), used primarily by year-round, sedentary agropastoralists. The transhumant herders then traveled in the late summer/early fall from 50 km to 150 km to Elassona, or the Plains of Thessaly, where they lived outside of large villages and towns and kept their herds of 250 to 1,000 animals on pastures and fodder from the lowlands. Although I was most interested in the ecological and economic constraints of long-distance pastoral transhumance, I observed and wrote about how the transhumant herders, by virtue of their membership in both upland and lowland villages, cultivated extensive and dense social networks with other transhumant herders, cheese merchants, and the political elite through patronage and kinship (Chang 1993).

In the mid-1990s I extended my analytical repertoire beyond the study of the material culture of pastoralism and began to consider the social and ideological reasons behind pastoral mobility. The ethnographic contrasts between Greek village herders of Didyma, where flocks ranged from under 100 head to 300 head of sheep and goats, to the Kazakh herdsmen, where a mixed herd of sheep, goats, cattle, and horses might include over 1,000 head, changed my understandings of pastoral mobility. Animals are also provided with fodder in the winter, especially when the snow is deep, and only horses can kick off the snow layers to reach native vegetation. Furthermore, the Talgar alluvial fan where our archaeological research on agropastoralism was conducted was located at the base of the Tian Shan Mountains, large glacier-peaked mountains. Within less than an 80-m area, the verticality of the environment ranged from high alpine pastures at about 2,600 to 1,300 m above sea level (asl) to the lowland alluvial fan, gently sloping from 1,100 m asl to 550 m asl. The alluvial fan flattens out to a semiarid steppe and desert. Kazakh pastoral mobility can range from vertical transhumance of 20 km to longitudinal mobility between north and south areas across steppes and valleys of up to 1,000 km. In contrast, the Greek countryside of mountains, plateaus, and valleys is heavily dissected, providing pockets of grazing and farming land. In Greek village pastoralism of the northeastern Peloponnesos, flocks of sheep and goats were grazed on green fodder, agricultural stubble, and fallow as well as maquis shrubland. The range of mobility may be from 5 km to 10 km on a seasonal round. Yet the Koutsovlach and Vlach herders of the Grevena region could move with their large flocks of sheep and goats up to 150 km or more between summer and winter villages.

Since variability is considerable within both Mediterranean and Eurasian steppe pastoralism, my main objective was to examine how herders used a particular landscape and whether we could trace their patterns of mobility. The visible pastoral traces on the Greek landscape included animal folds, corrals, milking pens, watering troughs, and shrines. For Kazakh herders the same facilities existed, such as encampments with yurts, pens, and corrals for the summer settlements, while stone or mud brick facilities were used during the winter months. For the Iron Age nomadic confederacies in southeastern Kazakhstan, the most visible landscape feature was earthen burial mounds, not the pastoral facilities or settlements of ancient agropastoral peoples. How were these mounds then related to distribution of grazing territories and other valuable resources? These new questions often led to a return to the essential factors of pastoral mobility, such as (1) how pasturelands were distributed and by whom (village government, rental relations, the state, or the collective); (2) the availability of grazing lands, fodder, and animal feed; (3) the articulation between cereal farming and animal herding (more closely articulated in the Greek case and less so in the Kazakh case, where cereal farming was often divorced from pastoralism in terms of labor and land management); and (4) the means by which the pastoral products were funneled into either household or kin networks and marketed or exchanged to broader networks, including urban centers.

In 1994, when I began archaeological study of the Iron Age and medieval period nomadic pastoralists of the Talgar region in southeastern Kazakhstan, two perplexing ideas about pastoralism dominated my Kazakhstani colleagues' notions about the Iron Age nomadic settlements we were excavating. They believed the first millennium BC nomadic pastoralists (Saka-Scythians, Wusun, Yuezhi) did not practice agriculture and had no semi-permanent architecture or substantial settlements. I tended to disagree with both notions because of the views of agropastoralism and Bronze Age settlement I had developed as a result of our ethnoarchaeological research in Greece. First, I was convinced that the "multi-resource strategies" of all nomadic pastoralists, as defined by Salzman (1980, 2004), needed to be better understood by archaeologists and ethnographers—especially any who wished to trace the origins, history, or evolution of specialized pastoralism (nomadic or transhumant). After all, if pastoralists had to engage in relations with an "outside world," it stood to reason that either they participated in these other economies themselves (cultivating crops, foraging, or fishing) or they had to develop trade and exchange relations with others who did practice those economies (Khazanov 1984).

The second problem that troubled me and countless other researchers was pastoral mobility. In my case, I thought the village pastoralists I had studied in the late 1970s were certainly mobile, even though they could be categorized as settled village herders. Most households moved between summer and winter pastures, and while the distances they traveled seemed minimal next to those of classic long-distance transhumant pastoralists or nomadic pastoralists, they did travel daily and seasonally on a pastoral round. Furthermore, when long-distance transhumance is practiced, such as among the Koutsovlach and Kutpatshari of the Grevena area in northern Greece, far more is involved than just the usual economic and ecological aspects of obtaining pasture, water, and other necessary resources. Pastoral mobility, in the case of these ethnic groups of Mediterranean herders, was also shaped by political and ideological strategies necessitating the maintenance of social networks in both lowland and upland communities. The meaning of mobility for the pastoralists themselves was as important as the economic and ecological reasons that necessitated the seasonal or daily movement of pastoral groups (Chang 1993). Frederik Barth (1986 [1961]:153), in discussing the Basseri of south Persia, attempted to grapple with what appeared to be a paucity of actual ritual practices among these nomadic people, stating, "If one grants the possibility, on the other hand, it becomes very reasonable to expect the activities connected with migration to have a number of meanings to the nomads, and to be vested with value to the extent of making the whole migration the central rite of nomadic society."

The Soviet-trained archaeologists I worked with on the surveys and excavations of Iron Age pastoral sites dating from about 700 BC to AD 100 based their notions of the evolution of steppe pastoralism on (1) textual accounts derived from the Greek historian Herodotus, Persian inscriptions, and Chinese chronicles; (2) ethnographic analogies with historic groups of nomadic pastoralists, such as the Hsiung-nu or the Mongols; and (3) Marxist theories on the evolution of primitive societies (Masanov 1966). In the mid-1990s, as our excavation and surveys of the Talgar area along the northern flanks of the Zailiisky Alatau (a section of the Tian Shan Mountains) progressed, I began to challenge the models and ideas of my Kazakhstani colleagues by adding two new perspectives to their research agenda: (1) a systematic investigation of paleo-ethnobotanical and zoo-archaeological data at Iron Age settlements that might determine the extent to which both farming and herding took place in this area of the Eurasian steppe, and (2) the idea that the "nomadic pastoralists" had semi-sedentary villages or settlements where evidence for both subsurface and aboveground architecture of a more enduring nature than the traditional yurt-type structure existed on the alluvial fans just north of the Tian Shan Mountains. In other words, my Kazakhstani colleagues and I began to amass archaeological data that addressed two fundamental issues previously unexplored in the history of nomads of the Eurasian steppes in the first millennium BC: the range of economic activities, including the existence of crop cultivation among the so-called nomadic populations, and the nature of pastoral mobility as revealed in the archaeological record (Chang et al. 1999; Chang et al. 2003).

In this chapter I outline some of the problems I, as a Western anthropologist, have encountered while researching the social evolution of communities on the Eurasian steppe. The cultures we study have been labeled the Saka-Scythians or Wusun of the first millennium BC, yet such labels are derived from textual accounts naming the various nomadic confederacies of the Semirech'ye (Seven Rivers area along the Ili River Basin) (Chang and Grigoriev 1999). The historiography of such groups is in and of itself quite complicated, dependent on such sources as Herodotus and Chinese chroniclers.

To clarify some of the conceptual divides (political, ideological, and intellectual) my Kazakhstani colleagues and I have had to cross to find common ground in our research interests, I have had to resort to tracing the history of Russian and Soviet studies of Eurasian pastoral nomads. In the next section I lay out the basic chronological divides in the historiography of the Russian and Soviet studies of Eurasian pastoral nomads, specifically the Kazakh-Kirghiz.[2]

A BRIEF OUTLINE OF RUSSIAN AND SOVIET PERSPECTIVES ON EURASIAN STEPPE PASTORALISM

Beginning in the eighteenth century, Russian explorers, diplomats, and intellectuals attempted to chronicle the pastoral nomads they encountered on the far eastern borders of Imperial Russia. The Russian tradition of documenting these peoples' folklore, customs, beliefs, languages, and physical characteristics produced a solid record of nineteenth- and early-twentieth-century accounts of such groups as the Kazakh-Kirghiz (the horse-riding pastoral nomads who occupied the newly independent states of Kyrgyzstan and Kazakhstan). After the 1917 revolution, these national groups came under the direct jurisdiction of Lenin's and Stalin's policies pertaining to the preservation of national and ethnic groups. Soviet historical science—a broad category including the fields of history, ethnology, and archaeology—then became the guiding model for research on Eurasian steppe nomads. Because of the stringent policies of collectivization, Soviet ethnologists most often had to rely on pre-revolutionary accounts of Kazakh social organization and customs. The seventy-year period of Soviet rule produced an intellectual cadre dedicated to the aims and goals of historical science and Marxist-Leninist formulations used to describe these nomadic groups.

Since the breakup of the USSR and the independence of the Soviet Republics at the end of 1991, a new synthesis of nomadic studies has emerged in the Republic of Kazakhstan. This synthesis combines elements of Soviet historical science, influences from Western ethnography and social anthropology, and the emerging nationalism of the Kazakh elite. I leave a more detailed discussion of the nature of such changes over the past decade and a half to others (Khazanov 1996; Yessenova 2005; Liu 2011:123–24). For the sake of clarity, I distinguish in this chapter between what I label broadly Soviet research (which includes Russian, Soviet, and post-Soviet perspectives) and Western research (which includes European and American perspectives). As with any cross-cultural comparison, whether made about scholarship or customs and traditions, there are bound to be broad generalizations that obscure the nuances and variations within either categorical group.

What sets Eurasian steppe pastoralism apart from other regional forms of pastoral nomadism is the dependence on horse riding and the herding of sheep, goats, cattle, horses, and in some cases camels and yaks on the vast grasslands of the mountain, steppe, and desert regions. The Russian term for pastoralists is *kochevniki* (livestock herders) and is often used in contemporary language as a derogatory term for ethnic groups whose origins belie a predominantly pastoral orientation. While kochevniki are distinguished from farmers, it is well understood by Soviet ethnographers that herdsmen often engage in agrarian pursuits such as the cultivation of wheat, barley, and millet in rich oasis regions and along fertile floodplains (Khazanov 1984).

When Soviet scholars describe pastoral mobility, their classification system approximates what Western anthropologists and historians have labeled "transhumance," or the movement of herds between two or more places on a seasonal basis. The two major kinds of mobility are (1) long-distance longitudinal movement between summer pastures, found in the northern steppe grasslands, and winter pastures, found in the southern steppe grasslands, and (2) short-distance vertical movement between summer pastures found at high elevations (1,000 m or above) and winter pastures found at lower elevations (below 1,000 m) (see Barfield 1993:141).

KEY CONCEPTS OF SOVIET HISTORICAL SCIENCE AND THE STUDY OF NOMADS: PATRIARCHAL FEUDALISM AND ETHNOGENESIS

Soviet and post-Soviet researchers are quite broad-minded in their geographic descriptions of various mobility strategies employed by pastoralists. They recognize that these two basic strategies can be combined and used interchangeably by any given pastoral group. There may be one central reason for this: what has dominated Soviet research is attention to placing pastoral nomadism into a scheme that follows Marx and Engels's formulation of the evolution of societal forms. In the instances where Soviet anthropologists expressed lip service or actual interest in employing Marxist models for the evolution and development of societies, nomadic pastoralism cannot be easily placed within any of the evolutionary stages such as foraging, agricultural, or industrial societies. The solution for ethnologists and historians studying the Kazakhs was to describe the pastoral means of production as a social form quaintly labeled "patriarchical feudalism." Patriarchical feudalism refers to the social organization of the Kazakh clan by which rich herdsmen owned herds (but did not own land privately) and therefore exploited their poorer relatives and even a "slave class" of herdsmen expected to maintain their rich relatives' vast herds and flocks (Tolybekov 1959). The Kazakhs, who have a conical clan social organization based on patrilineal descent, also have a superimposed political-military division into three major hordes. Patriarchical feudalism, or the fact that rich herders exploited poor relatives and had attached groups of herdsmen who were lower in status and similar to slaves, did not lead to true class divisions within the clan and lineage organization (Gellner 1984). Therefore, patriarchical feudalism, as an economic and social form of organization, did not lend itself to the classical stages of social evolution used by historical materialists.

This contrasted considerably with the Western model of nomadic pastoralism, which stressed the egalitarian, fiercely independent nature of herdsmen. Dyson-Hudson's (1972) herding model still seems applicable, especially with respect to any discussion of the human ecology of pastoralism. The model (ibid.:24–25) focused on these variables: (1) varieties of animals herded; (2) animal numbers in relation to carrying capacity and human labor; (3) the nature of herd makeup (age/sex composition) to human groups (household, lineage); (4) distribution and variety of grazing, browsing, and foddering that might affect stocking rates and herd management practice; (5) the human group's degree of dependence on and level of commitment to livestock; (6) ability to convert livestock to labor, land, food, cash, prestige, and personnel; 7) alternate possibilities for resource exploitation; and (8) the coincident nature of the human-livestock association. Throughout his discussion of this model, Dyson-Hudson interwove ideas about the nature of the pastoral experience, noting the social, political, and ideological conditions under which a herdsman might be linked to others within or outside the group, to other economies or markets, and to the state. Not only was the pastoral way of life characterized as highly variable and flexible, but pastoral societies themselves—far from isolated either in time or space—were the brokers, middlemen, and purveyors of goods, personnel, and ideas.

The other dominant feature of Soviet studies of pastoral nomadism on the Eurasian steppe has been attention to the concept of ethnogenesis. As I have come to understand the term, *ethnogenesis* refers specifically to the origins and development of an "ethnos," or a particular ethnic group (Chard 1963; Bromley 1979; Bromley and Kozlov 1989). The concept of ethnogenesis could be used in connecting archaeological cultures with historically known tribes (Chard 1963:541). Determining how an ethnic group comes into being is a decidedly Soviet preoccupation (Kohl 1998).[3] Western anthropology has tended to emphasize the nature of ethnic groups in maintaining boundaries and perpetuating political and cultural identities in national states rather than investigation of the actual origins and historical process that give rise to a new ethnic group. Recently, ethnogenesis as a process has been discussed by Western researchers employing a world-systems perspective to explain the role of frontiers and boundaries at the edges of core states, semiperipheries, or peripheries (Hall 2000:241). Hall (ibid.) argues that it is at frontiers where such processes as ethnogenesis and its opposite, ethnocide, can occur rapidly. Bromley and Kozlov (1989:433–35) also indicate that "evolutionary and transformational ethnic processes" could occur at frontiers as well as within geographically defined areas. This is especially true if one accepts the notion that ethnogenesis as used by the Soviets was an outgrowth of Marx and Engels's notions of the evolutionary development of society.

Soviet anthropology used the term *spiritual culture*, derived from German ethnology, to characterize the mentality, beliefs, and ideology of *ethnos* or *ethnikos*, which according to Bromley (1979) means specific common and stable cultural and psychic aspects. In their elaborate schema, Bromley and Kozlov (1989) explained that in the late 1940s, ethnographers used the Russian term *narodnost* (masses of people) to describe ethnic groups undergoing transformation (periods of slavery and feudalism) who had not yet formed *natsional'nost,* or nationality in the form of being part of a state.[4] Thus, the beliefs, personality characteristics, and moral nature of pastoral nomads were as much a part of what distinguished them from other ethnic groups as their economic dependence on livestock herding. Therefore, ethnogenesis, although an evolutionary process, should be understood as encompassing both the material and ideological aspects of culture.

In the post-Soviet society of Kazakhstan, the study of ethnogenesis has taken on a decidedly nationalistic bent, although to directly pinpoint this process of Kazakhification, a clear understanding is required of natsional'nost, a term that changed throughout the Soviet period, at times glossed as both nationality and ethnicity (Beardmore 2007).[5] The effort to prove the ethnic origins of the Kazakhs, whose history begins in the fifteenth century AD with the development of the Kazakh-Uzbek Kingdom, has spurred local historians and archaeologists to investigate even earlier periods with the expectation that there is a deeper antiquity for the nomadic Kazakhs, going back to the Saka period (Schatz 2000:496).[6] For a comparative case, one can examine Shnirelman's (1995, 2006) discussion of the claims made by North Ossetian authorities in the early 1990s that attempted to establish historical continuity between themselves and the Alans of the ninth–tenth centuries, a nomadic confederacy whose origins trace back to the Sarmatians-Scythians (thus Indo-Aryans).[7]

THE CASE STUDY: KAZAKH ETHNOGRAPHY

In the main body of this chapter I use the ethnographic and historical studies of the Kazakhs as an example of how Soviet researchers investigated pastoral nomadism. The main source of information for the examples of Kazakh ethnography and history presented here come from the book-length essay published in Russian in 1966 by leading Kazakh historian and ethnographer Edige I. Masanov. Over the last century the Republic of Kazakhstan has been an ideal "natural laboratory" for the study of nomads. The republic, geographically situated in the far eastern region of the former USSR, is bordered by China and Mongolia to the east and north, respectively. Since independence, official statistics indicate that more than 60 percent of the population is classified as ethnic Kazakh; the Kazakhs speak a Turkic language and underwent nominal conversion to Islam from the seventeenth century onward (Olcott 1987). For both Russian and Soviet ethnographers, the Kazakhs, who occupied the vast eastern steppe lands from the Caspian Sea to China and Mongolia, represented an "untouched" nomadic population who successfully adapted to the temperate climate of the Eurasian steppe. These horse-riding pastoralists keep mixed herds of sheep and goats, cattle, horses, and camels. Although other Central Asian and Middle Asian pastoralists are also known for their dependence on riding horses, the Kazakh and Kirghiz are distinguished by their practice of eating horse meat and drinking mare's milk (*kumiss*).

The Kazakhs' social organization is complicated by the formation of three military confederacies known as hordes but more correctly translated as "hundreds" (see Olcott 1987:12–13). The hordes were historical federations or groups of tribes without common ancestry. Olcott (ibid.:11) states that by the seventeenth century the current division into the Great Horde, Middle Horde, and Small Horde was established by geographic territories. The Great Horde occupied the area known as Semirech'ye of southeastern Kazakhstan (the Seven Rivers area, including the river valleys of the Chu, Talas, and Ili). The Middle Horde occupied the southwestern territory of the lower Syr Daria and held summer pasturelands in the central steppe area. The Small Horde originated in western Kazakhstan along the lower Syr

Daria and Ural Rivers and the steppe areas in northern Kazakhstan. These military and political groups were essential for maintaining vital summer and winter pasture areas. In Kazakhstan today, the horde divisions continue to play an important role in political alliances and the integration of large geographic regions among ethnic Kazakhs. Since independence, President Nursultan Nazarbaev, a member of the Great Horde, has continued to employ principles of horde alliance and loyalty among the elite ranks of the Kazakh leadership.

THE TSARIST PERIOD

Russian travelers, diplomats, translators, and military officers from the seventeenth through the nineteenth centuries collected ethnographic materials on the Kazakh-Kirghiz nomads (the Kazakh and Kirghiz nomads were often categorized as a single ethnic group by the Imperial Russian administration). These early ethnographers collected material on the Kazakhs' kinship structure, settlement and residence patterns, household organization and labor, religious beliefs, temperament, and material culture. These ethnographic materials were in many ways similar to the anthropological material collected by British colonial officials, military officers, and explorers who traveled to Africa, Asia, and the Pacific during the Victorian period (Stocking 1987). These early Russian ethnographers also developed interest in the folklore of the Kazakhs, collecting poetry, songs, legends, and folktales. Other anthropological studies included the physical anthropology, linguistics, and prehistoric archaeology of ancient cultures inhabiting the territory of Kazakhstan.

These Russian travelers noted the existence of the three Kazakh hordes. In particular, they were interested in the historical connection between the Kazakh hordes and the Golden Orda of the Mongol Empire. The Tsarist regime paid close attention to the political and military nature of the horde system, with an eye toward using this "indigenous political system" as a means of mandating Russian colonial and imperial control over the Kazakhs. During the second half of the nineteenth century, when Imperial Russia undertook a program of Russification, information on Kazakh history and ethnography was seen as vital. The four major areas of Tsarist interests identified by Masanov (1966:138) included (1) the political and administrative nature of the horde and the clan and lineage structure of Kazakh social organization, (2) the linguistic and cultural relationship of Kazakhs to the other Turkic-speaking groups of Central Asia, (3) the system of law and jurisdiction among the Kazakhs, and (4) the establishment of Kazakh schools in Russian territories such as Orenburg (Siberia).

The Tsarist interest in Kazakh history and ethnography that persisted well into the Soviet period and the current period of Kazakh independence was mainly in the role of the Kazakh language in uniting Kazakh people. As early as 1862, B. B. Grigoriev, a Russian colonial administrator in Orenburg, decided to transcribe the Kazakh-Kirghiz languages using the Cyrillic alphabet (the original spoken language had already been transcribed into Arabic). At the local Kazakh school in Orenburg, students were taught the written version of these Turkic languages using the Cyrillic alphabet. During the 1920s, under Soviet rule, the language question was brought up again when the suggestion was made that Kazakh should adopt the Latin alphabet. From 1926 to 1940, the Kazakh language was written in Latin letters (Olcott 1987:196). The choice of a Cyrillic or Latin alphabet over the original Arabic script had definite political and nationalist implications. After the breakup of the USSR in 1992, the independent republics of Kazakhstan and Kyrgyzstan again debated whether they should use Latin or the modified Cyrillic form. One might speculate that if the Latinized form of Kazakh and Kirghiz writing were adopted (as discussed in earlier Soviet periods; see Masanov 1966), it could reinforce a cultural and political allegiance with the modern state of Turkey and a connection with all Turkic cultures, signaling a clear break with the Russian colonial past and Soviet rule.

The leading historian and ethnographer of Kazakh history in the nineteenth century was Chokan Valihanov, a Kazakh intellectual trained in Orenburg. During the Soviet period, the Institute of History, Archaeology, and Ethnography at the Kazakh National Academy of Sciences bore Valihanov's name. During his most productive years of research (1855–59), Valihanov began to explore the question of Kazakh origins (ibid.:153). His investigations led to these major fields of research: (1) the study of ethnography and folklore as a means for reconstructing the origins of the Kazakh; (2) the study of architecture and archaeology (materials from burial mounds, graves, and mausoleums) to establish the Kazakhs' ancient history; and (3) the study of physical anthropology (craniological measurements of skulls) as a means of proving the Kazakhs' Mongol-Turkic origins.

Valihanov's nineteenth-century interest in ethnography, folklore, architecture and archaeology, and physical anthropology as a means for tracing the history and origins of the Kazakhs also parallels the development of American anthropology during the end of the nineteenth century and the beginning of the twentieth century. Nineteenth-century American ethnographers, physical anthropologists, linguists, and archaeologists also sought to establish the "origins" of indigenous American societies. In a broader historical sense, one might say that Edward Tylor's impact on questions of origin and the stages of humankind had a widespread impact on such distant intellectual traditions as Tsarist period ethnography on Kazakhs and American pre-Boasian and Boasian anthropology on American Indian cultures.

The other striking parallel between the Tsarist period ethnographic tradition and

Anglo-American anthropology was the notion that primitive people represented "the survivals" of earlier stages of humankind. Masanov (1966:206) notes in particular the work of P. E. Makovetskii who, writing in the 1890s, said that Europeans must try to understand contemporary Kirghiz not just for the sake of modern ethnography but also because the nomads of the steppe represented a stage in the history of the development of humankind. This notion of "cultural survival" lends itself well to the agenda of Soviet historical science.

Two late-nineteenth-century scholars of Kazakh history, Altinasarina and Arendenko, wrote specifically about the Kazakh practice of agriculture (ibid.:176, 190). They described the origins of agriculture and irrigation agriculture in Turkestan among Kazakh people that predated the period of contact with Slavic peasant populations. As scholars of nomadic pastoral populations know, the practice of agriculture in those societies is not only of interest to scholars but to national governments as well. The blanket assumption that nomads have no history of practicing agriculture usually gives national governments and those in power the rationale for resettling pastoralists and usurping their pasturelands. If the case is made that a nomadic pastoral group has a history of practicing agriculture, then it becomes much harder for the colonial or national regime to take land away from indigenous pastoralists. In fact, at the end of the nineteenth century, the Tsarist regime did implement land policies designed to settle the Kazakh nomads. These policies included the designation of specific winter and summer pasture areas within the districts granted to the Kazakh *auls* (groups of extended families who share winter or summer pasture areas) (Olcott 1981b:13–14). Olcott (1987:87) states that the most devastating aspect of the late-nineteenth-century Russian land policies levied against the Kazakhs was the 1891 Steppe Statute. In particular, Article 120 stated that land in excess of Kazakh needs was to be given to the Public Land Fund. Ironically, the Kazakh auls were to be settled by designating specific territories for their summer and winter pastures and by restricting the amount of land they were allowed as pasturelands using the same size land grants awarded to Russian immigrants who were peasant farmers. Furthermore, the state could then legitimately usurp Kazakh pastures under Article 120. As early as the end of the nineteenth century, it was clear that Kazakh nomads would be forced to give up land to Russian peasant farmers and the Tsarist state. Furthermore, if mobility was more than just an economic strategy and also one necessary for maintaining long-distance trade and communication routes as well as military alliances, all Russian policies directed toward settling the Kazakhs on land or at least limiting their mobility by "fixing" pasturelands would have the effect of weakening the Kazakhs both politically and militarily. Regardless of whether the Russian Tsarist state understood this consciously, it was certain that its land policies would also limit Kazakh mobility (see Kuznar, this volume, for a similar case regarding Navajo stock reduction in the 1930s, which also curbed long-distance Navajo pastoral mobility).

THE SOVIET PERIOD

After the 1917 revolution, Kazakh history and ethnography became an integral part of Lenin's policy toward the national people (natsional'nost) and ethnic groups. By preserving the national languages, customs, costumes, and folklore of the many minorities now under Soviet rule, Lenin hoped to ameliorate ethnic or nationalistic divisiveness in the new USSR. Ethnography in the service of outlining the geographic territories held by nationalities and ethnic groups and in tracing the history or ethnogenesis of such groups became the aim of Soviet historians and ethnographers. Masanov (1966) points out that during the 1920s and 1930s, ethnographers and historians working in regional and provincial museums drew the line between the bourgeois intellectual tradition of Tsarist Russia and that of the new Soviet ethnologists and historians who followed the historical perspectives of Marxist-Leninist formulations. During the most difficult period of Stalin's policies of forced Kazakh settlement and collectivization in the 1930s (when at least 2 million Kazakhs died as a result of collectivization, imprisonment, and starvation [Olcott 1981a]), a series of regional conferences led by Moscow- and Leningrad-trained ethnographers were held throughout Kazakhstan. The goal of these regional conferences was to introduce the Marxist-Leninist model of historical science into ethnographic and historical research on Kazakhs and other ethnic groups. Soviet historical science was supposed to take on a political agenda in which the "nationalist-bourgeois elements" and "foreign influences" were squelched in favor of a model for society that encouraged "the building or construction of socialism" among all nationalities and ethnic groups.

It was also during this time period that the questions of Kazakh origins and the placement of Kazakh nomadism into a larger evolutionary scheme became a central theoretical focus for Soviet historians and ethnographers. Pastoral nomadism could be seen as a form of patriarchal feudalism. Although there is no direct documentation, I believe the *biis* (clan or aul heads) could be labeled "bourgeois" since the biis could be seen as slave owners or, at the very least, as those who exploited the labor and resources of their poorer kinsmen. If the correct scientific method could be applied using the evolutionary models of historical science, then by implication the former patterns of Kazakh social and political organization could be seen as antiquated and primitive. Such ideas worked in favor of subduing the nomadic population under the banner of socialism.

In a similar vein, I believe many Soviet ethnographers and historians tended to emphasize the elements of pastoral mobility and the lack of agricultural practices among the Kazakh people as a way of justifying Soviet policies of forced collectivization. Even if I am incorrect about the details of how ethnography and history were used in the service of defending state policies, it is clear that those in power did fear the repercussions of Kazakh intellectuals and nationalists. Kazakh ethnography or history in the hands of a discontented populace could indeed have become a powerful weapon against an oppressive state. Those studying Kazakh history and ethnography did not become the "champions" or spokespeople for the implementation of more liberal policies toward the Kazakh pastoral nomads. Clearly, if they had done so, charges of bourgeois nationalism could have been easily leveled against them by the ethnographic cadre from Leningrad and Moscow.

Later ethnographic accounts of the nomadic Kazakhs, such as the important monograph by the Kazakh ethnographer Sergali Tolybekov (1959) in the late 1950s, relied on the pre-revolutionary accounts of Kazakh economy and social organization. In fact, the best ethnography of Kazakh nomadism depended on the Tsarist period accounts that described Kazakh life before the end of the nineteenth century, when Kazakh nomads were able to practice livestock herding without the interference of Tsarist or Soviet land policies. The historical materialism used by Soviet scholars showed an unabashed concern with the description of Kazakh society that could be used to place pastoral nomadism within a particular evolutionary stage and that addressed the question of ethnogenesis, or origins of the Kazakhs. This is particularly evident in the close alliance established between archaeology and ethnography in Soviet historical science after World War II.

Sergey P. Tolstov, a leading historian of Central Asia during this postwar period, attempted to answer the question of Kazakh origins. He proposed that the ethnogenesis of the Kazakh, Karakalpaki, Turkmen, Uzbek, and Kirghiz people took place earlier than the Middle Ages. Rather than link the origins of these Turkic nationalities to the Middle Ages, as most Soviet and pre-revolutionary ethnographers did, he said that their origins could be traced back to the ancient oasis cultures of the Sogdian, Bactrian, Chorezm, Massagetae, Usunei, Eftalitov, and ancient Turks (Masanov 1966:310). Tolstov's conjectures resulted in an increase in physical anthropological, ethnographic, linguistic, and folklore studies carried out on the Turkic groups of Central Asia and Kazakhstan. Archaeologists working on the excavations of graves and burial kurgans of nomadic populations such as the Scythians and Sarmatians from the first millennium BC began to speculate about the direct historical connections between these early Indo-Iranian nomads and contemporary Turkic groups. This theme has been revisited in the post-Soviet period by archaeologists, historians, and ethnographers, as Kazakh intellectuals attempt to establish scientifically the antiquity of Kazakh ethnogenesis. Clearly, a newly formed independent state, which prides itself on its Kazakh origins, must now use ethnography and history in the service of nationalism.

KAZAKH HISTORY AND ETHNOGRAPHY IN SERVICE OF THE STATE

It has become almost a scholarly axiom these days, especially in light of the influence of postmodernism and critical theory on the humanities and social sciences in the West, to state that all knowledge, even so-called scientifically objective knowledge, is shaped by the political and social context of the day (e.g., in archaeology; see Wylie 2000). It should be obvious from my brief outline of Kazakh history and ethnography during the Tsarist and Soviet periods that this chapter has only demonstrated the obvious: that history and ethnography can be used as a form of intellectual propaganda to justify, rationalize, and even disguise the agenda of national states (Trigger 1984). What is less clear is the extent to which the view of the "cultural other," whether a so-called primitive or an "exotic" in the eyes of the investigator, has been accurately described and documented by a historian or ethnographer. While intellectual propaganda and fashion serve up what the state or a group of intellectuals deem is the truth (and nothing but the truth), we return again to the most basic questions: Who are pastoral nomads, and can their way of life and their history be portrayed accurately?

THE DIVIDE BETWEEN SOVIET AND WESTERN STUDIES OF PASTORAL NOMADS

My motivations for writing this chapter are twofold. First, I wish to assert that even under the most pernicious regimes of any state or empire, informed and accurate research can and should be conducted on indigenous people, whether nomads, farmers, or foragers. In many ways Masanov's (1966) work and that of many other Kazakh ethnographers and historians, although seeming to serve the purpose of the state, also accurately describes the historiography of Kazakh ethnology. A Western scholar, not used to scholarship conducted under authoritarian regimes, must learn to read between the lines, where the truths are not hidden from discerning readers or fieldworkers (for a detailed discussion of this dilemma, see De Soto and Dudwick 2000). Second, the Eurasian pastoral nomads, especially in Kazakhstan, after having withstood the hardships of at least 200 years or more of ill-advised state policies, deserve an accurate accounting of their way of life, something Soviet ethnographers and historians recognized.

The problem of researching Eurasian nomadic steppe populations exists more in the area of theory than in the actual methods used by Tsarist and Soviet researchers.

The Tsarist researchers asked many of the same questions their European and American contemporaries did when they began to document the nature and character of indigenous societies. What were these peoples' beliefs, religious practices, folklore, economy, languages, and political and social organization? The Victorian notion of viewing primitive peoples as "evolutionary throwbacks" or primitive customs as "survivals" of earlier stages of humankind was not restricted to the European and Western ethnologists and ethnographers of the time but was also an interest of Tsarist period researchers. In Western anthropology, the study of craniology and physical measurements of indigenous groups provided the basis for biological determinism and the origins of scientific racism. In the nineteenth century, Valihanov hoped Kazakhs' physical measurements and craniology might be used to determine the Mongol-Turkic origins of these pastoral nomads. It can be said that perhaps Valihanov was also participating in a form of scientific racism; however, he was attempting to use craniology as a means of tracing ethnic origins, similar to the nineteenth-century racialism of the Victorian anthropologist.

Evolutionary models, whether based on Tylor's stages of socio-cultural evolution or Marx and Engels's historical materialism, need not be inherently wrong. In fact, concepts of socio-cultural evolution and historical materialism can be found in many contemporary anthropological studies. While dismissed by some contemporary anthropologists as examples of "crass materialism," the fact is that the "exotic, the primitive, and the savage" continues to inform the comparative ethnographic research carried out by contemporary anthropologists. We could say that the study of the other is by definition an exercise that attempts to marginalize the mainstream and glamorize the marginal. What might have been most pernicious about labeling Kazakh nomadism a form of patriarchal feudalism is the way this category was used to describe Kazakh nomadic society as inherently class-based and thus antithetical to contemporary Soviet socialism of the twentieth century. The truth is that Western anthropologists also suffer from the disease of evolutionary reckoning when it comes to the subject of pastoral nomadism. Who are the nomads? How can we define a true nomad, when most migratory livestock herders also participate in foraging, agriculture, trading, and raiding? How do pastoralists accumulate property if private ownership takes place in the form of owning animals and not land? Are pastoralists fiercely independent and egalitarian as Western anthropologists tend to portray them, or are they as Soviet ethnographers portrayed them—militaristic, hierarchical, even slave owners? If one considers the nature of these questions, there seems to be an evolutionary agenda in place. Are pastoral nomads really "primitives," or are they the "survivals" of an original egalitarian society or a clearly ranked feudal society?

As I have shown through my ethnographic studies of Greek pastoralists, important ecological-economic aspects of the pastoral way of life such as mobility have important ideological, social, and political dimensions. This is equally true when considering the function of mobility among the Eurasian steppe nomads. As Lattimore (1940:76–77) recognized more than seventy years ago in discussing the issue of wealth versus mobility in the nomadic pastoral societies of Mongolia: "In every possible combination the emphasis had to be either on wealth or on mobility. There was no ideal balance equally suitable to every region. The extreme of mobility was the mounted warrior; the extreme of invulnerability or inaccessibility was the camel rider of the Gobi; the extreme of wealth was the patch of intensively cultivated land in a watered valley isolated in the general expanse of steppe and exploited by nomad overlords."

Mobility was not merely determined by ecology and economics; for the Eurasian nomad it was a strategy selected for political, social, and ideological purposes. A warrior could use the threat of rapid military deployment and mobility to subdue his agricultural subjects or opponents. Mobility across the steppe was vital for trade and communication networks among dispersed peoples. And the inherent instability of agricultural and pastoral regimes in Mongolia resulted in fluctuations and cycles that vacillated between degrees and types of mobility.

KAZAKH HISTORY AND ETHNOGRAPHY IN THE SERVICE OF ARCHAEOLOGICAL STUDIES OF NOMADIC PEOPLES OF THE EURASIAN STEPPE

In the Republic of Kazakhstan there is a long history of scholarship known as paleo-ethnography, in which studies of material culture have been used as comparative databases for interpreting archaeological remains. In some ways the divide among ethnography, history, and archaeology has been less contentious than it has in the West, where ethnoarchaeology has been a long-debated subject. Paleo-ethnography in the Soviet tradition has been useful for our studies of Saka-Scythian materials in these areas: (1) outlining the possible strategies of seasonal movement between mountain and steppe areas; (2) a clear understanding of climatic conditions such as the phenomenon of *zhudt,* or the black ice that covers pasturelands and leads to starvation of animal herds, which must have devastated herds periodically in our region of Semirech'ye during the first millennium BC; (3) the description of Kazakh populations in oasis regions that participated in the cultivation of millet and wheat; (4) the descriptions of Kazakh housing and architecture that could have parallels in some of the Iron Age sites we have excavated (aboveground house floors with center ridges) and large, rectangular stone-walled or semi-subterranean *zimovki* (winter houses); (5) the description of kinship and horde organization of historical and contemporary Kazakhs, while separated in time, which suggests interlocking social organization of politico-military confederacies with household (aul) and lineage segments for earlier steppe confederacies.

It seems apparent even from this brief list that a direct relationship will always be drawn between the historical populations of Kazakh nomadic pastoralists and the archaeological evidence of earlier cultures of nomadic pastoralists. If this, then, is to be the "ghost in the machine," what needs to be made more explicit for ethnographers, historians, and archaeologists working on the issue of steppe nomadic pastoralism is the need to discard our evolutionary notions of nomadic pastoral adaptations and adopt models, methods, and theories that simply consider the multiple dimensions of the pastoral way of life.

CONCLUSIONS

For the purpose of this chapter, I have focused on the multi-resource nature of pastoralism and on pastoral mobility itself. Also, I have stressed the perceptions that nation-states and imperial powers might adopt regarding local nomadic pastoralists within their territories. Clearly, the Kutpatshari and Koutsovlach, the ethnic groups of Aroumani transhumant pastoralists in northern Greece, have been marginalized by the Greek nation-state, yet in other arenas their traditional identity gives them the right to hold local elections in the summer in their traditional "summer villages"; further, in many senses their traditional mobile way of live is celebrated, not denigrated (Chang 1993). Today the Kazakhs promote their own identity, language, traditions, and history in an independent post-Soviet state that they see as their "homeland." The herdsman is both a symbol of the fierce and independent individual surviving against all natural odds in a timeless primordial past and the landless victim of Soviet subjugation. The image of the pastoral nomad in post-Soviet times now permits the nomads to fulfill their natural destiny by reclaiming the promised homeland in a modern nation-state. Western and local scholars who study pastoral nomadism, or agropastoralism, are placed in the throes of a larger argument of "the invention of tradition" (Hobsbawm 1983). If indeed, as we have proven, the ancient Iron Age nomadic confederacies of Semirech'ye (Saka, Wusun, and Yuezhi) contained farmers as well as herders, does that heighten or undermine the official historical narrative of the primordial nomadic past of the modern nation-state?[8]

The study of pastoralism is beset with these two central issues: (1) that of mobility (how long, how far, and under what circumstances) and (2) that of identity and the level of social formation within larger social arenas (ethnic group, tribe, clan, state). These are the two central themes that have perplexed and vexed me throughout my professional career. Yet I continue to believe that insufficient notice is given to the rhythms and nuances of the everyday lives of livestock owners, the care with which they must keep, maintain, pasture, water, and feed their flocks and herds. This quotidian nature of the everyday lives of those who practice livestock husbandry today or who practiced animal husbandry in the deep historical past suggests that we have chosen to "imagine" the pastoral nomad using evolutionary blinders in the service of our own ideological and political agendas. The ethnographic and archaeological accounts of pastoralists continue to be informed by these analytical impulses to pigeonhole and classify pastoral groups in terms of indexes of mobility and evolutionary distancing from the "civilized" world. While I have come directly out of an intellectual tradition that focused on the ecological and economic nature of all pastoral adaptations, my own research has demonstrated that, of course, there are social and ideological factors that should be examined in all pastoral societies. It is not just ethnographers and political elites who use nomadic narratives to reinforce stereotypes; at times the pastoralists themselves also project a self-awareness and romantic image by which they fashion and negotiate their own tropes of identity in larger national arenas, so as to occupy a place in the dominant national discourse. Perhaps it is to their own fashioning of the pastoral rhetoric that we must now turn to view the world of ethnic identity from the inside and not as outside bystanders. The fact that academics, politicians, elites, and others have chosen to use pastoral narratives in service of ideological and political agendas merely suggests that our progress in the area of pastoral studies has been slow and that the future requires another re-visioning of the pastoral nomadic narrative.

ACKNOWLEDGMENTS

The research for this chapter was conducted in 1997 and 1998 and was supported by an NEH fellowship and a sabbatical leave from Sweet Briar College. The Archaeology Program of the National Science Foundation, Grant SBR–9603661, has funded the long-term excavations and surveys at Talgar. During this time I have been formally associated with the Institute of Archaeology, the Kazakh Academy of Sciences, in Almaty. I thank these institutions and my Kazakhstani hosts for their generosity and support of my scholarly activities. Many friends and colleagues in Kazakhstan, the United States, Germany, and France have listened to my sometimes misguided notions about Eurasian pastoral nomadism. They have listened, critically evaluated, and at times disagreed with my ideas. My intellectual debts include Martha Brill Olcott's thorough scholarship on Kazakh history, Edige I. Masanov's fascinating essay on the historiography of Kazakh history and ethnology, George Stocking's account of Victorian anthropology, and Rebecca Beardmore's insightful study of the role of archaeological research in

Kazakhstan state building in the early 2000s. Nick Kardulias and Mark Shutes provided invaluable advice in revising this chapter. I deeply regret that Mark Shutes did not live to see the fruits of his work.

NOTES

1. People from the Republic of Kazakhstan identify themselves according to their ethnic identity such as Kazakh, Russian, Ukrainian, Tartar, and so on. In this chapter I refer to any citizen of Kazakhstan, regardless of ethnicity, as Kazakhstani. See Suny (2001:883) for a detailed discussion as to why President Nazarbaev's nationality policy in the mid-1990s attempted to stress Eurasian, Kazakhstani, and Kazakh ethnic identities in service of state building and nationhood.

2. The Russians and Soviets studied many of the nomadic groups within and adjacent to their national borders. For the purposes of this chapter I focus on the research done on the Kazakh-Kirghiz, a Turkic group of pastoralists residing in the present-day territory of the independent state of Kazakhstan. Although the Kazakh-Kirghiz did not emerge as an ethnic group until the fifteenth century, historical and archaeological research on earlier Eurasian steppe groups is often included in the Soviet historiography of contemporary Eurasian steppe nomads such as the Kazakh-Kirghiz, for reasons discussed later.

3. The role ethnogenesis played in promoting the search for the "origins" of ethnic groups and thus promoting questions of nationality and territory is explained at considerable length by Kohl (1998:231–32).

4. Bromley and Kozlov (1989:431) trace the historical use of the term *narodnost* in the late 1940s in Soviet literature, which describes the moment in which an ethnic group had survived the disintegration of its tribal identity but had not become a nationality (capitalist relations are implied) under a nation-state. This is fascinating, if obscure, because it indicates the degree to which ethnographers attempted to link the process of ethnogenesis with evolutionary process and classical Marxism.

5. Beardmore (2007) discusses how later Soviet policy in the 1980s attempted on the one hand to encourage a folk identity of Kazakhs but on the other hand attempted to build a concept of nationality by which the Republic of Kazakhstan was a political administrative unit within the larger Soviet Union. By default, because Kazakhstan was a political administrative unit of over 100 different ethnic designations, this posed certain contradictions in Soviet policy and later in post-Soviet policy after independence.

6. Schatz (2000:496) describes informants who report that they were instructed by their director at the Institute of History and Ethnography to "find the roots of Kazakh statehood in the Sak period (the first millennium BCE)."

7. Shnirelman (2006) discusses how Ossetian leaders in the 1990s adopted the name Alania, obviously derived from the Alans.

8. Baipakov (2008), our collaborating Kazakh archaeologist, has written convincingly about the importance of Iron Age agriculture as the foundation for later medieval urban development along the Silk Route in Semirech'ye. His literature review demonstrates that an agropastoral origin for the Saka was previously posited during Soviet times, but archaeologists lacked the methodological means for recovering seeds and other remains that would have proved the existence of Early Iron Age agriculture.

REFERENCES

Akishev, Kemal A. 1978. *Kurgan Issyk: Iskusstvo Sakov Kazakstana* [Issyk Mound: The Art Saka in Kazakhstan]. Moscow: Iskusstvo.

Baipakov, Karl M. 2008. "Gorod I Step' v Drevnosti: Osledost' i Zemledelniye i Sakov I Usunei Zhetisu" [City and Steppe in Ancient Times: Settlement and Farming of the Saka and Wusun of Semirech'ye]. *Izvestia* (National Academy of Science, Republic of Kazakhstan), Seriya 1 (254): 3–25.

Barfield, Thomas J. 1993. *The Nomadic Alternative.* Englewood Cliffs, NJ: Prentice-Hall.

Barth, Frederik. 1986 [1961]. *Nomads of South Persia.* Prospect Heights, IL: Waveland.

Beardmore, Rebecca. 2007. "The Role of Archaeology in the Development of the Concept of Nation in Kazakhstan: A Case Study on the Engagement of Archaeological Objects in Semirechie, South-east Kazakhstan." *Acta Eurasica* 4 (38): 91–114.

Bromley, Julian, and Viktor Kozlov. 1989. "The Theory of Ethnos and Ethnic Processes in Soviet Social Sciences." *Comparative Studies in Society and History* 31 (3): 425–38. http://dx.doi.org/10.1017/S001041750001598X.

Bromley, Yu V. 1979. "Towards Typology of Ethnic Processes." *British Journal of Sociology* 30 (3): 341–48. http://dx.doi.org/10.2307/589912.

Chard, Chester S. 1963. "Soviet Scholarship on the Prehistory of Asiatic Russia." *Slavic Review* 22 (3): 538–46. http://dx.doi.org/10.2307/2492498.

Chang, Claudia. 1993. "Pastoral Transhumance in the Southern Balkans as a Social Ideology: Ethnoarchaeological Research in Northern Greece." *American Anthropologist* 95 (3): 687–703. http://dx.doi.org/10.1525/aa.1993.95.3.02a00080.

Chang, Claudia. 2000. "The Material Culture and Settlement History of Agro-Pastoralism in the Koinotis of Dhidhima: An Ethnoarchaeological Perspective." In *Contingent Countryside: Settlement, Economy, and Land Use in the Southern Argolid since 1700*, ed. Susan B. Sutton, 125–40. Stanford: Stanford University Press.

Chang, Claudia, Norbert Benecke, Fedor P. Grigoriev, Arlene M. Rosen, and Perry A. Tourtellotte. 2003. "Iron Age Society and Chronology in South-east Kazakhstan." *Antiquity* 77 (296): 298–312.

Chang, Claudia, and Fedor P. Grigoriev. 1999. "Tuzusai, a Late Iron Age Settlement in Southeastern Kazakhstan: A Report of the 1994–1996 Field Seasons." *Eurasien Antiqua, Band* 5: 319–410.

Chang, Claudia, and Perry A. Tourtellotte. 1993. "The Ethnoarchaeological Survey of Pastoral Transhumance Sites." *Journal of Field Archaeology* 20 (3): 249–64. http://dx.doi.org/10.1179/009346993791549192.

Chang, Claudia, Perry A. Tourtellotte, Karl M. Baipakov, and Fedor P. Grigoriev. 1999. "The Kazakh-American Talgar Project Archaeological Surveys in 1997 and 1998 in the Talgar Region." *Izvestia, Obshestvennye Nauk* 1 (219): 168–84.

De Soto, Hermine G., and Nora Dudwick, eds. 2000. *Fieldwork Dilemmas: Anthropologists in Postsocialist States*. Madison: University of Wisconsin Press.

Dyson-Hudson, Neville. 1972. "The Study of Nomads." In *Perspectives on Nomadism*, ed. William Irons and Neville Dyson-Hudson, 2–29. Leiden: E. J. Brill.

Gellner, Ernest. 1984. "Foreword." In *Nomads and the Outside World*, by Anatoly M. Khazanov, ix–xxv. Cambridge: Cambridge University Press.

Hall, Thomas D. 2000. "Frontiers, Ethnogenesis, and World-Systems: Rethinking the Theories." In *A World-Systems Reader: New Perspectives on Gender, Urbanism, Indigenous Peoples, and Ecology*, ed. Thomas D. Hall, 237–70. Lanham, MD: Rowman and Littlefield.

Hobsbawm, Eric. 1983. "Inventing Traditions." In *The Invention of Tradition*, ed. Eric Hobsbawm and Terence Ranger, 1–14. Cambridge: Cambridge University Press.

Khazanov, Anatoly M. 1984. *Nomads and the Outside World*. Cambridge: Cambridge University Press.

Khazanov, Anatoly M. 1996. *After the USSR Collapsed: Ethnic Relations and Political Process in the Commonwealth of Independent States*. Madison: University of Wisconsin Press.

Kohl, Philip L. 1998. "Nationalism and Archaeology: On the Constructions of Nations and the Reconstructions of the Remote Past." *Annual Review of Anthropology* 27 (1): 223–46. http://dx.doi.org/10.1146/annurev.anthro.27.1.223.

Lattimore, Owen. 1940. *Inner Asian Frontiers of China*. London: Oxford University Press.

Liu, Morgan Y. 2011. "Central Asia in the Post–Cold War World." *Annual Review of Anthropology* 40: 115–31.

Masanov, Edige I. 1966. *Ocherk Istorii Ethnographischeskogo Isycheniya Kasakhskogo Naroda v SSSR* [Essay on the History of Ethnographic Research on the Kazakh People in the USSR]. Alma-Ata: Nauka KazSSR.

Olcott, Martha B. 1981a. "The Collectivization Drive in Kazakhstan." *Russian Review* 40 (2): 122–42. http://dx.doi.org/10.2307/129204.

Olcott, Martha B. 1981b. "The Settlement of Kazakh Nomads." *Nomadic Peoples* 8: 12–23.

Olcott, Martha B. 1987. *The Kazakhs*. Stanford: Hoover Institution Press.

Salzman, Philip C. 2004. *Pastoralists: Equality, Hierarchy and the State*. Boulder: Westview.

Salzman, Philip C., ed. 1980. *When Nomads Settle: Processes of Sedentarization as Adaptation and Response*. New York: J. F. Bergin.

Schatz, Edward. 2000. "The Politics of Multiple Identities: Lineage and Ethnicity in Kazakhstan." *Europe-Asia Studies* 52 (3): 489–506. http://dx.doi.org/10.1080/713663070.

Shnirelman, Victor. 1995. "From Internationalism to Nationalism: Forgotten Pages of Soviet Archaeology in the 1930s and 1940s." In *Nationalism, Politics and the Practice of Archaeology*, ed. Philip L. Kohl and Clare Fawcett, 120–38. Cambridge: Cambridge University Press.

Shnirelman, Victor. 2006. "The Politics of a Name: Between Consolidation and Separation in the Northern Caucasus." *Acta Slavica Iaponica* 23: 37–73.

Stocking, George F. 1987. *Victorian Anthropology*. New York: Free Press.

Suny, Robert G. 2001. "Constructing Primordialism: Old Histories for New Nations." *Journal of Modern History* 73 (4): 862–96. http://dx.doi.org/10.1086/340148.

Tolybekov, Sergali E. 1959. *Obshchestvenno-Ekonomicheskiy Stroi Kazakhov v XVII–XIX Vekakh* [Socioeconomic Structure of the Kazakhs in the Seventeenth to Nineteenth Centuries]. Alma-Ata, Kazakhstan: Izdatelstvo AN KAzSSR.

Trigger, Bruce G. 1984. "Alternative Archaeologies: Nationalist, Colonialist, Imperialist." *Man* 19 (3): 355–70. http://dx.doi.org/10.2307/2802176.

Wylie, Alison. 2000. "Questions of Evidence, Legitimacy and the (Dis)Union of Science." *American Antiquity* 65 (2): 227–38. http://dx.doi.org/10.2307/2694057.

Yessenova, Saulesh. 2005. "'Routes and Roots' of Kazakh Identity: Urban Migration in Postsocialist Kazakhstan." *Russian Review* 64 (4): 661–79.

3

The Ecology of Inner Asian Pastoral Nomadism

Nikolay N. Kradin

The world of nomads had always been terra incognita for residents of settled agricultural civilizations, both scaring and intriguing them. The centaur, itself a mysterious creature, was a symbol of the steppe world. It is no coincidence that the former crusader William of Rubruck, who had seen much in his lifetime, included in the first chapter of his writings a description of a trip to visit the ruler of the Mongolian Empire (Rockhill 1900:52): "When I found myself among them it seemed to me of a truth that I had been transported into another century" (in translation by Jackson in van Ruysbroeck [1990:71]: "I really felt as if I were entering some other world"). In fact, the economy, culture, mode of life, and social-political organization of the inhabitants of steppe regions have all differed from the customary way of life of plowmen and townspeople.

The nomads made their life in the arid steppes and semi-deserts, where agriculture was practically impossible. However, they were able to breed grass-grazing animals, which proved an effective mode of existence in the native zones considered here. The nomadic lifestyle originated in Inner Asia approximately at the turn of the second to first millennia BC. However, this chapter emphasizes the period of the imperial confederations and empires of nomads (200 BC–AD 1600) because, first, the number of narrative and archaeological sources for the preceding period is greatly limited. Second, with the conquest of Mongolia by the Qing Empire and Russian campaigns to Siberia, the epoch of colonialism had started. From the very start of the modernization period, it was apparent that the nomads were incapable

DOI: 10.5876/9781607323433.c003

of competing with the industrial economy. The development of repeating (magazine) firearms and modern artillery gradually ended the military power of nomadic peoples. Later, attempts at the sedenterization of nomads, creation of the cattle-breeding market, and conversion of pastures to arable lands combined to upset the ecological balance of the region; this alteration is a subject for another study.

The term *Inner Asia* can be used in several ways. In the broadest sense, it indicates a region of Asia to the east of the Urals that excludes the settled agricultural civilizations, meaning without the Middle East, India, China, and the South Pacific (Sinor 1990:6). Sometimes the term *Inner Asia* is used as a synonym for Central Asia (e.g., it is characteristic of the Soviet geographic school). In a restricted sense, Inner Asia is the land to the east of Central Asia (Di Cosmo 2002:13) and includes territories of three modern countries: Russia (southern Siberia and the Trans-Baikal region), Mongolia, and China (Sinkiang, Inner Mongolia, northern Shensi, northern Shansi, Manchuria).

In this chapter I use the term *Inner Asia* in this last sense. It is important because the nomadic ecology of Western Eurasia (the steppes of Eastern Europe, the Middle East, and Central Asia) differs from that of Eastern Eurasia (Inner Asia) (see figure 3.1). In Western Eurasia, steppe prairies are only characteristic of the Kazakh and South Russian steppes. Most of the Middle Eastern territories, Afghanistan, and Central Asia (without Kazakhstan) are characterized by semi-deserts. Here, in the river valleys and near great water sources, there were oases suitable for agricultural occupation. As opposed to the inhabitants of the Mongolian steppes, the nomads and farmers in this region coexisted in close proximity. In Eastern Eurasia, the vast steppe prairies prevailed. To the south, China was isolated by impenetrable deserts. The spacious extent of the steppes enabled the establishment of the great nomadic empires (Fletcher 1986; Barfield 1992; Golden 2001).

The geographic and ecological boundaries between the north (steppe region) and south (China) have coincided with the cultural frontier (Lattimore 1940). To the Chinese, the north is known as the region of the nomads. The nomads are considered the powers of darkness (Yin) and are believed to have the hearts of wild beasts (Di Cosmo 2002). In the Chinese astrological system, the planet Mercury, associated with the winter and war, was assigned to the nomads. The frontier had become a base for cultural and political mobilization of the south and north, resulting in the establishment of the bipolar system between China and steppe nomads in Inner Asia.

CLIMATE AND PASTORAL ECONOMY

The climate of Inner Asia horrified the residents of the agricultural states. The cold winters and strong winds, as well as the heat and droughts of summer, rendered the

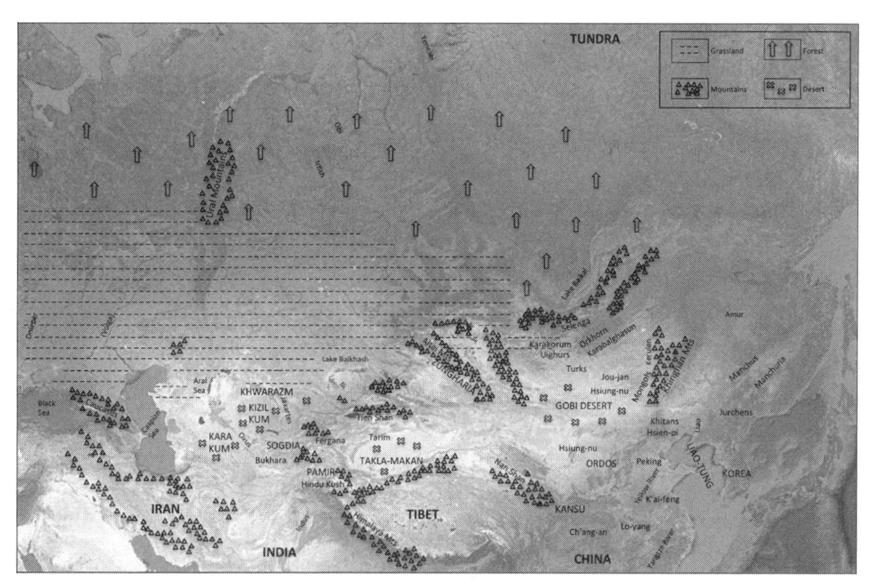

FIGURE 3.1. Map of Central Asia, with environmental zones and cultural groups indicated.

nomadic steppes practically unsuitable for agriculture. In Mongolia, only a small percentage of the territory was suitable for husbandry. Pastoral nomadism was shaped strongly by the cataclysms of nature and climate. Snowstorms (Mong. *dzut*), droughts, and epidemics were perpetual misfortunes for nomads. The Xiongnu experienced climate-related hardship at least every ten years (Kradin 2002:34). There are records of murrains among the Turks (Bichurin 1950:236), Uighurs (ibid.:334), and Mongols (Bichurin 1829:277, 279, 301; Munkuev 1977:417, 427–28).

The data for the colonial period yield more information. In Mongolia, about 20–30 percent of domestic animals perished (Maisky 1921:118). In difficult years, the Kazakhs lost more than 50 percent of their livestock (Tolybekov 1971:79–80, 542–43). In Tuva, about 15–17 percent and up to 50 percent of cattle herds were lost in years of snowstorms and epizootic outbreaks, respectively (Dulov 1956:68–70; Vainshtein 1972:52–53). Even in modern Mongolia at the turn of the twenty-first century, about 30 percent of livestock (approximately 11 million animals) were lost as a result of dzut (Legrand 2001; Janzen and Enkhtuvshin 2008). As a result, both poor and rich nomads could lose their means of subsistence in a single night.

Researchers have time and again suspected that major losses of cattle were characterized by particular cycles (Shakhmatov 1961; Krupnik 1989:128–40; Kradin 1992:54–55; Masanov 1995:100). This periodicity could possibly be related to the eleven-year cycles of solar variability and other natural cycles (Maksimov 1989). If so, one can attribute the mass cattle plagues among the nomads to the recurrence of extreme cold, snowstorms, droughts, and similar phenomena every ten–twelve years. Typically, about half of the entire herd would perish each cycle. Restoring the herd would have taken ten to thirteen years. On this basis, one can theoretically suppose that the number of livestock following the disruption of an ecological zone should have oscillated cyclically around this particular level. That number increased as a result of favorable conditions or decreased because of catastrophic natural factors. Meanwhile, the amount of livestock increased at a faster rate than the population. Therefore, animals had exhausted the grass before the human population could develop correspondingly. Thus, the number of domestic animals and the number of cattle breeders changed in accordance with a complex cyclical model similar to the Lotka-Volterra "predator-prey" model.

Undoubtedly, the economic system was fragile and dependent on the environment. To counteract natural crises, the nomads developed an efficient system of mutual aid. In a situation of livestock loss, fellow tribespeople would provide an individual with one or two heads of cattle, thus restoring a means of subsistence. But the aided group or person(s) was bound to reciprocate the same service to the benefactors if necessary.

Nomads could also lose livestock to raids by foreign tribes. These forays were a favorite recreation of nomads. The *Secret History of the Mongols* describes different stories related to the raiding ventures of the future Chinggis Khan. So in §§90–93 of this chronicle (de Rachewiltz 2004), a touching story of young Temujin's acquaintance with his future closest associate Bo'orču, who helped him strip those who stole his horses, is reported. Another section (§128) tells about the breach between Chinggis Khan and Jamuqa following the murder of Jamuqa's brother, Tačar, after he had stolen a herd of horses from one of Chinggis Khan's camps. One can find many other such examples.

Another important factor involving household welfare among the pastoral nomads was the marriage dowry. If a family had many sons, it stands to reason that a number of animals would have been lost as marriage gifts. If the number of sons in the family was very large, this could have deferred their marriage (Irons 1975:164–67).

The specificity of cattle breeding suggests a dispersed life. Concentration of great herds of animals in one place resulted in overgrazing, excessive grass trampling, and a higher risk of contagion by infectious diseases among the animals. It was impossible to accumulate cattle ad infinitum because their maximal number was determined by the steppe landscape's limited capability. As such, regardless of the stock keeper's nobility, his entire herd could be killed by dzut, drought, or disease. Therefore, it became more profitable either to apportion the livestock among poor kinsmen for pasture or to distribute them as gifts to raise one's social status. Thus, pastoral nomadism could not provide the stable food surplus needed to support large groups of people who did not participate in food production—the ruling aristocracy, officialdom, soldiers, priests, and similar groups.

In this connection, it is significant that only agriculture has provided the firm economic basis for the emergence of statehood. This thesis was examined by Andrey Korotayev (1991), who used cross-cultural methods. His study, based on George Murdock's (1981) *Atlas of World Cultures*, showed that among societies reaching statehood, none had the appropriate type of economy, and there were no pastoral ethnohistorical societies. Later, a close correlation was established among the type of economy (agriculture), population density, size of community, and the development of internal stratification and political centralization (Korotayev, Malkov, and Khalturina 2006:112–13; Kradin 2006).

What caused this correlation? The essential factor that influenced a society's adaptive capabilities in the course of the evolutionary process was its ability to store considerable stocks of food. The availability of storage technology allowed societies to overcome periodic food shortages and was of great importance in the creation of stratification and permanent authorities (see Claessen 1989:233). Cross-cultural studies show that if societies with early agriculture were able to sustain food stores for a long time, then they developed internal stratification and a political system of governance at a faster rate (Korotayev 1991:166–78, tables 18–28). This conclusion seems well-grounded. From the viewpoint of systems theory, the rise in the vital capacity of any system depends on the society's maximum ability to derive energy from the environment, to keep it and consume it with maximum utility (Odum and Odum 1976).

One popular explanation for the origins of nomadic empires focused on global climate change. The idea that increased aridity on the steppes was a catalyst for the invasions of nomads was suggested at the end of the nineteenth century to the early twentieth century by Ellsworth Huntington (1915) and Mikhail Bogolepov (1908). Later, Arnold Toynbee (1934) and Grigory Grumm-Grzhimailo (1926) were ardent supporters of this view. Lev Gumilev stressed the role of increased moisture rather than aridity, stating that hungry nomads on lean horses would have had difficulty waging devastating wars against agricultural civilizations. Beginning in the mid-1960s, Gumilev (1993) developed his ideas into a coherent theory. However, despite the array of paleo-climatic data Gumilev presents, the information does not conclusively demonstrate a strict correlation between cycles of humidity-aridity and periods of nomadic prosperity decline. While his reconstructions from works spanning a period of twenty years (Gumilev 1993:237–319) suggest that wetness was a driving factor, more recent data (Ivanov and Vasilyev 1995:tables 24, 25) indicate that nomadic empires could emerge in both humid and arid periods. The only important fact that should be recognized is that the existence of the two largest nomadic empires (Turkic khaganate and Mongolian Ulus) fell within the period of maximum humidity on the steppes of Inner Asia (Dinesman et al. 1989:204–5).

The other point of view related the genesis of nomadic empires to a global cooling. This approach suggests that in the period 1175–1260, a great drop in temperature occurred in Mongolia and other regions of the world. Jenkins (1974) suggests that these conditions could have spurred the unification of the Mongols. However, the actual historical process was subject to more complicated fluctuations. Comparison of the temperature rhythm curves with the economic data and political history of agrarian civilizations shows a lack of strict correlations. The prosperity of certain civilizations fell in periods of maximum warming, while others rose during periods of cooling (Turchin and Hall 2003:50–52). In addition, climate scientists do not always agree on the dating of periods of cooling-warming. According to established opinion, the thirteenth century, the time of Chinggis Khan's imperial foundation and the rule by Mongols in Eurasia, was a period of significant drop in temperature (Monin and Shishkov 1979; Yasamanov 1985). However, according to other data, the period between 1120 and 1280 witnessed a medieval maximum in terms of average annual temperature and solar activity (Turchin and Hall 2003:50).

In an effort to identify a relationship between climatic variations and activity of nomads, Chinese researchers systematically compiled information from the annals of natural disasters, migrations and raids of nomads, and changes of conditional boundaries between the steppe region and China from 100 BC through the beginning of the eighteenth century (Fang and Liu 1992). Analysis of their tables shows a clear correlation between a drop in the average annual temperatures by one degree and an increase in the number of floods, sandstorms, and other natural disasters on the one hand and an increase in migrations and raids by nomads on the other hand. However, it is not known what initiated military activity among nomads. This activity could have been caused by both deterioration of the pastoralists' living conditions and a reduction in deliveries of craft and agricultural products because of a crisis in China. These reasons could also act in parallel.

The other circumstance, not revealed by Fang and Liu, seems more important. The start of military activity by nomads did not become a cause for establishing the great nomadic empires. To the contrary: during periods of cooling, the nomads migrated closer to Chinese territory and established on its northern territory either small buffer states in the fourth and fifth centuries or great empires (T'opa Bei, Liao, Jurchen) with hybrid agricultural/stock-raising economies. It can also be assumed that the medieval Manchoo expulsion southward was dictated by the fact that the heart of the Chinese world-system from the tenth century onward gradually shifted to South China. Since this shift resulted in the displacement of commodity flows southward, nomads of the Mongolian steppes and their Manchoo neighbors (Khitan, Jurchen) were obliged to establish states in the territory of North China (Tabak 1996).

The time of the typical medieval nomadic empires (Turks, Uigurs) coincided with a period of warming (AD 550–1100). It is possible that data about ancient natural disasters are incomplete, but it can be expected that the periods in which the Xiongnu and Xianbei Empires existed are also not related to climatic stresses. However, many Chinese researchers believe the genesis of the Mongolian Empire occurred during a cold snap. But as a whole, large-scale variations in the average annual temperature did not exert great effects on the dynamics of nomadic empires and preindustrial macroeconomic trends. The phases of growth and decline of the great world civilizations (Frank and Gills 1993) are not directly correlated with global natural climatic variations (Turchin and Hall 2003). By all appearances, the finer cause-and-effect relationship involved the interaction of various factors. For this reason, when interpreting particular political events, it is more accurate to take into account not only large-scale climatic variations but also local cataclysms—droughts, floods, epizootic outbreaks, earthquakes, and similar natural disasters.

FIVE KINDS OF LIVESTOCK

The nomads' mode of life and culture did not change during preindustrial times. The famous treatise of the Chinese historian Ssu-ma Ch'ien, *Shih chi* (*Historical Records*), describes the composition of cattle herds characteristic of Xiongnu: "Most of their domestic animals are horses, cows, sheep, and they also have rare animals such as camels, donkeys, mules, hinnies and other equines known as *t'ao-t'u* and *tien-hsi*. They move about according to the availability of water and pasture, have no walled towns or fixed residences, nor any agricultural activities, but each of them has a portion of land" (quoted in Bichurin 1950:39–40; Watson 1961:129; Taskin 1968:34). A thousand years later, Mongols had the same herd composition: "The Tartars are quite rich in animals: camels, cattle, sheep, goats, and they have so many horses and mares that we did not believe there were that many in all the world, but they have few pigs or other animals" (Plano Carpini 1996:41; see also Lin and Munkuev 1960:137).

Around 100 years later the situation had not changed (Przevalsky 1875:141; Maisky 1921:33–35; Pevtsov 1951:112–13; Radloff 1989:130, 153–62, 168, 260, 335). Even the modern nomads of Inner Asia identify five major kinds of domestic animals. Mongols call this phenomenon *tavan khoshuu mal*, five kinds of livestock. Sheep constituted the most numerous species in nomads' herds (50–60%). Horses and cattle each comprised about 15–20 percent of the herd. The remaining portion was made up of goats and camels. It is significant that the nomads of Inner Asia had Bactrian camels rather than dromedaries (Khazanov 1975, 1994; Markov 1976; Cribb 1991; Dinesman and Bold 1992; Kradin 1992, 2002; Ivanov and Vasilyev 1995; Masanov 1995; Tortika and Mikheev 2001).

Of all the species, the horse was of major military and economic importance. It was no coincidence that in places where so-called horseman forces became widespread (Eurasia and North Africa; the camel fulfilled the role of the horse among Afro-Asian groups), the nomads experienced a significant increase in mobility compared with their settled neighbors and played a major role in the military and political history of preindustrial civilizations (Khazanov 1990:6; Pershits 1994:154–55, 161–63).

Mongolian horses were small but robust and adapted well to severe environmental conditions. They were used for riding and transporting goods. Reins were made of mane hair. Mongols invented a special technology of fattening horses, which made the animals more durable and allowed them to survive without food and water for eight to nine days (Bold 2001:38). The Mongolian horse has been known to be capable of covering a distance of 320 km in a week and 1,800 km in a period of twenty-five days (Hoang 1988).

Horses played an important role in the exploitation of pastures in winter. In the event of heavy snowfall covering the landscape, the horses were sent out so their hoofs would crush the compact snow cover, reaching through to the grass (*tebenevka*). For this reason, the ideal horse-sheep ratio in the herd was not less than 1:6, a ratio reflected in medieval sources of Mongolian history: "In their country, someone who possesses one horse should surely have six-seven sheep. Therefore, if [a man] has a hundred horses then he should without fail possess a flock of six-seven hundred sheep" (Munkuev 1975:69).

Mare's milk, or kumiss (Mong. *airag*), has a special significance in nomadic cultures. "To satisfy a hunger and slake a thirst [they] drink only mare's milk. Generally, the milk of one mare is sufficient to satiate three persons" (ibid.). As a whole, horses have played a major part in the economic and cultural life of nomads. They are symbolically embodied in folklore and legend. The status of pastoral people was determined by the number of horses they owned, and in the lore of townspeople and peasants, the mythologized image of the warlike nomad was often associated with the fierce centaur.

The nomads' cattle were not characterized by high productivity. A typical Mongolian cow yielded only 400–500 liters of milk a year, whereas Dutch cows could provide about 3,800 liters of milk a year (Simukov 2007:292–93, 733). However, Mongolian livestock was well adapted to the rigorous climate. There was no need to sustain cows with warm winter sheds and great reserves of hay. Cattle were also used as draft animals. A story from the *Secret History* (de Rachewiltz 2004:§§100–102) tells of the young wife of Temujin Borte, who was captured by Merkid because Borte lacked a horse and his wife drove in a closed sleigh pulled by a speckled cow. A bull could carry 200–250 kg of cargo for a distance of 15–20 km a day (Simukov 2007:399–400).

Sheep formed the majority of the nomads' herds in Inner Asia. "For the most part [Tatars] rear sheep and use [their meat] for food" (Lin and Munkuev 1960:139). Sheep required no special care. They could find and eat grass all year, drink muddy brackish water, and eat snow in winter. After the hunger of winter, sheep quickly restored their weight, experiencing about a 40 percent increase in weight during the summer. Sheep fertility typically reached about 105 lambs per 100 ewes. Sheep were the principal source of milk and meat for nomads, and mutton was considered the best meat in terms of taste and nutritional value. Everyday clothes were made of sheepskin, and felt was manufactured from sheep's wool. An average sheep's weight reached 40–50 kg, and the meat yield was about 20–25 kg. In summer, each sheep was sheared and yielded slightly more than 1 kg of wool (Kradin 2002:68–69).

The nomads of Inner Asia also kept a small number of goats (5–10% of the total herd). Goats were even easier to keep than sheep; in places where grass quality was poor, they took the place of sheep in their use by nomads. Goats were able to orient themselves well in varied terrains and led the way for the entire herd. Though goat's milk offered the highest fat content, it generally did not please the nomadic palate. Breeding goats was considered less prestigious than breeding sheep, and traditionally Mongols deemed that only poor people kept goats (Timkovsky 1824:79).

The camel was the last of the important livestock of Inner Asia nomads. It is known that medieval Mongols bred both dromedaries and Bactrians (Lin and Munkuev 1960:137). It is notable that among a camel's major qualities, its abilities to survive without water and food for up to ten days, as well as to drink water with a high salt content and eat plants unsuitable for other species, would have been especially important to the nomads. Other important advantages of the camel are its great strength, high speed, and large mass (200 kg of meat and 100 kg of fat). In Inner Asia, camels were used mainly to transport goods. With a load pack, a camel was capable of transporting up to 300 kg and with a sleigh, 500–600 kg. The usual length of a day's passage was 25–40 km (Masanov 1995: 70–71).

Camels also comprised an important source of wool and milk products. An individual camel could yield 3–6 kg of wool. In southern Mongolia, camels were as important as cattle. Because camel's milk is very fatty, it was employed as an equivalent to kumiss, cheese, and other products. In addition, dried camel excrement was an important fuel source in the Gobi Desert. Nevertheless, camels had drawbacks;

they could not travel well on slippery roads, they tired quickly, could not withstand cold and dampness, and bred slowly (about every 2–3 years) (Masanov 1995).

NOMADIC DIET

All nomadic interests were invested in livestock, their source of life and indicator of success. The Chinese immigrant, eunuch Chung-hang Yüeh, described the life of nomads: "According to Xiongnu custom, people eat the meat of their animals, drink their milk, and wear [clothes made with] their hides" (quoted in Taskin 1968:46). More records exist of the Mongol diet in the Middle Ages. In the headquarters of chiefs and khans, beef, mutton, and horsemeat dishes were essential dietary components. Most people, however, only used the meat of slaughtered animals when entertaining guests. For most cattle breeders, diet consisted mainly of the milk products of mares, sheep, and camels; stews; occasional cereals; and "Mongolian" tea (Munkuev 1975:69; Plano Carpini 1996:53; Juvaini 1997:21). Mongols call such food *white*. In modern Mongolian, white is a synonym for dietary nutrition. The greatest reliance on dairy food occurred during the spring-summer period, when winter meat reserves were depleted and the cattle had not yet recovered their weight.

Cattle were normally slaughtered in late fall after fattening and when the average daily air temperature was below zero, which enabled herdsmen to store meat for a long period of time, until early spring. Possibly, in instances of meat shortage, poor nomads may have used fallen animals for food. Horses were only slaughtered during high holidays (Lin and Munkuev 1960:139). In this case, horsemeat and mutton belonged to so-called hot (Mong. *haluun*) meals.

Mare's milk was a favorite delicacy of Mongolian nomads:

> They drink mare's milk in great amounts. If they should happen to have it, they drink the milk of sheep and cows, goats and even camels. The Tartars do not have wine, ale, or mead unless it is sent by other people or given to them. In winter, unless they are rich, they do not have mare's milk. They cook millet with water, which they make so thin that they can drink rather than eat it. And each one of them drinks one or two cups in the morning, and then nothing more during the day. In the evening however each is given a very small amount of meat, and drinks meat broth. In summer because they have enough mare's milk, they rarely eat meat unless it happens to be given to them or they take some animal or bird by hunting. (Plano Carpini 1996:53)

Because protein was generally absent from the diet of common and poor nomads, Mongols compensated for this deficit by hunting. In the *Secret History,* different kinds of Mongolian hunting are repeatedly mentioned: hunting of wild ungulate animals (de Rachewiltz 2004:§§12–13, 200), waterfowl (ibid.:§§54, 189), and different species of rodents (Lin and Munkuev 1960:139). Tarbagan (*Pharaoh's Rat,* in Marco Polo 2001:77) was considered a delicacy. Mongols prepared it as a delicacy by cooking it in hot stones (*boodog*). The meat was said to be very succulent and tender. The hunting of fur-bearing animals—sables and squirrels (de Rachewiltz 2004:§§109, 182)—was also of great importance. Cattle breeders could exchange animal pelts for much-needed goods and agricultural products.

The *battue* (beating woods and bushes to flush out game) had a significant role. It was performed by a large quantity of people forming a long chain. This chain would gradually transform into a kilometers-long ring, which continually contracted until all game animals were bunched up and enclosed. At that time, the hunters would begin to fell their quarry. This kind of hunting required a sophisticated level of coordination by the horsemen. It also became a form of military training for nomads (Juvaini 1997:27–28). A favorite pastime among nomads was falconry. The legendary Bodonchar used this sport for subsistence, and, later, Chinggis Khan himself engaged in it as an enjoyable leisure activity (de Rachewiltz 2004:§§31, 232). Nomads also foraged and fished as a means of supplementing their diet. However, such activities were not considered honorable. After the death of Yisugei Ba'atur, his children were forced to survive by fishing to avoid starvation (ibid.:§75).

The nomads of Eurasia were also familiar with agriculture (Khazanov 1975:11–12, 117, 150–51; Markov 1976:159, 162–67, 209–10, 215–16, 243; Kosarev 1991:48–53; Masanov 1995:73–76). However, nomads disapproved of the sedenterization that farming required. It was an offense to the nomadic mentality and customs. Thus, it is no coincidence that Tartars had a proverb warning "stay put as a Christian and sniff [your] own stink" (Mekhovsky 1936:213n46). Therefore, sedentary nomads considered their static status temporary and, at the first opportunity, returned to mobile pastoralism, thereby demonstrating flexibility (Tolybekov 1959:335–38; Markov 1976:139–40, 163, 165, 243–44; Kosarev 1991:46–50; Khazanov 1994:83–84).

For these reasons, nomads preferred to acquire agricultural products by means of war or trade. At the time of the nomadic empires of Xiongnu and Xianbei, nomads established conquered farming settlements as part of their territory (Davydova 1968; Hayashi 1984; Taskin 1984:80; Kradin 2002). In the nomadic empires of Uigurs, Kitans, and Mongols, rulers built fortresses and towns and relocated captured farmers and craftsmen to these enclosures. Garrisons were also quartered in these towns, and functionaries, tradesmen, and monks resided there. The nomadic elite also seasonally occupied these settlements (Kiselev 1957, 1965; Ivliev 1983; Hayashi 2003; Rogers, Erdenebat, and Gallon 2005; Kradin and Ivliev 2008). However, the stark Mongolian steppe offered few chances for sedentary life.

Agriculture is only effective where there is at least 400 mm of annual rainfall or a nearby branch of a river network. A large portion of Mongolia fails to meet these conditions (Murzaev 1948:192, 207, 220–33). Ultimately, only 2.3 percent of Mongolian lands are suitable for husbandry (Yunatov 1946).

How many nomads could subsist on the Mongolian steppes? Based on the approximate number of animals possessed by one family (100–130 sheep) and the estimated strength of Mongolian nomads in the thirteenth to fourteenth centuries (Munkuev 1970), Dinesman and Bold (1992:179) assumed that total livestock numbers should not exceed 28 million–30 million head. These data appear to be overestimates. More likely, the numbers would have been comparable to those in present-day Mongolia. Bat-Ochir Bold used finer calculation methods. His work was based on the known quantity of horses in the hands of Mongols in the early thirteenth century (1.4 million head), and, in accordance with the traditional herd structure, he calculated the approximate total number of domestic animals to be 15.1 million head (Bold 2001:40–41). Based on his data, each household had 17.7 head of domestic animals. In this case, the total population of Mongols is taken into account, including fighters who were under arms in the territory of China.

These conclusions are more consistent with those of other researchers. Namio Egami (1963:353–54) has calculated that the ancient Xiongnu had about 19 head of animals per capita. According to the thorough statistical calculations of Maisky's (1921:67, 124) 1918 expedition, each person in Mongolia had 17.8 head of all breeds of livestock. In comprehending these data, Maisky advanced a key idea about the fundamentally limited possibilities of extensive nomadic economic development: "I hesitate to advance a fully categorical opinion because it is impossible to present any reliable data in its confirmation, but my general impression is such that the Autonomous Mongolia with its system of primitive pastoralism was not in a position to feed the cattle number exceeding essentially its present one. Maybe, its grass resources would be sufficient for the numbers of livestock which is 1.5 times more than the present number in case of austerity economy . . . but not greater" (ibid.:134).

Is it too much or too little? Tortika and colleagues (1994) proposed a formula for the nomads' food supply index (FSI). They used an averaged equivalent of 36 sheep per capita as a basis. This factor was empirically derived from known data about the correlation between the numbers of people and animals in different pastoral societies. It means that a herd of 36 head of sheep is necessary for the survival of a single person. Then, a formula is as follows:

FSI = number of cattle per capita in nominal sheep: 36 nominal sheep.

If this index is more than 1, then a household has adequate resources to support life (true, if it is much more than 1, it threatens to place an excessive load on pastures). If the index is less than 1, then a household is in a situation of stress that requires either involvement of additional sources (e.g., agriculture, hunting, war) or the development of client relations with well-heeled cattle owners (ibid.; Tortika and Mikheev 2001).

Using this formula, one can calculate the FSI for any nomadic society. Let us consider this problem using the example of the medieval Mongols. As noted, Bold concluded that among the Mongols there were 17.8 animals per capita. However, the total population of animals is considered here, so it should be transformed into "nominal sheep." Because the herds of nomads in Eurasia included 50–60 percent sheep, we will consider that small cattle comprised about ⅔ of this value (12 head). Then the number of great cattle and horses will be about 6 head.

There are different variants of calculating the nominal (conventional) equivalent for different kinds of livestock. Rudenko (1961:5) defined some nominal equivalent of the livestock (300 nominal sheep or 25 horses) needed for the minimal self-contained existence of a five-member family. According to his data, 1 horse corresponds to 6/5 head of horned stock and 6 sheep or goats. In Mongolia, a conditional unit (*bodo*), which is equal to half a camel or 1 horse or 1 head of great cattle or 7 sheep or 14 goats, was taken for convenience (Murzaev 1948:48). A similar system of cost relationship between different kinds of livestock was accepted by the Russian administration in Kazakhstan in the nineteenth century (Kosarev 1991:37). Based on this system of calculations, one can calculate that every Mongol had about 50 ($6 \times 6 + 12$) nominal sheep. Even granting the relativity of the calculations, it is evident that the number of cattle among medieval Mongols exceeded the minimal standard of FSI.

"IN SEARCH OF GRASS AND WATER . . ."

Because animals constantly require new pastures, nomads were forced to move from one place to another several times a year. In Chinese accounts, a stereotyped formula asserts that the Xiongnu nomads "move about according to the availability of water and pasture, [and] have no walled towns or fixed residences, nor any agricultural activities" (Bichurin 1950:40; Watson 1961:129; Taskin 1968:34). Similar descriptions were recorded concerning Turkish people (Bichurin 1950:229), Uigurs (ibid.:216), Mongols (Kafarov 1866:286, 288), and other nomads.

By virtue of this mobile way of life, nomads lived unpretentiously, occupying light, collapsible dwellings (yurts, tents, marquees). Their household necessities were few. Dishes were commonly made of unbreakable materials (wood, leather). Clothes and footwear were sewn, as a rule, of leather, wool, and fur. The yurt was a major architectural invention of the nomads. I became familiar with it during my expeditions in Mongolia. Its circular form allows the greatest amount of interior living space. The yurt protects against cold in winter and against heat in summer. It has optimal aerodynamic qualities, rendering it tolerant of strong winds and hurricanes:

> Tartar homes are round and prepared like tents made cleverly of laths and sticks. In the middle of the roof there is a round window through which light comes in and smoke can leave, because they always have a fire in the center. The walls and the roof are covered by felt and even the doors are made of felt. Some huts are large and some are small, depending upon the wealth or poverty of the owners. Some are taken apart quickly and put back together again and carried everywhere; some cannot be taken apart but are moved on carts. The smallest are put on a cart drawn by one ox, the larger by two or three or more [oxen] depending upon how large it is and how many are needed to move it. Whenever they travel, whether to war or other places, they always take their homes with them. (Plano Carpini 1996:41)

The yurt, without its timber floor, weighs about 200 kg, 75 percent of which is the weight of the felt. Yurts could be assembled within one hour. Felt carpets could be used for up to five years, while a timber frame lasted more than twenty years.

For mobility's sake, Mongols placed yurts on special carts: "On the carts, the rooms [are arranged so that] one can sit and lie. They are called 'carts-marquees' (Mong. *ger-tergen*). Into the four corners of the cart, sticks or planks are driven and they connect crosswise above" (Lin and Munkuev 1960:137–38; Polo 2001:77). Nomads in earlier times probably used similar carts.

Vladimirtsov (1948) suggested that prior to the Chinggis Khan Empire, Mongols moved from place to place in large groups. For safety, they encircled their camps with a ring formed by their carts (Mong. *kuren*). It was not until later that they began to migrate in single family groups (Mong. *ail)*. However, the historical and ethnographic data show that these two variants of wandering cannot be considered in an evolutionary context. Their existence at one time or another was related to different ecological, socio-economic, and political factors (Khazanov 1975:10–13; Markov 1976:57–58, 240–41; Kradin 1992:48–49; Masanov 1995:114–30).

In Mongolia, there is no one universal way of migrating, even today. No fewer than ten variants of seasonal roaming from place to place are known. The majority of Mongols relocate with their livestock an average of two to four times a year. However, frequency and distance of migration vary considerably based on the pastures' productivity. In the fruitful Khangai steppes, nomads wander within a radius of 2–15 km. The Khangai cattle breeders' summer camps are usually located in wide river valleys, while the winter camps are established atop outer valleys for wind protection. In the semiarid regions of the Gobi Desert, this travel radius is much greater (50 km to 70 km). There, summer pastures are located in the open plains and winter camps are established in the hilly and sub-montane areas, where the settlement is protected from cold winter winds. On the Onon River, nomads spend winters in quiet submontane valleys or in mountains, and in summer they descend into the wide, productive river valleys. The largest migrations—covering 100–200 km—are made by the Mongols of the Ubur-khangaisky and Bayan-khongorsky aimaks. The number of annual migrations in these aimaks can reach fifty or more (Murzaev 1948:48–49; Dinesman and Bold 1992:193–94; Bold 2001:54–55; Simukov 2007:274–75).

In the Middle Ages, Mongol migration was also subject to seasonal rhythms. In summer, Mongols relocated to places where pastures were fresh and there were plentiful water sources. In winter, they moved to warmer places open to the sun (Kafarov 1866:586; Polo 2001:76–77). After the conquest of the Eastern Europe steppes, the Mongols would relocate to the south and in the summer would move to cooler locations to the north (ibid.:22–23). Routes of seasonal migration were more or less well-established. As William of Rubruck argued, "Every captain, according as he hath more or less men under him, knows the limits of his pasture land and where to graze in winter and summer, spring and autumn. For in winter they go down to warmer regions in the south: in summer they go up to cooler [areas] towards the north. The pasture lands without water they graze over in winter when there is snow there, for the snow serveth them as water" (quoted in Rockhill 1900:53).

The Mongolian Khans spent all year making seasonal journeys (Rashid ad-Din 1960:41–42). Shiraishi reconstructs this seasonal route as follows: In spring, Ögödei Khagan traveled to the north of Karakoram where, in the neighborhood of Doityn balgas, he enjoyed practicing falconry. At the beginning of summer he relocated to the south, where the Khangai Mountains provided cool shelter. With the advent of winter's cold, Ögödei traveled to the south, nearer the Gobi. In early spring he returned to Karakoram to settle affairs of state. The total length of this route was about 450 km (Shiraishi 2004:113–15).

Güyüg and Möngke also engaged in this mobile lifestyle. Khubilai spent half the year, from September to February, in the new capital of Khanbalyk. For the next three months the khan hunted, and from the beginning of May to the end of August he resided in his bamboo palace in Shàngdū (Polo 2001:124–27).

The continual movement of such large masses of people sparked genuine awe among the farmers of the settled agricultural states. When moving from one place to the other, the clusters of yurts, carts, and other elements of the caravans stretched for many kilometers over the steppe. According to Lin and Munkuev (1960:138), "When migrating, the carts move in one row of five each. [When preparing for migration, they], as the strings of ants, as the fibers for plaiting of cord, reach [one place] from the right and left at a [distance] of fifteen *li* (about six kilometers). When (a column of moving carts) becomes straight and half [of them] reach the water source a [column] makes a stop."

POLITICAL ECOLOGY OF THE STEPPE FRONTIER

Possibly the most intriguing question in the history of the Great Steppe is, what compelled nomads to undertake mass migrations and destructive campaigns against agricultural civilizations? Regarding this question, a great many diverse opinions have been proposed. These opinions can be classified as follows: (1) diverse global climatic changes (drying according to Arnold Toynbee [1934] and Grigorii Grumm-Grzhimailo [1926]; humidification according to Lev Gumilev [1993:237–340]); (2) the nomads' warlike and greedy nature; (3) overpopulation of the steppe; (4) growth of productive forces and class struggle, with concomitant weakening of the agricultural societies as a consequence of feudal division (Marxist conceptions); (5) the need to replenish an extensive cattle-breeding economy by means of raids on more stable agricultural societies; (6) settled peoples' unwillingness to trade with nomads (the cattle breeders had nowhere to sell their surplus products); (7) personal property of rulers of the steppe societies; (8) ethno-integrating impulses (*passionarity* according to Gumilev [1989]).

The majority of the factors listed here have a certain rationality of their own. However, the importance of some has been overestimated. So, the present paleogeographic data do not conform to a strict correlation between periods of the steppe drying (humidification with periods of decline) and the prosperity of nomadic empires (Ivanov and Vasilyev 1995:tables 24, 25). The Marxist class struggle thesis concerning nomads has proved to be erroneous (Markov 1976; Khazanov 1994; Kradin 2003a). The role of demography is not entirely known because the numbers of livestock increased faster than the human population. An increase in livestock catalyzed grass depletion and a crisis of the ecosystem. Nomadic life can naturally contribute to the development of certain military characteristics. But farmers outnumbered the nomads many times over, and they also had an ecologically complex economy, reliable fortresses, and a more powerful handicraft-metallurgical base.

It seems to me that several important factors should be taken into account:

1. Ethnohistorical studies of the present pastoral people of Asia and Africa show that the ecology of pastoralism, the extensive nomadic economy, the low population density, and the absence of a settled way of life do not require any sort of legitimated hierarchy. Thus, one can assume that a state system has not been intrinsically necessary for nomads (Lattimore 1940; Bacon 1958; Tolybekov 1971; Markov 1976; Irons 1979; Fletcher 1986; Barfield 1992; Khazanov 1994; Masanov 1995).
2. The degree of centralization among nomads is directly proportional to the extent of neighboring agricultural civilizations. In terms of world-system analysis, nomads have always occupied the status of a "semiperiphery" (Hall 1991;

Chase-Dunn and Hall 1997; Kardulias 1999; Chase-Dunn, Hall, and Turchin 2007), which has consolidated different regional economics into a common space (local civilizations, "world empires"). In each local regional zone, the political structuring of the nomadic semiperiphery was in direct proportion to the size of the core. That is why, in order to trade with oases or attack them, the nomads of North Africa and the Near East united into tribal confederations of chiefdoms, with the Eastern European steppe nomads living on the margins of the Ancient Rus'–established "quasi-imperial" state-like structures, while in Inner Asia, for example, the "nomadic empire" became such an important mode of adaptation (Lattimore 1940; Khazanov 1975, 1994; Irons 1979; Barfield 1981, 1992, 2001; Fletcher 1986; Golden 1992, 2001; Kradin 1992, 2002; Masanov 1995; Hall 2005; Kradin and Skrynnikova 2006; Rogers 2007).

3. Thus, the imperial and quasi-imperial organization of Eurasian nomads first developed after the axial age ended (Jaspers 1949), from the middle of the first millennium BC, at the time of the mighty agricultural empires (Ch'in in China, Maur in India, Hellenistic states in Asia Minor, the Roman Empire in Europe)—first in regions where large spaces favorable to nomadic pastoralism were available (e.g., regions off the Black Sea, Volga steppes, Khalkha-Mongolia) and second where the nomads were forced into long and active contact with more highly organized agricultural urban societies (Scythians and old oriental and ancient states; nomads of Inner Asia and China; Huns and the Roman Empire; Arabs, Khazars, and Turks and Byzantium).
4. It is possible to trace a synchronism between the processes of growth and decline in agricultural world empires and in the steppe semiperiphery. The Han Empire and Xiongnu power appeared over one decade. The Turkish Khaganat appeared just as China was consolidated under the dominion of the Sui and T'ang dynasties. Similarly, the steppe region and China entered into subsequent periods of anarchy over a short period of time. When the sedition and economic crisis started in China, the systematic remote exploitation of nomads ceased to work, and the imperial confederation collapsed into separate tribes until peace and order were reestablished in the south (Barfield 1992, 2001; Chase-Dunn and Hall 1997; Chase-Dunn, Hall, and Turchin 2007).

Nomadic empires were organized in the form of "imperial confederations" (Barfield 1981, 1992). From the outside, these confederations appeared autocratic and state-like (they were created to withdraw the surplus products outside the steppe), but from the inside they were consultative and tribal. The stability of steppe empires depended directly on the supreme power's skill at organizing the production of silk, agricultural products, handicraft articles, and delicate jewels in settled territories. As these products could not be produced under the conditions of a cattle-breeding economy, obtaining them by force and extortion was the priority of the ruler of a nomadic society. As the sole intermediary between China and the steppe, the ruler of a nomadic society had the opportunity to control the redistribution of plunder from China and thereby strengthen his own power. This allowed him to maintain the existence of an empire that could not survive on an extensive pastoral economy alone.

We can identify six signs of nomadic empires: (1) multistage hierarchical character of social organization affected at all levels by tribal and super-tribal genealogical ties; (2) dualistic (into the wings) or triadic (into the wings and center) principle of administrative division of the empire; (3) military-hierarchical character of social organization of the empire's center, more often on the decimal principle; (4) horse relay messenger service (*yam*) as a precise way of organizing the administrative infrastructure; (5) a particular system of power inheritance (empire is a property of the whole khan clan, institution of co-government [*kuriltai*]); and (6) specific nature of relations with the agricultural world (Kradin 1992, 2000, 2003b).

The majority of nomadic empires did not often exist for more than 100–150 years. Researchers have repeatedly cited reasons that could have caused the decline and collapse of nomadic empires. Among them are (1) natural phenomena (drying of steppe, short-term climatic stresses and epidemics); (2) foreign policy factors (invasion of enemies, delayed wars, cessation of outside incomes, crisis of nearby agrarian civilizations); and (3) internal causes (demographic outburst, loss of internal unity and separatism, gigantic size and weakness of the administrative infrastructure, class struggle, internal wars of khans and civil wars, political rulers who lacked talent).

Natural phenomena as a causative factor constituted a popular explanation in earlier years, but more recent data have thrown doubt on this suggestion. The paleogeographic data of the 1990s suggest a lack of direct relation between global cycles of drying/moistening of the steppe and periods of collapse/rise of steppe empires (Ivanov and Vasilyev 1995). A thesis of nomadic class struggle proved erroneous because the phenomenon was not observed (Markov 1976; Kradin 1992; Khazanov 1994). However, the majority of the other causes cited earlier had a real impact on the success of the steppe powers, and comparative-historical analysis often shows that the collapse of nomadic empires was often the product of multiple simultaneous factors. As a rule, misfortunes never came alone. The internal internecine wars could be accompanied by both local ecological catastrophes (Xiongnu, Uighurs) and invasions of enemies (Rourans, Uighurs).

At the same time, some events potentially contributed to the structural instability of nomadic empires: (1) external sources of surplus products, which integrated economically independent tribes into a unified imperial confederation; (2) mobility and armament of nomads, which forced the supreme power of an empire to restore balance between different political groups; (3) the specific *tanistrial* system of power inheritance, by which each descendant of the ruling lineage (mothered only by main wives) was prioritized according to age for administrative promotions, including the right to a throne; and (4) polygamy among the highest elite of nomads. This last factor has had an ominous role. Even the theoretical argumentation needs to be reconsidered. Let us assume that each member of the elite had 5 sons by his main wife. Hence, the khan theoretically should have had 5 sons, 25 grandsons, and 125 great-grandsons. In reality, the number of descendants was much higher. Chinggis Khan, for example, had about 500 wives and concubines, Jochi 114 sons, and Khubilai 50 sons; one descendant of Chinggis Khan was named comically "the commander of a hundred soldiers" because he had 100 sons. In such a progression, a competition for inheritance would generally cause murderous internal wars, resulting in civil war and ending with the slaughter of most competitors or the collapse of the polity (Kradin and Skrynnikova 2006:484–86). Ibn Khaldun wrote about this trend many centuries ago. By analyzing the occurrences of new titles for nobility, I revealed such a pattern in the history of the Xiongnu Empire (Kradin 2002). Similar processes have taken place in other nomadic empires as well (Turchin 2003:32–137).

As a result, the fortune of the nomadic empire always depended on the extent to which its ruler was able to solve these problems, to redirect the energy of his numerous relatives and brothers-in-arms outside the polity. Therefore, the steppe element placed specific strains on candidates for rulers of the nomadic empires. In addition to strong personal superiority, an aspirant needed acute talent in policy and war. He needed to be an extraordinary person to attain control over the pastoralist tribes and chiefdoms that had depended on his predecessor, to force these tribes to submit absolutely, and to show generosity and magnanimity when dividing spoils and distributing gifts. If he failed to do so, the nomadic empire was doomed to collapse and historical oblivion.

CONTEMPORARY PASTORAL NOMADISM

In the period of modernization, pastoral nomadism has failed to compete with the industrial economy. The emergence of firearms and artillery gradually eroded nomads' military dominance over farmers. Nomads no longer play dominant roles in world history. The political status of the steppe societies has changed. They have become the colonial periphery of the industrial world-system. In Inner Asia, nomads became the vassals of two great world empires, the Chinese and the Russian. If in the nineteenth century the practice of indirect rule of nomads was normal, with the establishment of socialism, direct rule has prevailed.

There is a vast literature in which a long list of issues is discussed: the causes of nomadic collapse at the present time (environmental crises, inability of nomads to be modernized, strong pressure of industrialism), the nature of nomadic society (irrationality of behavior) and attempts to predict its response to various decisions, and the consequences of the active intervention of the modern state into nomads' traditional life. Many studies emphasize the need to restrict the influence of modernization on nomads and to offer them a chance to carry on traditional nature management (Salzman and Galaty 1990; Ginat and Khazanov 1998; Salzman 2004; Ikeya and Fratkin 2005; Janzen and Enkhtuvshin 2008; Scholz 2008). The peculiarities of the transformation of mobile pastoralism in the post-socialist societies have been the object of thorough investigation since the mid-1990s (Humphrey and Sneath 1996, 1999; Sitniansky 1998; Kradin 2004; Khazanov and Shapiro 2005; Finke 2011). Important investigations devoted to studying the stock-raising business transformation during the post-socialist period in Inner Asia (Mongolia, Sinkiang, and Autonomous Mongolia of the People's Republic of China; Tuva, Buryatia) were undertaken. However, many issues remain unsolved.

The experience of studying the societies of nomads-pastoralists shows that many projects geared to transforming their socio-economic structure inflicted considerable damage because the governments of different countries and functionaries at various levels have not taken into account the peculiarities of the economy and culture of nomads and imposed on them some variants of modernization that were not appropriate (Khazanov 2008). The socialist method of pastoralist reformation has resulted on the one hand in intensification of the economy, breeding of new varieties, and increase in delivery of meat, wool, and milk. A process of sedenterization was carried out, and pastoralists' living conditions improved. Houses, schools, and shops were constructed for them, and electrical equipment was installed in the seasonal sites. On the other hand, the sedenterization of the population and increase in the number of animals resulted in the degradation of pasture, while plowing the steppes contributed to erosion. By the end of the Soviet era, most pastures suffered from overgrazing.

I give just one example, the Buryats of the Agin steppes in East Transbaikalia (Kradin 2004). During the socialist years, more than 28 percent of the Autonomous area lands were plowed up. Because the arid climate of the Agin steppe is unfavorable for agriculture, this was a very serious mistake. Instead of heavy yields, a mass erosion of lands ensued. The Soviet state also rapidly implemented sedenterization. By the early 1970s, practically all Agin Buryats had been settled. The population of the region almost doubled over what it had been in 1917, reaching 70,000. After the settled habitations were established, several kinds of matrimonial arrangements were formed. In one case, the cattle raiser, with his wife and infants, lived at the site while their school-age children were boarded out. In other circumstances, parents with adult children who had their own families could reside at the site. Disjointed marriages also existed when the family of a collective farmer lived in the village while the herdsman stayed with the kolkhoz herd in the winter and summer pastures.

As a result of the transformation of pastoralism, the numbers of horses, goats, and great cattle were considerably reduced, and camels almost disappeared. In turn, the sheep population sharply increased. In the 1970s and 1980s, it reached 700,000 to 900,000 animals. Local people called one new breed "golden" because the wool was very expensive, and herdsmen were adequately rewarded for their work. The salary of a conscientious herdsman was higher than that of a university professor. The informants told me that party chiefs near the end of the Soviet era wanted the sheep population to reach a million head. By doing so, they hoped to obtain prestigious state awards and new job titles. However, environmental pressure produced an ecosystem crisis. The excess number of sheep placed a great burden on pastures. The sheep rapidly crushed the topsoil and ate and trampled the grass.

The intense degradation of pastures was observed almost everywhere. However, following the breakup of the USSR, the crisis was more economic than environmental. A similar scenario played out practically everywhere. The post-Soviet privatization, rise of energy prices, and reduction in planting acreage resulted in the decline of fine-wool sheep breeding. The crisis of wool–sheep breeding, however, was also dictated by the reduction in prices for wool all over the globe. Undoubtedly, it has most strongly affected those economies in which sheep breeding was central (Khazanov and Shapiro 2005:521).

The numbers of domestic animals have been drastically reduced. A naturalizing of the stock-raising business and nomadization of a considerable part of the population took place. In post-socialist Russia (Buryatia), the sheep population dropped by a factor of five and was below 500,000 head in the early twenty-first century. In the Ust-Ordynski Buryat district (the Irkutsk region), the number of sheep decreased from 280,000 to 19,000 during that time. The number of sheep among the Aginsky Buryats dropped from 771,000 to 168,000 head. In Tuva, the number of sheep and goats went from 1.2 million to 617,000 head. In Khakassia, the sheep population decreased by a factor of nine and in 1999 was slightly more than 150,000 head. The population of domestic animals in Kazakhstan, Kirghizia, and Turkmenistan also decreased significantly (Sitniansky 1998; Kradin 2004; Khazanov and Shapiro 2005; Finke 2011). Herd reduction was one major method of resolving the post-socialist pastoralist problem.

Mongolia adopted another major approach to overcome the crisis. It proved more viable, and perhaps the Russian specialists should, in some measure, take the experience of their nearest southern neighbor into account in their planning. During the first half of the 1990s, the country was in decline as a result of the destruction of collective farms, privatization of livestock, and the transition to a market economy. The crisis was accompanied by inflation. However, after the first five years of crisis conditions, the situation flattened out. The number of individual farms increased, and total livestock numbers grew by 17 percent, reaching 33 million head in the first ten years of the post-socialist era. Several intense oscillations in herd size between 1999 and 2003 strongly affected the nomads' well-being, but conditions resumed a regular pattern thereafter. By 2010, the total livestock numbers had reached 44 million head. Since around 2005, Mongolia has shifted to cashmere export, a very profitable business. In the composition of herds, the number of sheep decreased sharply while the number of goats increased, which resulted in an increased load on pastures (Bruun and Odgaard 1996; Humphrey and Sneath 1996, 1999; Janzen and Enkhtuvshin 2008; Janzen 2009). The world economic crisis also affected Mongolia. In 2008 the price for cashmere fell by half.

The third approach has a pronounced "intensive" nature. It was realized as a result of the economic expansion of the People's Republic of China, which served as a catalyst for the qualitative transformation of subsistence pastoralism into an industrial form of livestock husbandry. Under the control of the Chinese government, the purposeful policy of nomadic sedenterization, accompanied by widespread mechanization and a changeover of livestock to a stable nursing system, was fulfilled within the past several decades. As a result, total livestock numbers increased from 10 million in 1985 to 74 million head in 2005. This can be considered a gigantic upswing in the economy. However, a considerable portion of the pastures in Inner Mongolia is on the edge of serious degradation (Williams 2002). The future of this way of transforming nomadic pastoralism is related to significant environmental problems in the region under consideration.

REFERENCES

Bacon, Elizabeth E. 1958. *Obok: A Study of Social Structure of Eurasia.* New York: Wenner-Gren Foundation.

Barfield, Thomas J. 1981. "The Xiongnu Imperial Confederacy: Organization and Foreign Policy." *Journal of Asian Studies* 41 (1): 45–61. http://dx.doi.org/10.2307/2055601.

Barfield, Thomas J. 1992 [1989]. *The Perilous Frontier: Nomadic Empires and China, 221 BC to AD 1757.* Cambridge: Blackwell.

Barfield, Thomas J. 2001. "The Shadow Empires: Imperial State Formation along the Chinese-Nomad Frontier." In *Empires: Perspectives from Archaeology and History*, ed. Susan E. Alcock, Terence N. D'Altroy, Kathleen D. Morrison, and Carla M. Sinopoli, 10–41. Cambridge: Cambridge University Press.

Bichurin, Nikita Ya. 1829. *Istoriia Pervykh Chetyrekh Khanov iz Doma Chingisova* [History of the First Four Khans from the Dynasty of Chinggis]. Saint Petersburg: Echatano v tipografii Karla Kraia.

Bichurin, Nikita Ya. 1950. *Sobranie Svedenii o Narodakh, Obitavshikh v Srednei Asii v Drevnie Vremena* [Collected Information about the Peoples of Inner Asia in Ancient Times], vol. 1. Moscow-Leningrad: Academy of Sciences of the USSR Press.

Bogolepov, Mikhail A. 1908. *O kolebaniiakh klimata Evropeiskoy Rossii v istoricheskuiu epokhu* [On Climate Fluctuation of European Russia in the Historical Period]. Moscow: Kushnerov and Co. Publishing House.

Bold, Bat-Ochiryn. 2001. *Mongolian Nomadic Society: A Reconstruction of the "Medieval" History of Mongolia*. New York: Curson.

Bruun, Ole, and Ole Odgaard. 1996. *Mongolia in Transition: Old Patterns, New Challenges*. Richmond, Surrey: Curzon.

Chase-Dunn, Christopher, and Thomas D. Hall. 1997. *Rise and Demise: Comparing World-Systems*. Boulder: Westview.

Chase-Dunn, Christopher, Thomas D. Hall, and Peter Turchin. 2007. "Systems in the Biogeosphere: Urbanization, State Formation and Climate Change since the Iron Age." In *The World System and the Earth System*, ed. Alf Hornborg and Carole Crumley, 132–48. Walnut Creek, CA: Green Press.

Claessen, Henri J.M. 1989. "Evolutionism in Development: Beyond Growing Complexity and Classification." In *Kinship, Social Change and Evolution: Proceedings of a Symposium Held in Honour of Walter Dostal*, vol. 5, ed. Andre Gingrich et al., 231–47. Horn, Wien: Verlag Ferdinand Berger and Söhne.

Cribb, Roger. 1991. *Nomads in Archaeology*. Cambridge: Cambridge University Press.

Davydova, Antonina V. 1968. "The Ivolga Gorodiscche—a Monument of the Hiung-nu Culture in the Trans-Baikal Region." *Acta Arcaeologica Hungariacae* 20: 209–45.

de Rachewiltz, Igor. 2004. *The Secret History of the Mongols: A Mongolian Epic Chronicle of the Thirteenth Century*. Trans. with a historical and philological commentary by Igor de Rachewiltz. London: Brill.

Di Cosmo, Nicola. 2002. *Ancient China and Its Enemies: The Rise of Nomadic Power in East Asian History*. Cambridge: Cambridge University Press. http://dx.doi.org/10.1017/CBO9780511511967.

Dinesman, Lev G., Nina K. Kiseleva, and Alexander V. Kniazev. 1989. *Istoriia Stepnykh Ekosistem Mongolskoy Narodnoy Respubliki* [History of Steppe Ecosystems of the Mongolian Peoples Republic]. Moscow: Nauka.

Dinesman, Lev G., and G. Bold. 1992. "Istoriia Vypasa Skota i Razvitiia Pastbishchnoy Digressii v Stepiakh Mongolii" [History of Pasture Livestock and Development of Grassland Digression in the Mongolian Steppes]. In *Istoricheskaia Ekologiia Dikikh i Domashnikh Kopytnykh: Istoriia Pastbishchnykh Ecosystem*, ed. Lev G. Dinesman, 172–216. Moscow: Nauka.

Dulov, Vsevolod I. 1956. *Sotsialno-Ekonomicheskaia Istoriia Tuvy (XI –Nachalo XX v.)* [Socio-economic History of Tuva (Nineteenthth–Beginning of Twentieth Centuries)]. Moscow: Academy of Sciences of the USSR Press.

Egami, Namio. 1963. "The Economic Activities of the Xiongnu." In *Trudy XXV Mezdunarodnogo kongressa orientalistov*, Vol. 5. Moscow: Academy of Sciences of the USSR.

Fang, Jin-gi, and Guo Liu. 1992. "Relationship between Climatic Change and the Nomadic Southward Migrations in Eastern Asia during Historical Times." *Climatic Change* 22: 151–69.

Finke, Peter. 2011. "Problems of Post-Socialist Transformation in the Pastoral Regions of Central Asia: Comparative Lessons from Mongolia, Kazakstan and China." In *Nomadic Civilizations in Cross-Cultural Dialogue*, ed. B. Enkhtuvshin, 375–82. Ulaanbaatar, Mongolia: International Institute of the Study of Nomadic Civilizations.

Fletcher, Joseph. 1986. "The Mongols: Ecological and Social Perspectives." *Harvard Journal of Asiatic Studies* 46 (1): 11–50. http://dx.doi.org/10.2307/2719074.

Frank, A. Gunder, and Barry K. Gills, eds. 1993. *The World System: Five Hundred Years or Five Thousand?* London: Routledge.

Ginat, Joseph, and Anatoly M. Khazanov, eds. 1998. *Changing Nomads in a Changing World*. Brighton, UK: Sussex Academic Press.

Golden, Peter B. 1992. *An Introduction to the History of the Turkic Peoples: Ethnogenesis and State Formation in Mediaeval and Early Modern Eurasia and the Middle East*. Wiesbaden: Otto Harrassowitz.

Golden, Peter B. 2001. *Ethnicity and State Formation in Pre-Činggisid Turkic Eurasia*. Bloomington: Department of Central Eurasian Studies, Indiana University.

Grumm-Grzhimailo, Grigori G. 1926. *Zapadnaia Mongoliiia i Riankhaiskii Krai* [Western Mongolia and the Uiriankhai Country], vol. 2. Leningrad: Russian Geography Society.

Gumilev, Lev N. 1989. *Etnogenes i Biosfera Zemi* [Ethnogenesis and the Biosphere of the Earth], 2nd ed. Leningrad: L'eningrad University Press.

Gumilev, Lev N. 1993. *Ritmy Evrasii* [Cycles of Eurasia]. Moscow: Ekopros.

Hall, Thomas D. 1991. "The Role of Nomads in Core/Periphery Relations." In *Core/Periphery Relations in Precapitalist Worlds*, ed. Christopher Chase-Dunn and Thomas D. Hall, 212–39. Boulder: Westview.

Hall, Thomas D. 2005. "Mongols in World-Systems History." *Social Evolution and History* 4 (2): 89–118.

Hayashi, Toshio. 1984. "Agriculture and Settlements in the Xiongnu." *Bulletin of the Ancient Orient Museum* 6: 51–92.

Hayashi, Toshio. 2003. "The Role of Sedentary People in the Nomadic States: From the Xiongnu Empire to Uigur Qaganate." In *Urbanization and Nomadism in Central Asia—History and Problems*, ed. M. Abuseitova, 117–34. Almaty: Daik-Press.

Hoang, Michel. 1988. *Gengis-khan*. Paris: Fayard.

Humphrey, Caroline, and David Sneath. 1999. *The End of Nomadism?* Durham, NC: Duke University Press.

Humphrey, Caroline, and David Sneath, eds. 1996. *Culture and Environment in Inner Asia*, vol. 1–2. Cambridge: White Horse.

Huntington, Ellsworth. 1915. *Civilization and Climate*. New Haven: Yale University Press.

Ikeya, Kazunobu, and Elliot Fratkin, eds. 2005. *Pastoralists and Their Neighbors in Asia and Africa*. Senri Ethnological Studies 69. Osaka: National Museum of Ethnology.

Irons, William. 1975. *The Yomut Turkmen: A Study of Social Organization among a Central Asian Turkic-Speaking Population*. Ann Arbor: University of Michigan, Museum of Anthropology.

Irons, William. 1979. "Political Stratification among Pastoral Nomads." In *Pastoral Production and Society*, ed. L'Equipe Écologie et Anthropologie des Sociétiés Pastorales, 361–74. Cambridge: Cambridge University Press.

Ivanov, Igor V., and Igor B. Vasilyev. 1995. *Chelovek, Priroda i Pochvy Ryn-peskov Volgo-Uralskogo Meshdurechya v Golocene* [Ryn-Sands Country during the Holocene: Man and Nature]. Moscow: Intellect.

Ivliev, Alexander L. 1983. "Gorodishcha Kidaney" [Towns and Forts of Kitans]. In *Materialy po Bkymmyh u Srednevekovoy Arkheologii Yuga Dalnego Vostoka SSSR i Smezmykh Territoriyi*, ed. V. D. Lenkov, 120–33. Vladivostok: Far-Eastern Scientific Center of the Soviet Academy of Sciences.

Janzen, Jorg. 2009. "Changes in the Mongolia Pastoral Economy during the Transformation Period." *Nomadic Studies Bulletin* 19: 7–16.

Janzen, Jorg, and Batboldyn Enkhtuvshin, eds. 2008. *Proceedings of the International Conference Dialogue between Cultures and Civilizations: Present State and Perspectives of Nomadism in a Globalizing World*. Ulaanbaatar, Mongolia: Admon.

Jaspers, Karl. 1949. *The Origin and Goal of History*. New Haven: Yale University Press.

Jenkins, Gareth. 1974. "A Note of Climatic Cycles and the Rise of Chinggis Khan." *Central Asiatic Journal* 18 (4): 217–26.

Juvaini, 'Ala-ad-Din 'Ata-Malik. 1997. *Genghis Khan: The History of the World-Conqueror*. Trans. J. A. Boyle. Manchester: Manchester University Press.

Kafarov, Palladiy. I. 1866. *Starinnoe Mongolskoe Skazanie o Chingiskhane* [Ancient Mongolian Legends about Chinggis Khan], vol. 4. Saint Petersburg: Trudy Chlenov Rossiyskoy Dukhovnoy Missii v Pekine.

Kardulias, P. Nick, ed. 1999. *World-Systems Theory in Practice: Leadership, Production, and Exchange*. Lanham, MD: Rowman and Littlefield.

Khazanov, Anatoly M. 1975. *Sotsial'naia Istoriia Skifov* [Social History of the Scythians]. Moscow: Nauka.

Khazanov, Anatoly M. 1990. "Ecological Limitations of Nomadism in the Eurasian Steppes and Their Social and Cultural Implications." *Asian and African Studies* 24: 1–15.

Khazanov, Anatoly M. 1994 [1984]. *Nomads and the Outside World*, 2nd ed. Madison: University of Wisconsin Press.

Khazanov, Anatoly M. 2008. "Pastoralism in the 'Age of Globalization.'" In *Proceedings of the International Conference Dialogue between Cultures and Civilizations: Present State and Perspectives of Nomadism in a Globalizing World*, ed. Jorg Janzen and Batboldyn Enkhtuvshin, xiii–xxxviii. Ulaanbaatar, Mongolia: Center of Development, Admon Press.

Khazanov, Anatoly M., and Kenneth H. Shapiro. 2005. "Contemporary Pastoralism in Central Asia." In *Mongols, Turks, and Others: Eurasian Nomads and the Sedentary World*, ed. Reuven Amitai and Michal Biran, 503–34. Leiden: Brill.

Kiselev, Sergey V. 1957. "Drevnie Goroda Mongolii" [Ancient Towns of Mongolia]. *Sovetskaia Arkheologiia* 2: 91–101.

Kiselev, Sergey V., ed. 1965. *Drevnemongolskie Goroda* [Ancient Cities of Mongolia]. Moscow: Nauka.

Korotayev, Andrey V. 1991. "Nekotorye Economicheskie Predposylki Klassoobrasavaniia i Politogenesa" [Some Economics Preconditions of Origins of the State and Classes]. In *Arkhaicheskoe Obshchestvo: Uslovye Problem Sociologii Rasvitiia*, ed. Andrey V. Korotayev and Viacheslav V. Chubarov, 136–91. Moscow: Institute of History of the USSR, Soviet Academy of Sciences.

Korotayev, Andrey V., Artemy Malkov, and Daria Khalturina. 2006. *Introduction to Social Macrodynamics: Secular Cycles and Millennial Trends*. Moscow: KomKniga.

Kosarev, Mikhail F. 1991. *Drevniia Istoriia Zapadnoy Sibiri: Chelovek i Prirodnaia Sreda* [Prehistory of Western Siberia: Man and Environment]. Moscow: Nauka.

Kradin, Nikolay N. 1992. *Kochevye Obshchestva* [Nomadic Societies]. Vladivostok: Dal'nauka.

Kradin, Nikolay N. 2000. "Nomadic Empires in Evolutionary Perspective." In *Alternatives of Social Evolution*, ed. Nikolay N. Kradin, Andrey V. Korotayev, Dmitri M. Bondarenko, V. de Munck, and Paul Wason, 274–88. Vladivostok: Far-Eastern Branch of the Russian Academy of Sciences.

Kradin, Nikolay N. 2002. *Imperiia Hunnu* [The Xiongnu Empire], 2nd ed. Moscow: Logos.

Kradin, Nikolay N. 2003a. "Nomadic Empires: Origins, Rise, Decline." In *Nomadic Pathways in Social Evolution*, ed. Nikolay N. Kradin, Dmitri M. Bondarenko, and

Thomas J. Barfield, 73–87. Moscow: Center for Civilizational Studies, Russian Academy of Sciences.

Kradin, Nikolay N. 2003b. "Ernest Gellner and Debates on Nomadic Feudalism." *Social Evolution and History* 2 (2): 162–76.

Kradin, Nikolay N. 2004. "The Transformation of Pastoralism in Buryatia: The Aginsky Steppe Example." *Inner Asia* 6 (1): 95–109. http://dx.doi.org/10.1163/146481704793647234.

Kradin, Nikolay N. 2006. "Archaeological Criteria of Civilization." *Social Evolution and History* 5 (1): 89–108.

Kradin, Nikolay N., and Alexander L. Ivliev. 2008. "Deported Nation: The Fate of the Bohai People of Mongolia." *Antiquity* 82: 438–95.

Kradin, Nikolay N., and Tatiana D. Skrynnikova. 2006. *Imperiia Chngis-Khana* [Chinggis Khan's Empire]. Moscow: Vostochnaya Literature.

Krupnik, Igor I. 1989. *Arkticheskaia Etnoecologiia* [Ecological Anthropology of the Arctic Zone]. Moscow: Nauka.

Lattimore, Owen. 1940. *Inner Asian Frontiers of China*. New York: American Geographical Society.

Legrand, Jacques. 2001. "The Mongolian 'Zud': Facts and Concepts from the Description of a Disaster to the Understanding of the Nomadic Pastoral System." In *International Symposium on Dialogue among Civilizations: Interactions between Nomadic and Other Cultures of Central Asia*, ed. Batboldyn Enkhtuvshin, 14–30. Ulaanbaatar, Mongolia: International Institute of the Study of Nomadic Civilizations.

Lin, Chiun-yi, and N. Ts. Munkuev. 1960. "Hei-Ta shih-lue (Kratkie Svedeniia o Cchernykh Tatarakh)" [Hei-Ta shih-lue (Brief Notes on the Black Tatars) by P'eng Ta-ya and Hsu T'ing]. *Problemy Vostokovedeniia* 5: 132–58.

Maisky, Ivan M. 1921. *Sovremennaia Mongolia* [Mongolia Today]. Irkutsk: GIZ.

Maksimov, Anatoly A. 1989. *Prirodnye Stykly* [Natural Cycles]. Moscow: Nauka.

Markov, Gennadii E. 1976. *Kochevniki Asii* [Nomads of Asia]. Moscow: Moscow University Press.

Masanov, Nurbulat E. 1995. *Kochevaia Civilizatsiia Kazakhov* [Nomadic Civilization of the Kazaks]. Moscow: Gorizont and Sotsinvest.

Mekhovsky, Matthew. 1936. *Traktat o Dvukh Sarmatiiakh* [Treatise about Two Sarmatians]. Moscow: Academy of Sciences of the USSR Press.

Monin, Andrey S., and Yuri A. Shishkov. 1979. *Istoriia Klimata* [History of Climate]. Leningrad: Gidrometeoizdat.

Munkuev, Nikolay Ts. 1970. "Nekotorye Problemy Istorii Mongolov XIII v. po Novym Materialam. Issledovanie Yuznosunskikh Istochnikov" [Some Problems of Mongolian History in the Thirteenth Century on the New Dates: The Study of South Sung

Sources]. Dr. Sc. thesis, Institute of Oriental Studies of the Academy of Sciences of the USSR, Moscow.

Munkuev, Nikolay Ts. 1975. *Meng-ta Pei-lu* [Full Description of the Mongols and Tatars]. Moscow: Nauka.

Munkuev, Nikolay Ts. 1977. "Novye Materialy o Polozenii Mongolskikh Aratov v XIII–XIV vv" [New Materials about Mongolian Nomadic Peasants in the Thirteenth–Fourteenth Centuries]. In *Tataro-Mongoly v Azii i Evrope*, 2nd ed., ed. S. L. Tikhvinsky, 409–46. Moscow: Nauka.

Murdock, George Peter. 1981. *Atlas of World Cultures*. Pittsburgh: University of Pittsburgh Press.

Murzaev, Eduard M. 1948. *Mongolskaya Narodnaia Respoblika: Ficiko-geograficheskoe Opisanie* [Mongolian People's Republic: Physiographic Description]. Moscow: Gosudarstvennoe Izdatelstvo Geograficheskoy Literatury.

Odum, Howard T., and Elisabeth C. Odum. 1976. *Energy Basis for Man and Nature*. New York: McGraw-Hill.

Pershits, Abram I. 1994. "Voyna i Mir na Poroge Tsivilizatsii" [War and Peace before Civilization]. In *Voina i Mir v Rannei Istorii Chelovechestva*, ed. Abram I. Pershits, Yuri I. Semeniv, and Viktor A. Shnirelman, 129–244. Moscow: Institute of Ethnology and Anthropology, Russian Academy of Science.

Pevtsov, Mikhail N. 1951. *Puteshestviya po Kitaiu i Mongolii* [Travels in China and Mongolia]. Moscow: Geografizdat.

Plano Carpini, Giovanni. 1996. *The Story of the Mongols Whom We Call the Tartars*. Trans. E. Hildinger. Boston: Branden.

Polo, Marco. 2001. *The Travels of Marco Polo*. Trans. M. Komroff. New York: Modern Library.

Przevalsky, Nikolay M. 1875. *Mongiliia i Strana Tangutov* [Mongolia and Tanguts Country], vol. 1. Saint Petersburg.

Radloff, Wasiliy W. 1989. *Iz Sibiri* [From Siberia]. Moscow: Nauka.

Rashid ad-Din. 1960. *Sbornik Letopisei* [A Collection of Chronicles], vol. 2. Moscow: Academy of Sciences of the USSR Press.

Rockhill, William W. 1900. *The Journey of William of Rubruck to the Eastern Parts of the World, 1253–55, as Narrated by Himself, with Two Accounts of the Earlier Journey of John of Pian de Carpine*. Trans. William W. Rockhill. London: Hakluyt Society.

Rogers, Daniel. 2007. "The Contingencies of State Formation in Eastern Inner Asia." *Asian Perspectives* 46: 249–274.

Rogers, J. Daniel, Ulambayar Erdenebat, and Mathew Gallon. 2005. "Urban Centres and the Emergence of Empires in Eastern Inner Asia." *Antiquity* 79: 801–18.

Rudenko, Sergey I. 1961. "Voprosy o Formakh Skotovodcheskogo Khoziaistva i o Kochevnikakh" [On the Question of the Forms of Pastoral Economy and of the Nomads]. *Doklady Geograficheskogo Obshchestva: Materialy po Etnografii* 1 (Leningrad): 2–15.

Salzman, Philip C. 2004. *Pastoralists: Equality, Hierarchy, and the State*. Boulder: Westview.

Salzman, Philip C., and John Galaty, eds. 1990. *Nomads in a Changing World*. Naples: Istituto Universitario Orientale.

Scholz, Fred. 2008. *Nomadism: A Socioecological Mode of Culture*. Ulaanbaatar, Mongolia: International Institute of the Study of Nomadic Civilizations.

Shakhmatov, Viktor F. 1961. "O Vliianii Solnechnoy Aktivnosti na Selskoe Khoziaystvo" [About the Influence of Solar Cycles on the Activity of Peasant Economy]. *Istoriia, Arkheologiia, Etnografiia* 2: 43–53.

Shiraishi, Noriyuki. 2004. "Seasonal Migrations of the Mongol Emperors and the Peri-Urban Area of Kharakhorum." *International Journal of Asian Studies* 1: 105–19.

Simukov, Andrey D. 2007. *Trudy o Mongolii i dlia Mongolii* [Works about Mongolia and for Mongolia], vol. 1. Senri Ethnological Reports 66. Osaka: National Museum of Ethnology.

Sinor, Denis, ed. 1990. *The Cambridge History of Early Inner Asia*. Cambridge: Cambridge University Press. http://dx.doi.org/10.1017/CHOL9780521243049.

Sitniansky, G. Yu. 1998. *Selskoe Khoziaistvo Kirgizov: Traditsii i Sovremennost* [Rural Economy of the Kirgiz: Tradition and the Current Situation]. Moscow: Institute of Ethnology and Anthropology.

Tabak, Faruk. 1996. "Ars Longa, Vita Brevis? A Geohistorical Perspective on Pax Mongolica." *RE:view* 19 (1): 23–48.

Taskin, Vladimir S. 1968. *Materialy po Istorii Sunnu* [Materials on Xiongnu History], vol. 1. Moscow: Nauka.

Taskin, Vladmir S. 1984. *Materialy po Istorii Drevnikh Kochevykh Narodov Gruppy Dunkhu* [Materials on the History of the Ancient Nomadic Peoples Tung-hu]. Moscow: Nauka.

Timkovsky, Egor F. 1824. *Puteshestvie v Kitay cherez Mongoliiu v 1820 i 1821 godakh* [Travel to China through Mongolia in the Years 1820 and 1821], vol. 1. Saint Petersburg: Tipografiia Med. Departamenta MVD.

Tolybekov, Sergali E. 1959. *Obshchestvenno-Ekonomicheskiy Stroi Kazakhov v XVII–XIX Vekakh* [Socioeconomic Structure of the Kazakhs in the Seventeenth to Nineteenth Centuries]. Alma-Ata, Kazakhstan: Izdatelstvo AN KAzSSR.

Tolybekov, Sergali E. 1971. *Kochevoe Obshchestvo Kazakhov v XVII Nachale XX Veka: Politiko-Ekonomicheskii Analiz* [The Nomadic Society of the Kazakhs in the Seventeenth to the Beginning of the Twentieth Centuries: A Political-Economic Analysis]. Alma-Ata, Kazakhstan: Nauka.

Tortika, Alexander A., Vladimir K. Mikheev, and R. I. Kurtiev. 1994. "Nekotorye Ecologo-Ekonomicheskie i Sotsialnye Problemy Istorii Kochevykh Obshchestv" [Some Ecological, Demographic, and Social Problems in the History of Nomadic Societies]. *Etnograficheskoe obozrenie* 1: 49–62.

Tortika, Alexander A., and Vladimir K. Mikheev. 2001. "Metodika Ecologo-Demograficheskogo Issledovaniia Traditsionnykh Kochevykh Obshchestv Evrazii" [Methods of Ecological and Demographic Studies of Traditional Eurasian Nomadic Societies]. *Arkheologiia Vostochnoevropeiskoy Lesostepi* 15 (Voronez): 141–61.

Toynbee, Arnold. 1934. *A Study of History*, vol. 3. London: Oxford University Press.

Turchin, Peter. 2003. *Historical Dynamics: Why States Rise and Fall.* Princeton: Princeton University Press.

Turchin, Peter, and Thomas D. Hall. 2003. "Spatial Synchrony among and within World-Systems: Insights from Theoretical Ecology." *Journal of World-System Research* 9: 37–64.

Vainshtein, Sevyan I. 1972. *Istoricheskaia Etnografiia Tuvintsev* [Historical Ethnography of Tuvinians]. Moscow: Nauka.

van Ruysbroeck, Willem. 1990. *The Mission of Friar William of Rubruck: His Journey to the Court of the Great Khan Möngke, 1253–1255*. Trans. Peter Jackson. London: Hakluyt Society.

Vladimirtsov, Boris Ya. 1948. *Le Regime Social des Mongols: le Feodalisme Nomade*. Paris: Andrien Maisonneuve.

Watson, Burton. 1961. *Records of the Grand Historian of China*. Trans. Burton Watson, from *The Shih chi of Ssu-ma Ch'ie*, vol. 1. New York: Columbia University Press.

Williams, Dee M. 2002. *Beyond Great Walls: Environment, Identity, and Development of the Chinese Grasslands of Inner Mongolia*. Stanford: Stanford University Press.

Yasamanov, Nikolay A. 1985. *Drevnie Klimaty Zemli* [Ancient Climates of Earth]. Leningrad: Gidrometizdat.

Yunatov, Alexander A. 1946. *Izuchenie Rastitelnosti Mongolii za 25 Let* [Study of Mongolian Flora over Twenty-Five Years], vol. 2. Ulaanbaatar, Mongolia: Trudy Komiteta nauk MNR.

4

Agropastoralism and Transhumance in Hunza

Homayun Sidky

The people of Hunza, the Hunzakutz, live in a resource-scarce high-mountain desert environment in northern Pakistan. To survive, they have had to adopt a range of complementary subsistence strategies, which include intensive irrigation agriculture, cattle husbandry, and transhumance. During the isolation of the past, success in this enterprise—which was based on regulating the interactions among crops, livestock, humans, and the environment—depended upon careful scheduling of various economic activities by means of state-sponsored rituals. This study provides a general overview of how this system enabled the Hunzakutz to create a sustainable economic system in the harsh environment of the Karakoram Mountains. The discussion is restricted to traditional practices prior to the massive changes that have taken place in the Hunza Valley since the late 1970s. My approach is materialist in orientation, based on the premise that "sociocultural systems adjust themselves in patterned and predictable ways to ecological and demographic constraints" (Lett 1987:91).[1]

BACKGROUND

The former princely state of Hunza is situated in an area where the Hindu Kush, Karakoram, and Himalayan Mountains merge to form a massive network of glaciers, peaks, and secluded valleys (Kreutzmann 2005:41). This region has the most immense concentration of mountains anywhere on earth. Aside from the massive

DOI: 10.5876/9781607323433.c004

mountains, the expanse also has the largest system of glaciers, surpassed only by the planet's polar regions.

Covering approximately 7,900 km^2, Hunza's territory is hemmed in by Afghanistan, Russia, and China to the northwest, north, and northeast, respectively. To the east and southeast lie the valleys of Baltistan and Kashmir, while the valleys of Gilgit, Yasin, and Ishkoman are situated to the south and southwest (see figure 4.1). In the past, formidable geological barriers made travel to Hunza extremely difficult. Travelers had to contend with snowstorms, hazardous mountain passes, landslides, and a narrow and precarious trail meandering along dangerous sheer cliffs and gorges. Accounts provided by adventure seekers during the nineteenth and twentieth centuries detail the dangers and hardships of the journey to and from Hunza (Knight 1893:97–98; Shor 1955:275; Stephens 1955:155; Clark 1956:155; Mons 1958:74–85; Thomas and Thomas 1960:95–97; Hamid 1979:7–11, 1992:13; Staley 1982). As a result of the difficulties of travel, Hunza remained relatively isolated from the outside world until 1978, when the all-weather Karakoram Highway was completed (1,284 km in length), connecting Pakistan and China by way of the Hunza Valley (Ahmed 1988; Kreutzmann 1991).

HISTORY

Very little is known about the history of Hunza prior to the eighteenth century (Crane 1956:442). Some scholars have suggested that Hunza is mentioned in seventh-century Tibetan chronicles and works by Buddhist pilgrims (Poucha 1960; Stein 1972:35; Ali 1982:18; Hauptmann 2005:21). These sources, however, are extremely ambiguous and provide little useful data about Hunza's ancient past (Sidky 2003:310). Oral tradition offers a broad picture of Hunza's history. Such accounts relate that for many centuries Hunza was an independent agrarian principality governed by a hereditary ruler who went by the title Thum or Mir. The legitimacy of Hunza's ruler is said to have been based on a heavenly mandate, his magical ability to make rain, and his special relationship with spirits called *pari*, believed to reside in the surrounding mountains (Biddulph 1880:30; Knight 1893:330; Lorimer 1979:295; Hauptmann 2005:21–25).

Moreover, oral history details the ethnogenesis of the Hunzakutz people and gives a broad view of the events leading to the evolution of the Hunza State in the late eighteenth and early nineteenth centuries (see Sidky 1993a, 1996, 1997, 1999). We know that by the start of the nineteenth century the rulers of Hunza had established political control over the territory extending from the village of Maiyon to the south to the Kilik Pass, along the Chinese border, to the north—forging a powerful centralized polity with a defined territory, a monopoly over violence, the

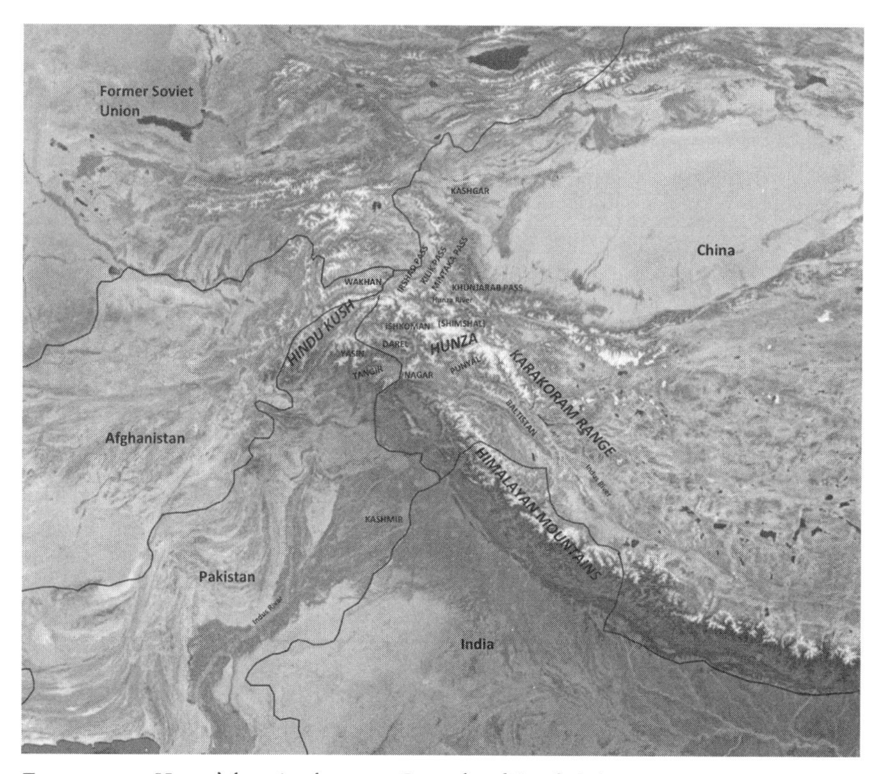

FIGURE 4.1. Hunza's location between Central and South Asia.

ability to raid neighbors, and the power to extract taxes in-kind and as compulsory labor (Staley 1969:229; Sidky 1993a). I consider Hunza a state rather than a chiefdom following Sanderson (1999:56), who notes that "a state is a form of sociopolitical organization that has achieved a monopoly over the means of violence within a specified territory."

By the end of the nineteenth century, the British, fearful over the Russian push into Central Asia and the false impression of a possible invasion of India by Russian troops, initiated a project to consolidate India's northern frontier. Younghusband (1896) and Durand (1899) provide firsthand accounts of these events, which are referred to as "the Great Game." In 1891 the British invaded Hunza and incorporated it into the Gilgit Agency within the Jammu and Kashmir State (Knight 1893:330–37).

Hunza's political fortunes continued unaltered until 1974, when the government of Pakistan relinquished the privileges of the princely states. Hunza was thus incorporated into Pakistan's Northern Areas. This was followed by the completion of the

Karakoram Highway in 1978. These two events had a major impact on traditional life in the Hunza Valley. As Kreutzmann (2005:71) has pointed out, the completion of the Karakoram Highway not only opened up "a new world of communication . . . [but] the close watch and control executed by the hereditary ruler lost its strength, [and] more opportunities and personal liberties for local enterprising people emerged."

THE HUNZA VALLEY

The people of Hunza live along a valley bisected by the Hunza River. In the past this valley was occupied by two political entities: the Hunza State on the north bank and the principality of Nagar on the southern bank. Three major ethnolinguistic groups live along the north bank of the valley. The lower part is the home of the Shin (Jettmar 1961:81; Lorimer 1979, 1980). This territory extends from the village of Nomal to the village of Hini. The second ethnic group is the Wakhi, immigrants from Afghanistan and former vassals of the Thums of Hunza. They live in upper Hunza, a stretch of land that extends from Gulmit village to Passu (Shahrani 1979:55–86; Buddruss 1985:27–29). The central portion of the Hunza Valley, which stretches from Murtazabad to Nazimabad, is the home of the Hunzakutz.

Hunzakutz farmers have traditionally operated an economy that combines the cultivation of cereal crops, vegetables, and fruit and nut trees with cattle husbandry and transhumance (cf. Bianca 2005:11). Because of the ecological constraints and limitations I describe here, the animal husbandry and pastoral components of the Hunzakutz's economy was crucial for the sustainability of their food production system as a whole.

ECOLOGICAL CONSTRAINTS ON AGROPASTORAL PRODUCTION

The specific climatic and geophysical conditions of Hunza's high-mountain environment are among the keys to understanding important aspects of Hunzakutz traditional adaptive strategies (Sidky 2003, 1996). Southwest monsoon rains do not affect this trans-Himalayan region; consequently, the amount of rainfall in the valley floors at an elevation of roughly 2,000–2,400 m—where farming communities are situated—seldom exceeds 130–200 mm per year (Kreutzmann 2000:93–96). This is not enough for dry farming. The Hunza River, which carries glacial meltwaters, in many places runs nearly 100 m below farms and is thus unusable as a source of moisture for natural vegetation or for irrigation farming without pump technology (Sidky 1996:30).

The areas around most Hunzakutz villages are categorized as "desert-steppes." These are biomes[2] where moisture is the primary limiting factor on biological productivity. Only species of drought-resistant *Artemisia* (sagebrush) grow here (Kreutzmann 1988:243). Otherwise, natural vegetation and animal life in these zones is very scanty. Higher up, rainfall increases, enabling the growth of a patchy juniper and cedar forest belt (*Juniperus macropoda, J. semiglobosa, Cedrus deodara*). This forest belt extends from around 2,800 m to 3,000 m (Eberhardt, Dickore, and Miehe 2006:109–21). Finally, grassy alpine pastures range from 3,000 m up to the snowline, at around 4,000 m (FAO 1961, 1978). At altitudes above 6,000 m, precipitation in the form of snowfall, approximately 2,000 mm a year, feeds the massive glaciers. These glaciers are the only source of water for the agricultural communities in the desert-steppe zones below (2,000–2,400 m).

Typically, the length of the growing season and agricultural output diminish with altitude (cf. Guillet 1983:563; Groetzbach 1988:25). Altitude also affects the types of livestock that can be maintained. As a general rule, according to Whiteman (1985:25), there is roughly a one-month decrease in the length of the growing season with every 500-m increase in elevation. This is clearly evident in a broad scale as one travels from Gilgit (1,454 m), the administrative center of the Northern Areas, to the Khunjarab Pass (4,733 m), roughly 275 km to the north. That journey takes us through double- to single-cropping zones (see Kreutzmann 2005:42).

In the lower double-cropping zone (up to 2,000 m), the growing season is a little more than nine months, with adequate time to plant maize as a second crop after wheat and barley. Here livestock include goats, sheep, cattle, donkeys, and horses. Traveling further brings us to the upper double-cropping zone (roughly 2,000 m to 2,400 m). The growing season here is seven to seven-and-a-half months. The principle cultigens include wheat and barley, followed by fast-maturing millets and buckwheat. Livestock at these elevations comprise sheep, goats, and cattle. In the single-cropping zone (2,400 m and higher), the growing season is about four months. At this altitudinal zone there is barely enough time to grow a single annual crop. Livestock herds here include sheep, goats, and yaks, beasts that thrive in high altitudes.

Aside from the limitations set by extreme aridity and altitude, other conditions in the Hunza Valley are more favorable for farming. One of these is the amount of solar radiation, which ranges between 65 percent and 70 percent of the maximum possible hours of sunshine (Whiteman 1985:19). In addition, more radiant energy is available for plant growth because the atmosphere at these altitudes is thin and low in humidity. There is therefore high potential for photosynthesis, which means high plant production (Whiteman 1988:73).

On the negative side, because the thin atmosphere cannot retain heat, the excessive number of overcast days contributes to lower temperatures. The growth, development, and maturity of crops are directly related to the total amount of warmth during a growing season of a particular length (cf. Duckham and Masefield 1970:29–31;

Tivy 1990:30). In Hunza, crop production can be adversely affected by too many cloudy days. The period May through August represents the thermal growing season in Hunza, with the greatest number of sunshine hours (more than 200) and warmest temperatures. This is the best time of the year for agricultural production. Agronomic yields depend in part both on the amount of sunlight received and the efficiency with which it is used by plants (ibid.:21). Through centralized decision-making and careful management and scheduling, Hunzakutz farmers attempted to utilize the maximum amount of solar radiation for plant growth. This was done by ensuring that their fields were covered by as much vegetation as possible as quickly as possible during the most favorable time of the year so sunlight did not go to waste by striking bare ground (cf. Langer and Hill 1991:337).

How quickly plants grow and produce a sufficient crop canopy is based on adequate quantities of nitrogen (N) and water (Langer and Hill 1991:344). By carefully scheduling planting and harvesting dates to correspond with the most favorable environmental conditions, when there was a supply of nitrogen-rich fertilizer and sufficient water for irrigation, crop canopies could be established quickly, resulting in photosynthetic effectiveness and dependable harvests (cf. ibid.:324). Moreover, because the growth and maturation of wheat and barley, the main cereal crops, is shorter than the seven-and-a-half-month growing season in central Hunza, with precise scheduling the remaining sunny days of the farming year could be used efficiently. This phase of the farming season occurred during October, when the number of sunshine hours drops to approximately 150; under ideal conditions this is sufficient to allow a fast-maturing crop such as millet or buckwheat to ripen.

As Whiteman (1988:65) has pointed out, this cropping pattern far exceeded the upper limits for double-cropping practiced elsewhere in the region. This is the case because during the isolation of the past, decision-making and the coordination of all aspects of agropastoral production were handled by the government, resulting in a level of management rarely found in conjunction with economies of this type (cf. Whiteman 1985:41).

By November, the number of sunshine hours drops to around 100. During the second half of the month, solar radiation receipts and daily temperatures decrease significantly. However, the conditions are just warm enough at the beginning of the month to allow an autumn wheat crop to be planted with sufficient time for the seedlings to reach a frost-resistant stage before the onset of winter. This gives the crop a head start when the growing season begins again. From December through February, the number of sunshine hours falls to around 90. The ground is frozen and plant growth terminates altogether. In March and April, sunshine increases once again, averaging about 100 hours per month; warmer temperatures mark the start of the growing season.

IRRIGATION, DEMOGRAPHIC PRESSURES, AND INFRASTRUCTURE FOR AGROPASTORAL PRODUCTION

To establish a sustainable food production system, the Hunzakutz have altered the terrain by transforming steep slopes into level fields or terraces and channeling water into the desert-steppe environmental zones. The primary source of water for irrigation farming in central Hunza is Ultar Glacier. To convey water from the glacier to the fields, the Hunzakutz, under the auspices of the Thum, who financed the project, constructed a complex network of canals and sediment settling tanks during the late eighteenth and early nineteenth centuries (Sidky 1997; Kreutzmann 2000). These canals are many kilometers long, and their construction transformed areas that could not otherwise be used for farming into arable lands. These fields belonged solely to the Thum as builder of the hydraulic system (see Sidky 1996, 1997). To use them required the payment of taxes in-kind and in labor to the state, which increased the Thum's prestige and coffers. The Thum also adjudicated all water rights disputes (see Schmid 2000).

Prior to the development of the hydraulic system, the Hunzakutz were confined to three small, overpopulated villages with limited resources because of a lack of water for farming. There is evidence, therefore, that the construction of the hydraulic works was initially driven by demographic pressures as a result of a scarcity of arable lands. Once the canal system was in place, it allowed for the establishment of numerous new settlements in Hunza and led to further increases in population and intensification of production (Sidky 1996; Kreutzmann 2005:67).

Giving land grants in the newly irrigated areas to select individuals (as well as livestock and seed to jump-start production on the new farms) resulted in increased political stratification and the emergence of a relatively wealthy landholding class. A hierarchical structure linking villages to the court emerged as well.

The establishment and operation of a large-scale hydraulic system had significant socio-political ramifications because irrigation is a strategy for the expansion and intensification of agricultural production that can serve as a potent source of wealth and power. The emphasis here, therefore, is more on the economic rather than the "organizational" dimensions of irrigation stressed by Wittfogel's (1956:155) hydraulic hypothesis (see Sidky 1997).

The evolutionary sequence in Hunza that I have identified ethnographically (Sidky 1996:49–74) accords with the observations by archaeologists Johnson and Earle (2000:305) regarding the rise of the state in the archaeological record: "Increasing population density requires a level of agricultural intensification that ultimately can be achieved only by large capital improvements such as irrigation systems . . . irrigation systems even on a fairly small scale permit economic control by elites who exchange access to irrigated areas for labor or shares of produce. States based on the control of productive technology are typically financed through staples produced on state-controlled improved lands."

Canal irrigation and population growth, however, did not result in the development of specialized fully nomadic pastoralism as predicted by Lees and Bates (1974:187). The data from Hunza do not support their hypothesis that "we see selective pressures for specialization in nomadic pastoralism arising from the practice of canal irrigation as a technique of agricultural intensification" (ibid.). One reason for this may be the geomorphology and ecology of the Hunza Valley—vertically situated and limited pasturelands, sparse vegetation, and extreme aridity in the zones surrounding settlements and beyond.

According to oral tradition, the Thum deployed the resources he thus acquired to continue other irrigation building projects that enabled the establishment of new settlements. He founded new villages in central and lower Hunza. He also built settlements in upper Hunza to be colonized by the Wakhi people he invited from Afghanistan. Once upper Hunza came under the control of the Thum, the Hunzakutz began to raid caravans plying the trade routes between Yarkand and Badakhshan and Yarkand and Leh, much to the vexation of the British in India and the authorities in China (Biddulph 1880:28–29; Knight 1893:329). The raiders used horses, maintained at great cost by the Thum, and their mobility and ability to retreat into the inaccessible recesses of their valley allowed them to attack caravans with impunity. The spoils of these raids quickly become another significant source of income for the Thum and met the increasingly extravagant tastes of the emerging class of wealthy landowners. Raiding by nomadic and semi-nomadic pastoral groups is not uncommon. For example, elsewhere, such as in Saudi Arabia, Bedouin pastoralists often raided other pastoralist encampments and sedentary villages to obtain animals (Irons 1965; Sweet 1965). However, this seems not to have been the primary objective of the Hunzakutz raiders, who were more interested in obtaining luxury goods rather than just livestock.

In addition to the geographic expansion, the Thum fortified his strategic villages and strengthened his military to defend against raiders from Nagar, across the river. This transformed the Thum's original domain, which initially consisted of three small overpopulated villages—Altit, Baltit, and Ganesh, with a population of 3,000 or more—into the nucleus of an extensive centralized agrarian polity. As noted, the establishment of production infrastructure resulted in additional population increases. This is in accord with Kreutzmann's (2005:65) observation: "The impact of this population growth has found its spatial expression in the expansion of settlements within the valley and the establishment of extra-territorial migrant colonies." It is estimated that the population of Hunza in the early to mid-1980s was approximately 32,000 people, spread out in fifty-two villages (Khan 1983:130; Charles 1985:19–26; Kreutzmann 2005:69).

ECOLOGICAL CONSTRAINTS ON AGROPASTORAL PRODUCTION

Since the completion of the Karakoram Highway in 1978, Hunzakutz society has undergone numerous changes. These changes, according to Nüsser and Clemens (1996:117), include a reduction in the use of alpine pastures after a major decrease of the male workforce as a result of access to the outside world, out-migration for wage labor and other opportunities in Pakistan, tourism, the construction and operation of hotels, the introduction of cash crops and fodder cultivation, and associated changes in cropping patterns. The discussion here, to reemphasize, is restricted to the analysis of traditional practices prior to the changes mentioned here.

Farming settlements are situated in those relatively few places where workable tracts of land and irrigation water occur together. Only about 90 km², or 1.2 percent, of the total 7,900 km² of Hunza's territory meet these criteria (Kreutzmann 1988:244, 2005:42–43). Surrounded by a barren landscape, Hunza's environmentally circumscribed settlements have been described as "mountain oases" (Whiteman 1985; Emerson 1990:105–8). These mountain oases posed both opportunities and limitations for sustainable food production. Although referred to as "mountain oases," these villages did not exist as closed self-sufficient systems but were integrated into the administrative structure of the Hunza State and linked together by the hydraulic system.

Hunzakutz farmers in these settlements had to make careful use of resources located at various elevations available to them by adopting a range of complementary strategies. Such verticality is a recurrent strategy in similar environments elsewhere in the world (cf. Rhoades and Thompson 1975). Village resources generally included its arable tracts of land, the alpine pastures to which it had traditional grazing rights, the desert-steppes that briefly provided forage for goats each spring,[3] and the upland forests where villagers gathered firewood for cooking and heating.[4]

Hunzakutz villages are located on glacial outwash fans (Saunders 1984:44–46; Whiteman 1985:28; Searle 2006). Mostly sandy loams, the soils here contain very little clay (about 15 percent), possess little structure, and have low water-holding capacity. But the heterogeneous origins of these soils means that they contain sufficient amounts of macro-nutrients, such as phosphorus and potassium, and most trace elements necessary for crop production (Whiteman 1985:28–30). However, these soils have an extreme deficiency of nitrogen (N), mainly because desert-steppe zones lack any significant quantities of organic materials, such as decomposed plants and animal remains. While variations exist between villages, in general, the soils in central Hunza contain about 1 percent organic matter and 0.5 percent nitrogen (ibid.:30).

Rapid crop cover and efficient photosynthesis depend upon the amount of nitrogen. This nutrient is therefore a crucial limiting factor for the production of crops (cf. Brady 1974:161; Frissel 1977). To maintain a sustainable food production system in the oasis villages, farmers tackled this limiting factor by relying on a number of interrelated strategies involving their livestock and particular land-use patterns. During the isolation of the past, when factory-made fertilizers were unavailable, the solution was to add animal manure to the fields.[5] The availability of sufficient quantities of manure depended upon ruminant livestock: local species of sheep, goats, and cattle (Saunders 1984:41). Ruminant livestock were central to Hunza's economy because they can digest plant materials high in cellulose. These animals are able to consume agricultural by-products, as well as vegetation from marginal areas where for various reasons farming is unfeasible. Thus, ruminant species not only contributed to the Hunzakutz's traditional diet by converting low-quality vegetation into high-quality milk and meat protein, they were also an indispensible source of manure used to maintain sufficient levels of soil nitrogen in the fields.

The problem farmers faced, however, was the dearth of vegetation in the terrain surrounding villages where livestock could graze and produce manure. Moreover, even after the expansion, there was a shortage of arable land in the villages that might be used to grow sufficient amounts of fodder crops and not affect the food-crop production necessary to feed the farmer and his family, as well as to pay taxes to the state. A careful balance had to be struck between producing crops for human consumption and growing and collecting fodder for the livestock. This factor restricted how many animals a farmer could sustain throughout the year.

In the past, farmers had to constantly struggle with the scarcity of natural forage and fodder resources to keep their animals alive. All agricultural by-products, such as straw and other residues generated from cereal cultivation, were fed to the livestock. Fresh barley and wheat stalks, which farmers cut when thinning their grain fields, were also fed to the cattle. Moreover, many households had access to *toqs*, terrain immediately above the villages where there is some water available naturally but not sufficient for cereal production. The grasses in these areas were either cut for storage or grazed (Sidky 1996:99). In addition, a segment of each household's arable fields, out of necessity, had to be allocated for the production of lucerne (*Medicago sativa*), which made up a portion of the needed fodder supplies. This, however, put an upper limit on how much food could be produced for human consumption. Livestock were also fed fresh leaves from privately owned fodder-trees—poplars (*Populus spec.*) and willows (*Salix spec.*)—along with mature and dead leaves from fruit and fodder-trees. In addition, farmers had to accumulate a sufficient stockpile of fodder to sustain their herds during the winter months, an extremely labor-intensive enterprise (cf. Dyson-Hudson and Dyson-Hudson 1980:17).

CATTLE HUSBANDRY

As Barfield (1993:137) put it in another context, the number and types of livestock that predominate in a herd are linked to local ecological circumstances (cf. Bates and Rassam 1983:111). These factors clearly influenced Hunzakutz herding patterns with respect to the numbers and types of animal species they managed. The least numerous of the ruminants were cattle. Each Hunzakutz farmer kept only one or two milk cows. Some families might also own one or two bulls. These animals are small local breeds. (An adult cow weighs approximately 300 to 400 kg.) Every household managed its own cattle. Typically, young boys and girls were assigned the task of tending the cattle and collecting their manure. As studies in highland environments elsewhere have shown, this labor pattern has energy conservation functions because cattle herding requires only light to moderate physical exertion, and children by age twelve can accomplish the task but use approximately 30 percent fewer calories than adults performing identical chores (Thomas 1976).

Cattle remained in the village year-round. In summer and part of autumn (approximately 180 days), cattle grazed along the edges of waterways and fields and sometimes underneath fruit trees. In winter and early spring (about 180 days), cattle scavenged around the village by day and were stall-fed at night from stockpiles of straw, dry lucerne, and grass gathered and stored by farmers during the previous growing season.

Cattle are energetically costly to own because of their high fodder requirement, which at roughly 6 km dry matter (DM) a day is almost three times that needed by an adult sheep or goat. Moreover, because Hunzakutz households managed their own beasts, cow herding involved duplication of effort in comparison to herding sheep and goats, which were managed by a few professional village shepherds.

The contribution cattle made to the Hunzakutz diet was negligible because the animals are poor milk producers, yielding typically no more than 170 liters over a seven-and-a-half- to eight-month lactation period. This is well below the milk output of the Hunzakutz's goats. Cow's milk was churned into butter or made into cheese. As a source of meat protein, cattle are also impractical because of the great amount of labor and energy invested in their rearing and maintenance. Beef was eaten very infrequently, usually when an animal was hurt beyond recovery or was no longer productive. Finally, in terms of reproduction, cattle are also inefficient in comparison to sheep and goats. They reach sexual maturity at around eighteen months, have a nine-and-a-half-month gestation period, and seldom have multiple births.

Low productivity and high maintenance costs notwithstanding, Hunzakutz cattle were indispensable, both as a source of manure and as draft animals for plowing fields and threshing barley and wheat crops. Manure accumulated in the sheds and was also collected from grazing areas and added to the stockpiles. Farmers applied this manure to their fields before planting spring, summer, and autumn crops.

TRANSHUMANCE

The Hunzakutz may best be described as a transhumant society. There is considerable confusion in the literature regarding what exactly constitutes transhumance (cf. Dyson-Hudson and Dyson-Hudson 1980). For example, Bates (2001:93) characterizes transhumance as found in parts of the Middle East, Eastern Europe, Switzerland, and Central Asia as follows: "Transhumant nomads often camp together for extended periods of time in two major grazing areas: Summer pastures in the mountains and winter pastures in the valleys. During the migration between seasonal encampments the roads and trails are crowded with people and animals on the move." What Bates (ibid.) describes here is actually nomadic pastoralism. According to Jones (2005:359), such characterizations, shared by many anthropologists, obfuscate rather than contribute to our understanding of the economic system under consideration. Jones (ibid.:358) points out that the terms *transhumance* and *nomadic pastoralism* have often been erroneously used as simply "two words for the same things." However, he adds:

> Pastoral nomadism . . . is an economic system based primarily on animal husbandry, supplemented by trade . . . Confined to marginal environments, it is necessary for the entire community to move seasonally in order to ensure adequate grazing and water for the livestock . . . Since mobility of the human and livestock is essential for the well-being of both, pastoral nomads characteristically live in tents, yurts . . . or some other portable dwelling. This need for flexible mobility, combined with the relatively arid nature of the lands available to them, generally rules out any form of agriculture. It is by trading surplus livestock, hides, goat hair, butter, cheese, meat, wool, and other products that the pastoralists obtain goods which they themselves do not make. Far from being independent "wanderers," pastoral nomads rely heavily on trading relations with settled communities. (Ibid.)

Jones's characterization supports the observation that "pure nomads" who operate without reliance on agricultural commodities—traded or produced by themselves—do not exist (cf. Spooner 1973:19).

Johnson (1969) provides an accurate definition of a transhumant society that perfectly fits the pattern in Hunza:

> Transhumance is a term used to describe a spacially limited pattern of movement in mountainous areas . . . A village of permanent buildings occupied by all or part of the

> population all year, rather than a mobile tent camp[,] form the nucleus of a transhumant society. Although pastoral activities are one of the concerns of a transhumant community, agriculture nearly always remains the dominant interest. In other words, pastoral movements are limited in scale, usually take place in one valley system, and are undertaken by only a small proportion of the total population. None of these features are shared by pastoral nomads. (cited in Jones 2005:358)

Jones describes a transhumant society in northeast Afghanistan, which also fits the pattern in Hunza:

> Here in the region known as Nuristan the population resides in permanent villages around which are irrigated hill terraces for the production of wheat, maize, millet, and barley . . . The year begins in spring with the livestock (mostly goats, but also some sheep and cattle) being taken out of winter stables and herded up to the first pastures. The majority of the population resides in the village throughout the year. In autumn, when the harvest is complete, the livestock are brought back to the village area and go into winter quarters, where they are stall fed until spring. (Jones 1974:22–38)

Based on his fieldwork in Afghanistan, Dupree (1973:164) has pointed out another distinction between transhumance and pastoral nomadism. In transhumance, mobility is vertical over short distances. In contrast, pastoral nomads are herdsmen who move over long distances;often but not always, their movement is horizontal (ibid.). As Bates and Rassam (1983:111) have observed, the movement involves considerable distances because in the foothills and lowlands, pastures are situated further apart than they are in mountainous areas. The key factors that determine the number of animals herded and locations of encampments are the availability of pasture and water (ibid.). However, temperature is also an important consideration.

Among the Hunzakutz, the bulk of the population remained in the villages in the lower elevations, where they practiced irrigation farming. Only a small segment of the population moved with the livestock to the mountains. With this form of animal husbandry, herders did not need to enter into complex symbiotic politico-economic relationships with sedentary farmers to acquire agricultural foodstuffs, which were produced in their own villages, or to gain access to grazing grounds in territories controlled by others in a manner Barth (1956) has described for the Swat region of Pakistan (cf. Bacon 1954:46; Casimir 2003:85).

HUNZAKUTZ TRANSHUMANCE CYCLE

Sheep and goats are the primary focus of the Hunzakutz transhumance. These animals' flocking instincts make them well-suited to forage together and enable a single shepherd to handle roughly forty-five to fifty animals. Thus, only a few professional herdsmen were needed to tend the sheep and goats that belonged to an entire village. In spring, these herdsmen drove the animals to the upland pastures. This strategy averted the duplication of effort associated with cattle husbandry, which meant more efficient use of human labor.

Species of *Agropyron, Poa, Festuca,* and *Bromis* are the main vegetation in the alpine pastures. Among these, *Festuca* and *Agropyron* have higher nutritional value than the other plants (Van Swindern 1978). Growth commences in late March and increases in April and May, reaching its peak between mid-June and the first half of August. Pastures are productive only during the spring and summer seasons. However, these biomes have limited carrying capacity and are unable to support intensive and prolonged grazing (cf. Briggs and Courtney 1989:167–69). The reason is that the upland pastures are environments with thin and immature soils, inadequate biomass, and slow rates of plant growth.

To prevent grassland degradation through overgrazing, Hunzakutz shepherds, like herdsmen elsewhere (cf. Rhoades and Thompson 1975:540; Sutton and Anderson 2004:223), had to move their flocks over clearly defined successive elevations in the upland areas. The objective was to obtain maximum sustainable yields without destroying the fragile ecosystem. As Casimir (2003:81) put it: "Sustainability . . . would mean nothing more than managing flocks on a given pasture in such a way that the carrying capacity is not exceeded. The optimum strategy would be one that maximized animal production, without endangering the collapse of the system through over-grazing." Sheep and goats have different feeding habits. Sheep are grazers and goats are browsers. As Sutton and Anderson (2004:223) pointed out, "Grazers eat primarily grasses and low growing plants while browsers eat primarily the foliage from bushes and trees." For this reason, the two species utilize different biomes. Possessing remarkable agility, goats are able to browse on scrubs and bushes growing on the precipitous slopes and ledges. In contrast, sheep must remain at lower elevations of the pasturelands, where they graze on grassy patches. By adjusting their herd composition, shepherds were able to utilize natural foliage resources at different altitudes on the alpine pastures more efficiently (cf. Holmes 1980:171).

Typically, each household kept fifteen to twenty goats. Goats are local breeds and are genetically affiliated with the once numerous ibex (*Capra ibex*) that roamed the Karakoram Mountains (Mason 1981:73–75, 97–98, 104–5). Of all the livestock in central Hunza, goats are the best adapted to the harsh alpine environments and temperature extremes of the Karakorams. The most versatile feeders among ruminant species (Tivy 1990:118), goats can eat a vast array of shrubs, weeds, low trees, and thorny bushes—vegetation that would not sustain sheep or cattle foraging in the same parcel of land (cf. Wilkinson and Stark 1987:109–10).[6]

Adult goats weigh roughly 40–55 kg and consume about 2 kg DM per day. Because of their low fodder needs, a farmer could keep more animals per unit of fodder than he could of cattle. By maintaining many small-sized animals rather than one or two large ones, the farmer also minimized his investment risk if some of his goats died (cf. Bates 2001:91). Hence, goat herding was a more viable economic enterprise than cattle husbandry. Also, because farmers could keep larger herds of small ruminants, they could afford to periodically slaughter some of their animals (usually males) for meat without jeopardizing their stock.

Unlike cattle, goats are seasonal breeders and reach reproductive age at roughly thirteen months. The mating season, triggered by decreasing day lengths, begins at the start of September. At this time the animals were still in the upland pastures, well fed and in excellent condition for reproduction. The animals mate approximately six to eight weeks after they are driven back to the villages in late September. Gestation is about five months, and kidding takes place toward the end of winter and the start of spring. Goats usually give birth to one or two kids. During a single lactation period, a female goat can produce nearly 300 l of milk. This amount is roughly three times that of a sheep and twice that of a cow, making goats the primary dairy animals. In addition to their indispensible contribution to dairy production, goats also provided hair and hides traditionally used for weaving rugs and hats, making shoes, and fashioning vessels for carrying water and churning butter.

Hunzakutz farmers also kept about ten to fifteen head of sheep. Originally bred in foothill environments (Tivy 1990:118), sheep do not thrive as well as goats in the high-altitude environment of the Karakoram Mountains. Moreover, as noted, sheep are not versatile feeders and cannot be sustained on the types of low-grade vegetation eaten by goats. To survive, sheep were dependent on grass resources found in the upland pastures and produced on the toqs and in farmers' fields. Thus, the number of sheep maintained on a sustainable basis by any community was ultimately determined by the total amount of pastures to which it had grazing rights and the size of its parcels of farmland. An adult sheep weighs around 45 kg and requires about 2 kg DM of feed per day. Nevertheless, because of their small size, sheep offer farmers comparable economic benefits over cattle, as do goats.

Reproductively, sheep are comparable to goats, reaching sexual maturity in about thirteen months, and they are seasonal breeders. Their mating season and lambing are also similar to those of goats. While not prolific milk producers, one animal can produce about 100 l of milk per lactation; sheep's milk contains more fat and has a higher caloric value than milk produced by the other ruminants (cf. Pyke 1970:43; Kon 1972). Although sheep were not highly valued as milk producers, they were the primary providers of meat and wool.

Goats and sheep on the upland pastures foraged all day and were corralled at night because of the danger of attacks by carnivores, such as snow leopards (*Panthera unica*). The animals on the pastures quickly gained weight and soon reached peak lactation. In addition to their usefulness as dairy animals, sheep and goats on the pastures offered other advantages. Being on the pasture for five months of the year, during the spring and summer, Hunzakutz could feed and raise surplus animals. These additional animals could be used to replace unproductive beasts, which were slaughtered for meat.

Moreover, sheep and goats produced manure while in the upland meadows, which constituted an external input for fertilizing the fields in the villages below. Sheep and goats produce less manure than cattle. However, their manure has higher quantities of nitrogen, calcium, and potassium. Sheep and goat manure also contributes to quicker bacterial action and humus formation (Winterhalder, Larsen, and Thomas 1974:100; Price 1981:425). The precious goat and sheep dung was carefully gathered from the corrals by herdsmen, dried on the roofs of their huts, and periodically taken down to the villages.

Around the autumnal equinox, the livestock returned to the settlements below. Once in the villages, the goats and sheep grazed for about a month on the toqs, in orchards, and in the recently harvested grain fields. All the manure produced during this time was gathered and stockpiled for the spring planting. Having eaten all available vegetation and crop residues on village farmlands, the animals were thereafter confined to their stalls.

Just before the animals were stalled for the winter, the Hunzakutz observed a brief purification rite for their goats. They ritually fumigated the animal shed with smoke from the leaves of the juniper, a plant deemed sacred because of its association with the pari. Finally, prayers were uttered for the animals' safety and fertility.

During the time goats and sheep were stalled, they were maintained on household supplies of fodder carefully stockpiled during the previous agricultural year. Nevertheless, Hunzakutz could not provide their animals with a diet that was sufficient enough and had the requisite nutrients, so the animals lost weight. In the past, livestock-fodder imbalances were rectified by slaughtering a few sheep or goats during the Thumushelling festival, a nighttime torch-lighting festival held in honor of the Thum that was celebrated during the winter solstice. Additional livestock were killed as fodder supplies diminished during late winter and early spring, a critical period when household supplies of food would have dwindled as well.

In January, about three weeks after the Thumushelling was celebrated, the Hunzakutz observed a ritual called Gerakus. Herdsmen would visit people's goat sheds to see if any of the animals were pregnant. They then prayed to ensure the birth of many kids (Lorimer 1979:40–48). This officially concluded the annual transhumant cycle.

RITUALS AND RESOURCE MANAGEMENT

In the past, pastoral activities and agricultural operations were coordinated by state decision-makers. Lorimer (1935) and Müller-Stellrecht (1973) have provided descriptive accounts of the ritualized agropastoral calendar. Ali (1981, 1982:113) has analyzed some of these rituals from a sociological perspective, that is, in terms of their ability to express and reinforce social solidarity. However, the ecological-regulating functions of Hunzakutz pastoral rites, which I describe here, have largely been ignored.

The annual transhumant cycle started in early May and finished during late September or early October. The move to the uplands was marked by a ritual called the Odi. The date for the Odi was determined by the Wazir (prime minister) based on observation of the maturation of grasses on the pastures, assessment of temperatures, and the availability of glacial meltwaters (cf. Salzman 2004:3). The Thum, who was believed to have magical powers, gave his blessing to the enterprise (as he did for different phases of agricultural production). Herders were now permitted to drive the animals to specifically stipulated pastures as grass cover reached maximum yields. Those who did not follow the official timetable had to pay a stiff fine.

This schedule had resource management functions that safeguarded the pastures from early arrival of the animals and overgrazing. Barth (1964) documented a similar pattern among the Basseri pastoralists of Iran, where individuals possess their own livestock but the assignment of herds to particular pastures is the task of tribal chiefs. This kind of practice, according to Sutton and Anderson (2004:223), "served to prevent over exploitation of pastures by any one segment, as overgrazing could endanger the pasture system of the entire group. In addition to the regulation of grazing, pasture assignment functioned to control animal populations, which in turn helped to control the human population by limiting the food supply."

The Odi ritual had ecological management functions because it calibrated the nutritional requirements of lactating goats and sheep with optimal grass growth on the upland pastures and at the same time protected biodiversity. As Sutton and Anderson (ibid.:115) noted: "[Rituals] can function as a form of resource management in a number of ways. All groups believe in some sort of supernatural power that has control over the environment, such as gods that control rain or the movement of the sun. Rituals, including prayers, ceremonies, and even some art, are used to influence these deities so that the sun will rise and the rain will come. These rituals, then, serve to manage the environment."

Another ritual behavior with resource conservation functions was the Hunzakutz belief in the sacredness of the upland pastures and a reverence for the spirits dwelling in the mountains. Such beliefs are not stochastic, irrational ideas subject to their own internal dynamics but must be construed in terms of the ecological systems that gave rise to them (Sidky 2004:373). Traditionally, the Hunzakutz regarded their upland grasslands and the mountains beyond as hallowed places, the domain of the pari. The belief in these beings seems to predate Islam and is linked to other surviving indigenous non-Islamic beliefs associated with oracles and mediums (Sidky 1994). These capricious and lethal entities, which were said to have special ties to the Thums of Hunza—and thus were linked to the state's religious ideological superstructure—were believed to have the power to bring good luck or misfortune. If offended, they could harm humans, devastate livestock herds, and destroy crops. Shepherds living in the alpine pastures felt the presence of the pari most acutely, hearing their voices at night or seeing them in various guises (ibid.). Herdsmen who were careless in tending their flocks, who did not heed the Thum's mandates regarding the upland pastures, or who damaged the fragile ecosystem by neglect or overgrazing, it was believed, were sure to incur the fury of these supernatural beings (ibid.).

Another belief among the Hunzakutz was that cattle should not be taken to the upland meadows because of the idea that cattle are ritually polluting and offensive to the pari.[7] This belief is the exact opposite of Hindu beliefs in the sacredness of cattle. Ostensibly, beliefs regarding the impurity of cattle may have fulfilled important ecological functions in a way that was different from, but comparable to, the idea of the sacred cow in India (Harris 1966, 1974). Keeping the cattle in the villages had sound ecological underpinnings because these animals, which have higher consumption requirements, would have competed with goats and sheep for the meager vegetation cover in the alpine pastures. Moreover, as heavier beasts, cattle hooves could easily damage the thin grass cover and therefore pose a hazard to the fragile alpine ecosystem. Finally, cattle were needed in the villages for plowing and threshing.

Careful analysis of Hunzakutz rituals and ritual beliefs in terms of what is known about the physiological and environmental basis of agronomic yields and animal husbandry suggests that they fulfilled ecological-regulating functions because they were based on years of practical experience and a sound understanding of the ecosystem, different biomes, vertical resource clusters, and plant and animal physiology.

These beliefs taken together, one could argue, averted "the tragedy of the unmanaged commons" (cf. Sutton and Anderson 2004:292–94; Hardin 2007:105). This is similarly the case elsewhere in the western Himalayas, where in areas protected by governmental or spiritual sanctions (graveyards/shrines), overexploitation is averted in comparison to pastures treated as "no man's land" (Casimir 2003:100–101). The damage done to Hunza's delicate alpine meadows following the lifting of traditional restrictions when the Hunza State was abolished in 1974 is evidence of the efficacy of traditional managerial controls and supernatural beliefs regarding the upland pastures (cf. Sidky 1993b).

CONCLUSIONS

I have shown how Hunza's "oasis environment" posed severe limitations on human subsistence, focusing on how, during the isolation of the past, the Hunzakutz attempted to address these limitations and how transhumance and cattle herding were the keys to their success. These limitations, as we have seen, included:

1. Demographic pressure
2. Shortages of arable land
3. Length of the growing season
4. Shortages of soil nitrogen
5. Shortages of fodder
6. Short growing season
7. Overexploitation of upland pastures

Demographic pressure on resources was dealt with through the construction of a complex hydraulic system and establishment of new settlements. The shortages of arable land and length of the growing season were managed through intensive land use and a carefully scheduled strategy of double-cropping regulated by the state's decision-making protocols to make the best use of seasonal variations in temperatures and amount of sunlight available for plant growth. Reliance on manure from goats, sheep, and cattle compensated for the shortage of soil nitrogen. The availability of manure was linked to the lack of significant natural forage resources and water in the arid areas surrounding the villages. The shortage of arable land prevented farmers from devoting significant portions of their fields, on which they needed to grow food for their families, to the cultivation of fodder. The solution was found in a set of strategic decisions involving vertical and horizontal land-use patterns, reliance on cattle husbandry, transhumance, careful regulation of herd size and composition, movement of livestock over successive levels in the pastures, adherence to a ritualized state-sponsored agropastoral calendar, and a belief in mountain spirits as supernatural protectors of the alpine meadows, which facilitated the most efficient use of time and space and the preservation of biodiversity in the alpine pastures.

NOTES

1. This chapter directly draws and elaborates on topics and ethnographic materials first presented in Sidky (1996, 1997).

2. A biome is defined as a "broad ecological unit characterized by a set of climatic parameters and types of floral and faunal associations" (Moran 2000:340).

3. The carrying capacity of trans-Himalayan desert-steppe zones is extremely low, averaging roughly 0.06 livestock units (goats) per hectare. Nevertheless, Hunzakutz farmers deployed their goats to exploit the sparse vegetation that grew in the areas surrounding their oasis villages. The absence of water in these arid zones, however, necessitated bringing the goats back to the villages each night.

4. On the degradation of the forests following the end of state controls, see Schickhoff (2006).

5. On the importance of manure in highland environments elsewhere, see Winterhalder, Larsen, and Thomas (1974).

6. This is probably why goats traditionally had a special place in Hunzakutz cosmology. The Hunzakutz believe the mountain spirits especially favor goats because these animals resemble the ibex and *markhor* (*Capra falconeri*), the "pets of the mountain spirits." The traditional importance of these animals in Hunzakutz cosmology is evident from their copious representations on ancient petroglyphs near Ganesh village in central Hunza. Because the ibex stands for strength and prosperity, the Thum placed a wooden statue of an ibex on the roof of his castle in Altit village. For the Hunzakutz, the connection between their goats and the ibex and markhor made these animals sacred. Sheep, in contrast, were considered ritually neutral and hence could be grazed and stalled together with the goats.

7. As I have noted elsewhere (Sidky 1996:122), comparable beliefs are pivotal to the "pastoral ideology" of the Kalasha, a non-Islamic people in the Hindu Kush Mountains of northern Pakistan (Drew 1875:428; Biddulph 1880:37–38; Jettmar 1961:87–91; Staley 1982:178; Parkes 1987:647, 1992:38). The Kalasha believe goats are the descendants of the sacred ibex and markhor (*Capra falconeri*), originally bestowed on their ancestors by the mountain deities. Thus, goats represent a tangible link between the world of humans and the gods (Parkes 1992:38). A related belief was that females are not to come into contact with goats because they menstruate and are therefore ritually unclean (cf. Staley 1982:179). Similarly, goats must be kept apart from cattle because of the latter's mundane origins and association with farming. These beliefs are the underpinning of a stringent division of labor in which men tend goats and take them to the upland pastures, while women are responsible for managing the cattle and farming down in the villages. One might conjecture that a similar ideology may have once prevailed among the Hunzakutz before the advent of Islam, of which only fragments have survived.

REFERENCES

Ahmed, Akbar S. 1988. *Pakistan Society: Islam, Ethnicity and Leadership in South Asia*. Delhi: Oxford University Press.

Ali, Tahir. 1981. "Ceremonial and Social Structure among the Burusho of Hunza." In *Asian Highland Societies in Anthropological Perspective*, ed. Christoph von Fürer-Haimendorf, 231–49. New Delhi: Sterling.

Ali, Tahir. 1982. "The Burusho of Hunza: Social Structure and Household Viability in a Mountain Desert Kingdom." PhD diss., Department of Anthroplogy, University of Rochester, New York.

Bacon, Elizabeth. 1954. "Types of Pastoral Nomadism in Central and Southwest Asia." *Southwestern Journal of Anthropology* 10 (1): 44–68.

Barfield, Thomas J. 1993. *The Nomadic Alternative*. Englewood Cliffs, NJ: Prentice-Hall.

Barth, Fredrik. 1956. "Ecologic Relationships of Ethnic Groups in Swat, North Pakistan." *American Anthropologist* 58 (6): 1079–89. http://dx.doi.org/10.1525/aa.1956.58.6.02a00080.

Barth, Fredrik. 1964. *Nomads of South Persia: The Basseri Tribe of the Khamseh Confederacy*. New York: Humanities Press.

Bates, Daniel. 2001. *Human Adaptive Strategies: Ecology, Culture, and Politics*, 2nd ed. Needham Heights, MA: Allyn and Bacon.

Bates, Daniel, and Amal Rassam. 1983. *The Peoples and Cultures of the Middle East*. Englewood Cliffs, NJ: Prentice-Hall.

Bianca, Stefano. 2005. "Introduction: Reclaiming the Cultural Heritage of the Northern Areas as Part of Integrated Development Process." In *Karakoram: Hidden Treasures in the Northern Areas of Pakistan*, ed. Stefano Blanca, 11–20. Torino: Umberto Allemande.

Biddulph, John. 1880. *Tribes of the Hindoo Koosh*. Calcutta: Office of the Superintendent of Government Printing.

Brady, Nyle E. 1974. *The Nature and Properties of Soil*. New York: Macmillan.

Briggs, David J., and Frank M. Courtney. 1989. *Agriculture and Environment*. Harlow, Essex: Longman Technical and Scientific.

Buddruss, Georg. 1985. "Linguistic Research in Gilgit and Hunza: Some Results and Perspectives." *Journal of Central Asia* 8 (1): 27–32.

Casimir, Michael. 2003. "Pastoral Nomadism in a West Himalayan Valley: Sustainability and Herd Management." In *Nomadism in South Asia*, ed. Aparna Rao and Michael Casimir, 81–103. New Delhi: Oxford University Press.

Charles, Christian. 1985. "La Vallee de Hunza Karakorum." PhD thesis, Institut de Geographie, Universite de Grenoble.

Clark, John. 1956. *Hunza: Lost Kingdom of the Himalayas*. New York: Funk and Wagnalls.

Crane, Robert I., ed. 1956. *Area Handbook on Jammu and Kashmir State*. Chicago: Division of the Social Sciences, University of Chicago.

Drew, Frederic. 1875. *The Jammoo and Kashmir Territories: A Geographical Account*. London: E. Stanford.

Duckham, Alec N., and Geoffrey B. Masefield. 1970. *Farming Systems of the World*. London: Chatto and Windus.

Dupree, Louis. 1973. *Afghanistan*. Princeton: Princeton University Press.

Durand, Algernon. 1899. *The Making of a Frontier: Five Years' Experiences and Adventures in Gilgit, Hunza, Nagar, Chitral and the Eastern Hindu-Kush*. London: John Murray.

Dyson-Hudson, Rada, and Neville Dyson-Hudson. 1980. "Nomadic Pastoralism." *Annual Review of Anthropology* 9 (1): 15–61. http://dx.doi.org/10.1146/annurev.an.09.100180.000311.

Eberhardt, Einar, W. Bernhard Dickore, and George Miehe. 2006. "Vegetation of Hunza Valley: Diversity, Altitudinal Distribution and Human Impact." In *Karakoram in Transition: Culture, Development, and Ecology of the Hunza Valley*, ed. Hermann Kreutzmann, 109–22. Oxford: Oxford University Press.

Emerson, Richard M. 1990. "Charismatic Kinship: A Study of State Formation and Authority in Baltistan." In *Pakistan: The Social Sciences Perspective*, ed. Akbar Ahmed, 100–145. Karachi: Oxford University Press.

FAO (Food and Agricultural Organization, United Nations). 1961. *Agricultural and Horticultural Seed*. FAO Agricultural Studies 55. Rome: United Nations Food and Agricultural Organization.

FAO (Food and Agricultural Organization, United Nations). 1978. *Middle East Grassland Education and Training with Special Reference to Iran, Pakistan, and Afghanistan*. Rome: United Nations Food and Agricultural Organization.

Frissel, Michael. 1977. "Cycling Mineral Nutrients in Agricultural Ecosystems." *Agro-ecosystems* 4: 17–25.

Groetzbach, Erwin F. 1988. "High Mountains as Human Habitat." In *Human Impact on Mountains*, ed. Nigel Allan, Gregory W. Knapp, and Christoph Stadel, 24–35. Totowa, NJ: Rowman and Littlefield.

Guillet, David. 1983. "Towards a Cultural Ecology of Mountains: The Central Andes and the Himalayas Compared." *Current Anthropology* 24 (5): 561–74. http://dx.doi.org/10.1086/203061.

Hamid, Shahid. 1979. *Karakuram Hunza: The Land of Just Enough*. Karachi, Pakistan: Ma'aref.

Hardin, Garrett. 2007. "The Tragedy of the Unmanaged Commons." In *Evolutionary Perspectives on Environmental Problems*, ed. Dustin Penn and Iver Mysterud, 105–7. New Brunswick, NJ: Aldine Transaction.

Harris, Marvin. 1966. "The Cultural Ecology of India's Sacred Cattle." *Current Anthropology* 7 (1): 51–66. http://dx.doi.org/10.1086/200662.

Harris, Marvin. 1974. *Cows, Pigs, Wars, and Witches: The Riddles of Culture*. New York: Random House.

Hauptmann, Harald. 2005. "Pre-Islamic Heritage in the Northern Areas of Pakistan." In *Karakoram: Hidden Treasures in the Northern Areas of Pakistan*, ed. Stefano Blanca, 21–40. Torino: Umberto Allemandi.

Holmes, William. 1980. *Grass: Its Production and Utilization*. Oxford: Blackwell Scientific.

Irons, William. 1965. "Livestock Raiding among Pastoralists: An Adaptive Interpretation." *Papers of the Michigan Academy of Science, Arts and Letters* 50: 393–414.

Jettmar, Karl. 1961. "Ethnological Research in Dardistan 1958: Preliminary Report." *Proceedings of the American Philosophical Society* 105 (1): 79–97.

Johnson, Allen, and Timothy Earle. 2000. *The Evolution of Human Societies: From Foraging Group to Agrarian State*. Stanford: Stanford University Press.

Johnson, Douglas L. 1969. *The Nature of Nomadism: A Comparative Study of Pastoral Migrations in Southwestern Asia and Northern Africa*. Chicago: Chicago University Press.

Jones, Schuyler. 1974. *Men of Influence in Nuristan*. London: Seminar.

Jones, Schuyler. 2005. "Comment: Transhumance Re-Examined." *Journal of the Royal Anthropological Institute* 11 (2): 357–59. http://dx.doi.org/10.1111/j.1467-9655.2005.00240.x.

Khan, Mohammed A. 1983. "Rural Development in the Karakoram-Himalayan Region of Pakistan: Report of an International Seminar, January 17–19, 1983." In *Management for Rural Development in Pakistan*, ed. Mirza Beg, Salim M. Anwer, and Harold Jonathan, 123–35. Peshawar: Pakistan Academy for Rural Development.

Knight, Edward. 1893. *Where Three Empires Meet: A Narrative of Recent Travels in Kashmir, Western Tibet, Gilgit, and the Adjoining Countries*. London: Longman, Green.

Kon, Stanislaw. 1972. *Milk and Milk Products in Human Nutrition*. Rome: United Nations Food and Agricultural Organization.

Kreutzmann, Hermann. 1988. "Oases of the Karakorum: Evolution of Irrigation and Social Organization in Hunza, North Pakistan." In *Human Impact on Mountains*, ed. Nigel Allan, Gregory W. Knapp, and Christoph Stadel, 243–54. Totowa, NJ: Rowman and Littlefield.

Kreutzmann, Hermann. 1991. "The Karakoram Highway: The Impact of Road Construction on Mountain Societies." *Modern Asian Studies* 25 (4): 711–36. http://dx.doi.org/10.1017/S0026749X00010817.

Kreutzmann, Hermann. 2000. "Water Management in Mountain Oases of the Karakoram." In *Sharing Water: Irrigation and Water Management in the Hindu Kush–Karakoram–Himalayas*, ed. Hermann Kreutzmann, 90–115. Oxford: Oxford University Press.

Kreutzmann, Hermann. 2005. "The Karakoram Landscape and the Recent History of the Northern Areas." In *Karakoram: Hidden Treasures in the Northern Areas of Pakistan*, ed. Stefano Blanca, 41–76. Torino: Umberto Allemandi.

Langer, Reinhart H., and George D. Hill. 1991. *Agricultural Plants*. Cambridge: Cambridge University Press. http://dx.doi.org/10.1017/CBO9781139170284.

Lees, Susan, and Daniel Bates. 1974. "The Origins of Specialized Nomadic Pastoralism: A Systemic Model." *American Antiquity* 39 (2): 187–93. http://dx.doi.org/10.2307/279581.

Lett, James. 1987. *The Human Enterprise: A Critical Introduction to Anthropological Theory*. Boulder: Westview.

Lorimer, David L.R. 1935. *The Burushaski Language*, vol. 1–3. Oslo: H. Aschenhoug.

Lorimer, David L.R. 1979. *Materialien zur Ethnographie von Dardistan, Teil I, Hunza*. Graz: Druck-und Veriagsanstalt.

Lorimer, David L.R. 1980. *Hunza: Materialize zur Ethnographie van Dardistan, Teil II, Gilgit, Teil III, Chitral and Yasin*. Graz: Akademische Druck-und Verlagsanstalt.

Mason, I. L. 1981. "Breeds." In *Goat Production*, ed. C. Gall, 61–135. London: Academic.

Mons, Barbara. 1958. *High Roads to Hunza*. London: Faber and Faber.

Moran, Emilio. 2000. *Human Adaptability: An Introduction to Ecological Anthropology*. Boulder: Westview.

Müller-Stellrecht, Irmatrud. 1973. *Feste in Dardistan: Durstellung urid kulturgeschichtliche Analyse*. Wiesbaden, Germany: Franz Steiner.

Nüsser, Marcus, and Jürgen Clemens. 1996. "Impacts on Mixed Mountain Agriculture in the Rupal Valley, Nanga Parbat, Northern Pakistan." *Mountain Research and Development* 16 (2): 117–30. http://dx.doi.org/10.2307/3674006.

Parkes, Peter. 1987. "Livestock Symbolism and Pastoral Ideology among the Kafirs of the Hindu Kush." *Man* (new series) 22 (4): 637–70. http://dx.doi.org/10.2307/2803356.

Parkes, Peter. 1992. "Reciprocity and Redistribution in Kalasha Prestige Feasts." *Anthropozoologica* 16: 37–46.

Poucha, Pavel. 1960. "Bru-a-Burasaski?" *Central Asiatic Journal* 5: 295–300.

Price, Larry W. 1981. *Mountains and Man: A Study of Process and Environment*. Los Angeles: University of California Press.

Pyke, Magnus. 1970. *Man and Food*. New York: McGraw-Hill.

Rhoades, Robert E., and Stephen I. Thompson. 1975. "Adaptive Strategies in Alpine Environments: Beyond Ecological Particularism." *American Ethnologist* 2 (3): 535–51. http://dx.doi.org/10.1525/ae.1975.2.3.02a00110.

Salzman, Philip C. 2004. *Pastoralists: Equality, Hierarchy, and the State*. Boulder: Westview.

Sanderson, Stephen. 1999. *Social Transformations: A General Theory of Historical Development*. Lanham, MD: Rowman and Littlfield.

Saunders, F. 1984. *Karakorum Villages: An Agrarian Study of Twenty-Two Villages in the Hunza, Ishkoman, and Yasin Valleys of Gilgit District*. Gilgit, Pakistan: United Nations Food and Agricultural Organization.

Schickhoff, Udo. 2006. "The Forests of Hunza Valley: Scarce Resources under Threat." In *Karakoram in Transition: Culture, Development, and Ecology of the Hunza Valley*, ed. Hermann Kreutzmann, 123–44. Oxford: Oxford University Press.

Schmid, Anna. 2000. "Minority Strategies to Water Access: The Dom in Hunza, Northern Areas of Pakistan." In *Sharing Water: Irrigation and Water Management in the HinduKush-Karakoram Himalayas*, ed. Hermann Kreutzmann, 116–31. Oxford: Oxford University Press.

Searle, Mike. 2006. "Geology of the Hunza Karakoram." In *Karakoram in Transition: Culture, Development, and Ecology of the Hunza Valley*, ed. Hermann Kreutzmann, 7–11. Oxford: Oxford University Press.

Shahrani, M. Nazif Mohib. 1979. *The Kirghiz and Wakhi of Afghanistan: Adaptations to Closed Frontiers*. Seattle: University of Washington Press.

Shor, Jean Bowie. 1955. *After You Marco Polo*. New York: MacGraw-Hill.

Sidky, Homayun. 1993a. "Irrigation and the Political Evolution of the High-Mountain Kingdom of Hunza." *Asian Affairs* 24 (2): 131–44. http://dx.doi.org/10.1080/714041207.

Sidky, Homayun. 1993b. "Subsistence, Ecology, and Social Organization among the Hunzakutz: A High-Mountain People in the Karakorams." *Eastern Anthropologist* 46 (2): 145–70.

Sidky, Homayun. 1994. "Shamans and Mountain Spirits in Hunza." *Asian Folklore Studies* 53 (1): 67–96. http://dx.doi.org/10.2307/1178560.

Sidky, Homayun. 1996. *Irrigation and State Formation in Hunza: The Anthropology of a Hydraulic Kingdom*. Lanham, MD: University Press of America.

Sidky, Homayun. 1997. "Irrigation and the Rise of the State in Hunza: A Case for the Hydraulic Hypothesis." *Modern Asian Studies* 31 (4): 995–1017. http://dx.doi.org/10.1017/S0026749X00017236.

Sidky, Homayun. 1999. "Alexander the Great, the Graeco-Bactrians, and Hunza: Greek Descent in Central Asia." *Central Asiatic Journal* 43 (2): 232–48.

Sidky, Homayun. 2003. "Verticality, Multiple Resource Utilization and Subsistence Economy in the Karakoram Mountains: The Case of the Transhumant Hunzakutz." In *Nomadism in South Asia*, ed. Aparna Rao and Michael Casimir, 307–41. New Delhi: Oxford University Press.

Sidky, Homayun. 2004. *Perspectives on Culture: A Critical Introduction to Theory in Cultural Anthropology*. Upper Saddle River, NJ: Pearson/Prentice-Hall.

Spooner, Brian. 1973. *The Cultural Ecology of Pastoral Nomads*. Addison-Wesley Modules in Anthropology 45. Reading, MA: Addison-Wesley.

Staley, John. 1969. "Economy and Society in the High Mountains of Northern Pakistan." *Modern Asian Studies* 3 (3): 225–43. http://dx.doi.org/10.1017/S0026749X00002341.

Staley, John. 1982. *Words for My Brother: Travels between the Hindu Kush and the Himalayas*. Karachi: Oxford University Press.

Stein, Rolf A. 1972. *Tibetan Civilization*. Stanford: Stanford University Press.

Stephens, Ian. 1955. *Horned Moon: An Account of a Journey through Pakistan, Kashmir and Afghanistan*. Bloomington: Indiana University Press.

Sutton, Mark, and E. N. Anderson. 2004. *Introduction to Cultural Ecology*. Walnut Creek, CA: AltaMira.

Sweet, Louise. 1965. "Camel Raiding of North Arabian Bedouin: A Mechanism of Ecological Adaptation." *American Anthropologist* 67 (5): 1132–50. http://dx.doi.org/10.1525/aa.1965.67.5.02a00030.

Thomas, Lowell, and Tay Thomas. 1960. "Sky Road East." *National Geographic* 117 (1): 70–112.

Thomas, Richard B. 1976. "Energy Flow at High Altitude." In *Man in the Andes: A Multidisciplinary Study of High-Altitude Quechua*, ed. Paul T. Baker and Michael Little, 379–404. Stroudsburg, PA: Hutchinson and Ross.

Tivy, Joy. 1990. *Agricultural Ecology*. Harlow, Essex: Longman Scientific and Technical.

Van Swindern, H. 1978. *Terminal Report of the Pasture and Fodder Technical Report*. Kathmandu: United Nations Development Program.

Whiteman, Peter T. 1985. *Mountain Oases: A Technical Report of Agricultural Studies in Hunza, Ishkoman and Yasin Valleys of Gilgit District*. Rome: United Nations Food and Agricultural Organization.

Whiteman, Peter T. 1988. "Mountain Agronomy in Ethiopia, Nepal and Pakistan." In *Human Impact on Mountains*, ed. Nigel Allan, Gregory W. Knapp, and Christoph Stadel, 57–82. Totowa, NJ: Rowman and Littlefield.

Wilkinson, John M., and Barbara Stark. 1987. *Commercial Goat Production*. Oxford: BSP Professional Books.

Winterhalder, Bruce, Robert Larsen, and R. Brooke Thomas. 1974. "Dung as an Essential Resource in a Highland Peruvian Community." *Human Ecology* 2: 89–104. http://dx.doi.org/10.1007/BF01558115.

Wittfogel, Karl. 1956. "Hydraulic Civilizations." In *Man's Role in Changing the Face of the Earth*, ed. William Thomas, 152–64. Chicago: University of Chicago Press.

Younghusband, Francis E. 1896. *The Heart of a Continent: A Narrative of Travels in Manchuria, across the Gobi Desert, through the Himalayas, the Pamirs, and Hunza, 1884–1894*. London: John Murray.

5

Animals, Identity, and Mortuary Behavior in Late Bronze Age–Early Iron Age Mongolia

A Reassessment of Faunal Remains in Mortuary Monuments of Nomadic Pastoralists

Erik G. Johannesson

Ancient Chinese historical texts of the Han Dynasty (200 BC–AD 200), such as Sima Qian's *Shiji*, describe that in the third century BC the nomadic steppe tribes of what today is Mongolia were united into a powerful confederacy the Chinese referred to as the Xiongnu. Within a few years of the formation of this polity, its leader, Modun, sent a declaration to the Han court announcing his subjugation of the region and ascendance to sole hegemon of the northern steppes. All were now considered Xiongnu. Henceforth the Han would have to contend with this—according to Sima Qian—aggressive and confrontational entity, thus setting the stage for the next four centuries of intermittent war and diplomacy with the Xiongnu.

So important was this interaction that Sima Qian dedicated an entire book of his history to the Xiongnu. This narrative has since figured prominently in histories of the relationship between the Chinese and nomadic peoples on their northern frontier (Lattimore 1967; Grousset 1970; Frye 1996). Sima Qian's description of the Xiongnu as nomadic herders has typically been accepted as accurate, but it raises a number of questions that are pertinent to identifying herding practices in the past. If the Xiongnu truly relied on a nomadic pastoralist economy, how can that practice and its associated behaviors be identified in material culture attributed to the Xiongnu? If the majority of people living under the Xiongnu yoke were pastoralists, when was the practice adopted in Mongolia? What complicates these questions is that nomadic pastoralists leave ephemeral traces archaeologically, which has

DOI: 10.5876/9781607323433.c005

traditionally constrained archaeological research to focus on mortuary contexts (Cribb 1991). Indeed, with a few exceptions, the preponderance of archaeological material on the Xiongnu and preceding periods derives from graves. Graves, their contents, and accompanying monuments constitute very particular kinds of archaeological material (Parker Pearson 2000). Mortuary contexts represent deliberate human actions and constructs that are inherently symbolic (Härke 1997). Therefore, graves and funerary monuments are not a direct reflection or correlate of the society that produced them but instead reveal only the symbolic ideologies relevant or important to the individuals who constructed them. This has tremendous consequences for identifying pastoralism in the past and for discussing the ecology of pastoralism in prehistoric human societies. Here I argue that faunal remains recovered from mortuary contexts cannot directly be used to infer nomadic pastoralism, nor can they form the basis for assuming that the occupant of a grave or its builders practiced animal husbandry. Instead, I maintain that the adoption of nomadic pastoralist life-ways entails corresponding changes in other aspects of culture and worldview, in particular in relation to how humans see the world, their relationship with particular animals, and where they fit into the natural order. Increased reliance on animal husbandry, nomadic or otherwise, is therefore not accessible archaeologically in funerary contexts based on the presence of animals but rather on how those animals figured into predominant symbolic and ideological repertoires. Through a description of mortuary practices during the Late Bronze and Early Iron Ages in Mongolia (1500–300 BC), leading up to the formation of the Xiongnu polity, I then illustrate how the incorporation of faunal materials changed over time and how that change suggests an increase in animal husbandry in the region. Finally, I demonstrate that upon the formation of the Xiongnu polity, its elite were manipulating symbols and associated ideology in funerary practice to express political and cultural unity while restricting the ideological expression of local lineages of leadership. A facet of this strategy focused on the use of animals, specifically pastoral resources associated with nomadic herding, to incorporate local areas into a wider Xiongnu political economy.

The data presented herein are primarily derived from Baga Gazaryn Chuluu (hereafter BGC), just north of the Gobi Desert in central Mongolia, with comparative material presented from published reports from other parts of Mongolia and Siberia (see figure 5.1). BGC is a constellation of granite cliffs and ridges that rise out of the desert-steppe just north of the Gobi Desert in the Dundgovi province of Mongolia. The area consists of approximately 85 km^2 of rocky terrain surrounded by a vast expanse of desert-steppe. Between 2003 and 2008 the Baga Gazaryn Chuluu Archaeological Survey Project conducted a systematic intensive pedestrian survey of approximately 240 km^2 of BGC and its hinterland, in addition to

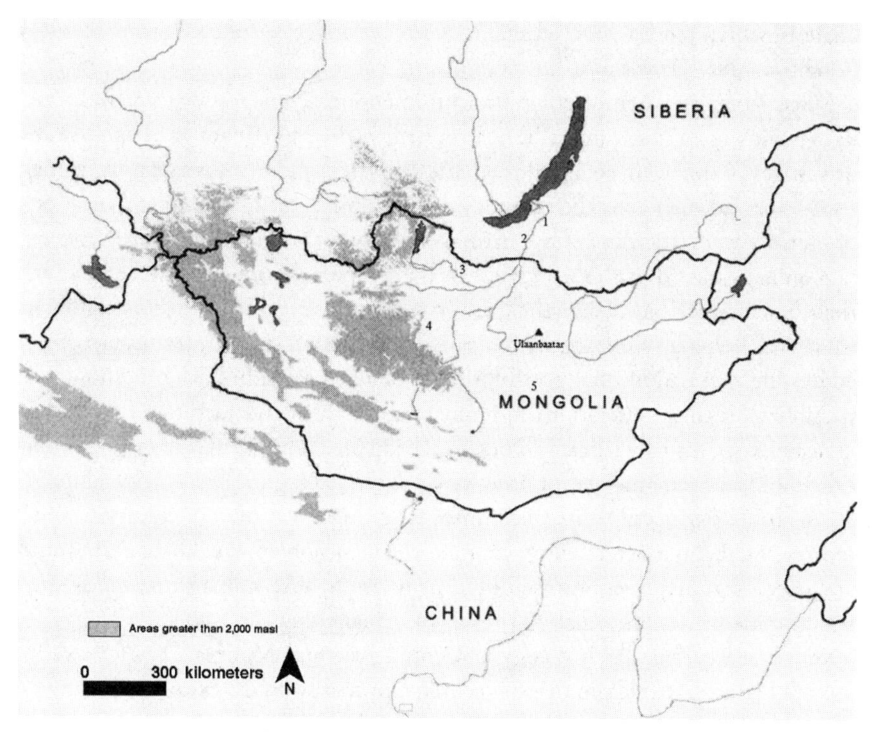

FIGURE 5.1. Locations of sites in Mongolia and Siberia mentioned in the text. 1. Ivolga, 2. Derestuy, 3. Egiin Gol, 4. Khanuy Valley, 5. Baga Gazaryn Chuluu.

extensive excavation. The purpose of the project was to identify, catalog, and document archaeological sites in the area and to create a comparative database of mortuary stone monuments in Mongolia. At the project's conclusion in 2008, roughly 1,750 archaeological sites had been identified, of which 812 could be assigned to the Bronze-Iron Age transition in Mongolia. Of these 812 monuments, 125 were excavated by the end of the project.

Before discussing the archaeology of Bronze and Iron Age Mongolia, it is necessary to mention a number of factors that complicate an assessment of animal use in Mongolia and, more broadly, Central Asia. First, archaeology in Mongolia is very much in its developmental phase and is complicated by a lack of high-resolution radiocarbon dates. Second, as mentioned, nomadic pastoralists leave ephemeral traces archaeologically, with the result that a predominance of archaeological materials recovered of nomadic pastoralists are derived from funerary contexts. Finally, mortuary contexts present a different set of parameters for the interpretation of faunal materials, upending traditional approaches to faunal analysis, which tend to view animal remains as derivatives of economic activities (Crabtree 1989; Driver 1990).

Since funerary assemblages represent intentionally deposited materials, faunal remains in burials cannot be interpreted in the same manner as spoil-heaps or vestiges of domestic economic activities (Pearson and Shanks 2001; Marciniak 2006). Instead, they must be recognized as symbolic and material manifestations of humans' understanding of the natural world order and reflective of the role certain animals played in the worldview of the people who incorporated them into their funerary customs. Nevertheless, a strong tendency remains in archaeology to view faunal remains as primarily indicative of economic activities without equally addressing these additional possibilities. In spite of these obstacles, it should be possible to identify alterations in animal exploitation and ideology from funerary contexts. This may be carried out by relying on multiple lines of evidence that examine faunal remains against other types of archaeological data and by adopting a diachronic perspective that traces material change over time. In addition, since mortuary remains are inherently symbolic, the inclusion of faunal remains in funerary assemblages represents a useful means to assess how animals may be used to transmit ideological information through mortuary practice.

In the last century, archaeology in Central Asia and Mongolia has followed a culture-historical paradigm that has centered primarily on the excavation of monumental burial mounds, or *kurgans* (Yablonsky 2000; Christian 2001; see Chang, this volume). Within this paradigm, faunal remains have rarely been the focus of systematic archaeological research and have played only a minor role in the interpretation of material change. A notable exception has been a large-scale focus on the domestication of the horse (Levine, Renfrew, and Boyle 2004; Olsen 2006; Anthony 2007). This general lack of consideration is conspicuous because of the fact that faunal remains are virtually omnipresent in Central Asian burial contexts. Moreover, most burials have been disrupted as a result of extensive pillaging activities, but faunal remains have often been left undisturbed because of their low economic value to grave robbers. As a result, faunal remains represent a largely untouched archaeological resource that may be used to interpret the life-ways of nomadic pastoralists.

Traditional zooarchaeological approaches to faunal remains have not been particularly informative in Mongolia. Part of the reason for this is the symbolic nature of mortuary contexts, which constitute the majority of archaeological sites. Age sets of domesticated animals, particularly of very young animals, have been used to identify the season in which a burial was constructed. However, identifying the time of year a burial and attendant funeral took place carries with it the assumption that such scheduling was indeed significant, which may not always have been the case (Parker Pearson 2000). Roger Cribb (1985) has demonstrated that there is no direct correspondence between species composition and age structure of animals from archaeological contexts and the herds from which they originally derived.

Determining the health and stature of animals in mortuary contexts is also problematic for the same reason. We simply cannot know if there were healthier, larger, or smaller animals to choose from. Furthermore, it is problematic to infer herd size or composition from mortuary contexts because it is unknown whether remains represent animals from one or multiple herds. This is a particularly important question, considering that funerals are often communal gatherings during which the living reaffirm or realign socio-political alliances. It is therefore possible that animals included in the faunal assemblages were brought by various people who attended the funeral rites. This, in turn, raises the question of whether faunal remains were the result of ritual feasting during the funeral itself or of the events that surrounded it.

Feasting has been suggested in both Central Asia and Mongolia, especially when faunal remains constitute the less meat-yielding parts of an animal such as the cranium, hooves, and cervical vertebrae (Kuzmina 2008; Miller et al. 2008). However, this may privilege the interpretation of faunal remains as inherently of nutritional and economic value and disregards the symbolic nature of mortuary contexts. Such assumptions may conceal other activities or functions of these particular elements, especially in the absence of other parts of the skeleton. The head and hooves are also the most portable portions of an animal's carcass. It is thus equally possible that animals were slaughtered and consumed elsewhere and the crania and hooves substituted as symbolic representations of each animal. Hence, demonstrated assumptions are made about the significance of faunal remains in mortuary contexts, particularly those that include livestock, that are in fact unknowns and which have significant impact on archaeological interpretation.

Perhaps the greatest assumption is that the presence of livestock suggests pastoral life-ways or animal husbandry on the part of the deceased or the builders of their graves. This need not be the case at all, since it is by no means certain from where animals in mortuary contexts were actually procured. The presence of livestock more accurately signals a preference for pastoral resources in mortuary practice. The significance of this observation is best demonstrated by the archaeological evidence itself. Xiongnu material culture, which is under consideration here, is commonly attributed to nomadic pastoralist groups based on the presence of livestock in mortuary contexts and the general absence of permanent architecture (Di Cosmo 2002; Honeychurch 2003). However, a handful of Xiongnu sites, such as Ivolga near Lake Baikal in southern Siberia, constitute settlement sites complete with fortifications and farming implements (Davydova 1995, 1996). Yet the burials at Ivolga exhibit the same prevalence of livestock in their funerary assemblages as Xiongnu tombs elsewhere in Mongolia and southern Siberia. Similarly, preliminary isotopic analysis to reconstruct and compare diets of Xiongnu populations at Baga Gazaryn Chuluu and Burkhan Tolgoi in northern Mongolia has demonstrated a divergence in subsistence strategies. Individuals at Burkhan Tolgoi were found to have a diet more consistent with a mixed economy of herding and hunting, whereas individuals at BGC had a primarily pastoral diet (Brosseder et al. 2011). Again, the burial practices in each locale are strikingly uniform, with livestock constituting the majority of animals represented. Given these observations, the presence of livestock in burials, regardless of their frequency, cannot be used to infer nomadic pastoralism.

Importantly, archaeological discourse is conceived of and framed within the context of our own contemporary understanding of the world (Holtorf 2007). For example, archaeologists often draw the distinction between wild and domesticated species because we believe that distinction to be important. However, it is not certain that it was equally important to prehistoric peoples, especially in the early stages of the domestication process. With this in mind, I suggest that the nature and implication of these categorical designations warrant further scrutiny. Bioarchaeology and zoo-archaeology constitute different sub-disciplines of archaeology, each with its own respective theoretical and methodological frameworks. This disciplinary separation reflects a modern understanding of the world in which humans are placed outside and above nature (Shepard 1996). Institutionalized differentiation between human and mammalian osteology often results in human and faunal remains being studied in exclusion of one another, even when derived from the same archaeological contexts. Therefore, a partition between the human world and the natural world exists and is continually reinforced through archaeological practice and organization. Humans are almost always privileged in these studies. Diet or pathology in human populations is interpreted as resulting from human choices and behavior, while diet and pathology in animals is also attributed to human behavior and exploitation. Humans are thus vested with an agency not attributed to animals. Animals instead become passive economic and dietary units that exist to better inform us about human behavioral practices. The focus tends to be not what animals were doing in prehistory but rather what humans were doing to them. This is a conceptual standpoint from which archaeologists rarely consider how humans' interaction with animals can affect how we perceive the world around us.

Humans' perception of animals is largely structured by how we choose to name, categorize, and organize the natural world. Modern classification systems follow a Linnaean taxonomy and seek to assign faunal remains to specific species, genus, family, order, class, division, and kingdom (Medin and Atran 1999). However, this obfuscates the existence of several other classification systems labeled folk taxonomies that are developed by local communities and that need not follow a Linnaean system at all (Berlin, Breedlove, and Raven 1973; Wapnish 1995; Marciniak 2006). Brian Smith (1991) has pointed out folk taxonomies in India, for example, that classify animals by their mode of procreation, whether they are domestic or wild or edible or inedible. Although a detailed reconstruction of prehistoric animal classification systems is hardly possible, it is likely they were more akin to folk taxonomies than our Western Linnaean taxonomy (Marciniak 2006). Archaeologists also draw distinctions between wild and domestic animals. Domesticated animals are further subdivided into livestock, pets, and feral categories, with the understanding that they all represent a separate class of animals in comparison to their wild counterparts (Shepard 1996). These distinctions are an important facet in understanding human-animal interaction, but like reconstructing prehistoric taxonomy, it is difficult to determine if people in the past adhered to similar frameworks of reference. Modern archaeologists are, of course, aware of the increasing compartmentalization of the discipline and of the impact of their own Western scientific paradigms on archaeological interpretation (Willis 1990). Nevertheless, zooarchaeological methodology continues to follow a Linnaean model that assumes and is partial to a Western understanding of nature.

Systematic categorization is not confined to the natural world. There is a long-standing tradition in anthropology and archaeology of classifying human societies according to their modes of subsistence, social organization, and language (to name but a few traits). Hence, nomadic pastoralism is a distinction drawn by archaeologists and ethnographers to distinguish between sedentary animal husbandry and herding strategies that are reliant on transhumance. However, the archaeological signatures used to differentiate between nomadic pastoralism and sedentary husbandry practices are not well-defined in Central Asian archaeology. Nomadic pastoralism is often implied in the absence of known settlement sites and on the presence in burials of horse trappings and livestock such as cattle, sheep/goats, and horses (Yablonsky 2000; Kuzmina 2008). This can be problematic, since it assumes that animals found in mortuary contexts are an accurate reflection of subsistence strategies among the living. Furthermore, it also presumes an overall dearth in settlements; hence, where habitation sites *are* found, they are more often than not assigned to different sedentary cultural groups. In addition, mobility and transhumance of prehistoric nomadic pastoralists are often assumed rather than demonstrated (although for an exception see Chang 2006). These factors combine to pose a significant obstacle to the interpretation of mortuary data in Central Asia and Mongolia.

To mitigate some of these problems, I approach faunal remains in mortuary contexts as one line of evidence among many that should be engaged in conjunction with other aspects of material culture to glean a more nuanced perspective of prehistoric life-ways and worldviews. I favor a synthetic approach that views faunal remains as a medium through which humans can change their way of life and mode of production but which in turn transforms them. Hence, I am not attempting to identify when nomadic pastoralism was adopted in Mongolia but rather how its adoption changed the ways pastoralists incorporated and used the animal world in their funerary practices. The fact that human domestication and exploitation of particular animals have resulted in the integration of those animals in ritual and cosmology is known from elsewhere. The domestication of bovids in the Mediterranean led to the emergence of symbolic representations of bulls as early as 10,000 years ago at Mureybet in the southern Levant (Cauvin 2000). At Çatal Hüyük and Haçilar in Anatolia, bull symbolism is evident in the bucrania, or cattle horns, embedded in architectural platforms and in representations of women giving birth to bulls' heads (Mellaart 1975). In Egypt, the Serapeum, a gallery of tombs built for sacrificial bulls, further attests to the existence of bull cults and the incorporation of that animal in ritual and religious beliefs (Grimal 1992). In many African pastoral societies, livestock also play an important role in communal rituals such as propitiation or naming ceremonies and serve an important role in the spiritual and psychological consciousness of groups such as the Tuareg and the Khoikhoi (Stenning 1959; Smith 1992). It is therefore reasonable to suppose that the adoption of pastoralism in Mongolia would also be accompanied by changes in the ritual use of livestock, as well as changes in worldview and understandings of the natural world order.

THE ARCHAEOLOGY OF BRONZE AND IRON AGE MONGOLIA (1500–200 BC)

The archaeology of the Bronze-Iron Age transition in Mongolia is underscored by the emergence of three roughly sequential types of mortuary stone monument called *khirigsuurs* (ca. 1500–800 BC), slab burials (ca. 800–300 BC), and Xiongnu ring tombs (300 BC–AD 200). The khirigsuurs appear relatively abruptly on the landscape where there were no previous stone monuments or ostentatious mortuary architecture (Wright 2006). These developments occur relatively late in Mongolia compared to elsewhere in Central Asia, where monumental earthworks and stone monuments are found as early as 2500 BC (Anthony 2007; Frachetti 2008; Kuzmina 2008). It thus appears that these traditions and the accompanying changes in economic and social aspects of culture, which involved the adoption of nomadic pastoralism, were not the result of endogenous processes in Mongolia but were likely incorporated from abroad. In the next section I describe the sequence of mortuary monuments of the Late Bronze–Early Iron Age in Mongolia and at BGC, focusing not only on how the construction and placement of these monuments change but also on alterations in funerary ritual and the symbolic integration of faunal remains.

FIGURE 5.2. Khirigsuurs, illustrating differences in size and shape of central mound and perimeter fence.

KHIRIGSUURS

Khirigsuurs consist of a central mound of stones surrounded by a circular or rectangular fence or alignment of rocks. Small circular stone features, or satellites, are sometimes located outside the perimeter fence in various arrangements (figure 5.2). The size of these monuments, including perimeter fence and accompanying satellites, can vary from 20 m to 50 m in diameter, but some are considerably larger (Allard and Erdenebaatar 2005). Khirigsuurs exhibit enormous variability in construction and placement. In some cases, khirigsuurs lack perimeter fences and satellites altogether, which can make it difficult to distinguish them from monument types from later periods. Khirigsuurs occur both individually and in clusters. There are instances where repeated building episodes in the same locale have produced extensive complexes in which the features of one or more khirigsuurs incorporate the features of another (Wright 2006).

The diversity in khirigsuur construction in conjunction with uneven taphonomic processes across Mongolia makes interpreting variation in form or function problematic. Considerable attention has been drawn to the inclusion of horse crania in the satellites of khirigsuurs in the Khanuy River Valley, suggesting exploitation and ritual inclusion of equids in mortuary practice (Allard and Erdenebaatar 2005). However, these results have not been replicated across Mongolia. In some cases, satellites are entirely absent and may represent local variation within a broader khirigsuur tradition. In spite of the high variability in khirigsuur construction, current research indicates that these were primarily funerary monuments, although they may have been the focus of other ritual activities as well (Wright 2006).

At BGC, 340 khirigsuurs have been identified and largely conform to observations made elsewhere in Mongolia. The extreme variability in khirigsuur construction is repeated at BGC, where khirigsuurs are found with both square and circular perimeter fences, including some without. Khirigsuurs are found throughout the research area and confer a significant visual impact on the landscape by taking advantage of the natural topography of the area. Excavation recovered primarily diagnostic ceramics, but no human remains were found either in the central mounds or in any of the satellites. Isolated osteological elements of horses and sheep/goat were identified, but none resembling the horse crania found in the satellites in the Khanuy River Valley. In 2007 the project also uncovered a quadrangular burial contemporaneous with khirigsuur construction, which indicates the existence of alternative burial practices taking place at BGC at this time. This burial is further significant in that it contained the remains of one horse, one bovine, a sheep, and a goat, species traditionally associated with animal husbandry (Johannesson and Machicek 2008; Machicek 2011). However, the presence of these species is not necessarily indicative of the practice of pastoralism but rather of access to pastoral resources.

Slab Burials

Khirigsuurs gradually begin to disappear by 800 BC, at which time they are replaced by a new type of monument. Slab burials, as the name suggests, are rectangular formations of large stone slabs raised around a shallow burial pit typically oriented east to west (figure 5.3). The average size of slab burials can range from 2 m to 4 m in length, but some can be substantially larger, with slabs reaching as high as 2 m aboveground (Wright 2006). These monuments are found throughout Mongolia and southern Siberia. Scholars have noted a spatial relationship between slab burials and khirigsuurs in which slab burials are often placed in close proximity to khirigsuurs (Honeychurch 2003; Tsybiktarov 2003; Wright 2006).

Around 287 slab burials have been identified at BGC, and, like khirigsuurs, they are found throughout the research area and typically occur in small linear clusters of three to nine graves. Although slab burials are found individually in BGC, these instances are relatively rare. The spatial relationship between khirigsuurs and slab

FIGURE 5.3. Slab burial at Baga Gazaryn Chuluu.

burials noted elsewhere in Mongolia is replicated at BGC where, in the same manner as khirigsuurs, slab burials confer a considerable visual impact on the landscape. Although numerous slab burials have been excavated at BGC, very little material has been recovered. This is not surprising, given the relatively shallow depth of these burials in addition to adverse taphonomic processes in the area, which include extensive pillaging episodes. However, slab burials excavated in other parts of Mongolia and southern Siberia where preservation is better rarely yield extensive funerary assemblages either. Objects typically found in slab burials include bronzes, ceramics, the occasional Chinese import, horse trappings, and faunal remains of both wild and domestic animals (Tsybiktarov 2003; Wright 2006). These types of materials have been found at BGC as well but usually constitute isolated finds rather than an entire assemblage of grave goods. A number of stone monuments excavated by the project and which were originally believed to date to other periods yielded radiocarbon dates, indicating that they were contemporaneous with slab burial construction. As in previous periods, this indicates that although an overarching architectural narrative seems to have directed funerary monument construction at BGC, there were alternatives to these regimes.

To summarize, the period immediately preceding the advent of the Xiongnu confederacy is characterized by a striking diversity in monument placement and construction. Both khirigsuurs and slab burials appear to reflect loosely defined criteria for construction and placement in which variation reflects particular choices by the builders of each monument. Khirigsuurs and slab burials require significant investments in labor, and this investment was directed at the monuments themselves rather than the accompanying burial assemblages. Given the visual component of both monument types and the diversity in their construction, it is likely that they were meant to be seen and experienced by people using the locales in which they were placed. The placement and view-sheds of these monuments also exhibit an intimate knowledge and understanding of the physical environment of BGC. This, in turn, suggests that the monuments built at BGC during this period likely belonged to local inhabitants and local lineages of leadership. However, in the case of both monuments, there were instances when people for reasons yet unknown were not buried in either monument type but were accorded different burials altogether.

THE XIONGNU

The appearance of Xiongnu material culture in the third century BC entailed sweeping changes in technology and mortuary practices that transformed the mortuary landscape throughout Mongolia and southern Siberia. Large tomb complexes of the Xiongnu elite have been documented in what is believed to be the core of the Xiongnu confederacy in central Mongolia in the Khanuy River Valley and in southern Siberia at Derestuy and Ivolga near Lake Baikal (Rudenko 1969; Konovalov 1976; Miniaev and Sakharovskaia 2006, 2007). These were large, square mounds oriented north to south, held together by stone walls that also often bisected the mound at right angles. An earthen ramp-like structure stretched out from the southern wall, but it is unclear whether these features served as practical access to the tomb during funerary rituals or if they denoted symbolic "funerary paths" (Miller et al. 2008). These tombs were truly monumental structures that could measure more than 25 m in length and which were often accompanied by several satellite burials (Miniaev and Sakharovskaia 2002). The superstructure covered a burial chamber containing a wooden coffin and a diverse mortuary assemblage. A number of these tombs contained a niche in the northern wall of the burial chamber that housed faunal remains. These remains typically consisted of crania of horses, goats, and cattle but also cervical vertebrae and the lower extremities of the legs (Miller et al. 2008). However, these tombs are the exception rather than the norm in Mongolia and southern Siberia and represent the uppermost echelon of the Xiongnu elite.

Instead, another monument type, the Xiongnu ring tomb, came to replace khirigsuurs and slab burials throughout Mongolia. These graves are marked on the surface by a broad band of stones between 4 m and 10 m in diameter that cover a burial shaft that ranges in depth from 1 m to 3 m (see figure 5.4). Xiongnu tombs typically occur in clusters or larger cemeteries and lack the visual prominence of khirigsuurs

FIGURE 5.4. Xiongnu ring tomb with the broad band of stones in the superstructure exposed.

and slab burials. In fact, relatively little investment appears to have been made in the superstructure of these monuments; instead, a great deal of resources were spent on the funerary assemblage itself. Xiongnu tombs typically include extensive mortuary assemblages consisting of ceramics, human and faunal remains, bone-plate bows, semi-precious stones and metals, beads, and imported long-distance luxury goods from China, Central Asia, and the Middle East.

At BGC, there is no evidence of the large square tomb complexes of the uppermost Xiongnu elite. Unlike slab burials and khirigsuurs, the 165 Xiongnu ring tombs identified in the area are not located within the central terrain of BGC but rather in clusters along its edges. Moreover, the placement of Xiongnu tombs in the area is significant in that they are found specifically in areas where preceding monuments are absent. Hence, there appears to be an intentional physical separation from previous mortuary traditions in the area. Xiongnu graves also exhibit a significant standardization in orientation, structure, and placement of objects within the burial. Standardization and possible regulation of ideologies on display in funerary ritual are also suggested by the absence of archaeologically visible alternative burial practices contemporary with the widespread appearance of Xiongnu

FIGURE 5.5. Crania of cattle, sheep, and goats buried in a niche in the northern section of Xiongnu tomb, Baga Gazaryn Chuluu.

material culture. At BGC, Xiongnu tombs also exhibit a specialized treatment of faunal remains. Similar to the large elite tomb complexes, the northern section of the grave typically contained a niche holding crania and hooves of goat, sheep, and cattle (figure 5.5). In addition, several tombs were accompanied by small secondary stone features located to the south of the grave, which consisted of a buried stone cist containing disarticulated remains of goat and sheep. These features were marked on the surface by an inconspicuous cluster of unaltered rocks.

These occurrences have been identified elsewhere in Mongolia and suggest a standardized treatment of faunal remains in funerary practice (Miniaev and Sakharovskaia 2002; Miller et al. 2008). Faunal remains were also found throughout the mortuary assemblage, and although they occasionally included dogs and deer, they were primarily composed of pastoral resources. Several elements, particularly teeth found in the funerary assemblage, exhibited evidence of burning. In addition, perforated astragali of wild and domestic goats were recovered from several graves. In comparison to faunal remains in slab burials and khirigsuurs, which include the occasional animal, the faunal assemblages in Xiongnu graves are extensive, often comprising more than two dozen animals.

There appears to be a conspicuous disconnect between Xiongnu ring tombs and previous mortuary traditions at BGC and elsewhere in Mongolia that can be summarized as follows:

1. The placement of Xiongnu graves in locations where other monuments are absent physically separates them from preceding mortuary monuments.
2. Xiongnu ring tombs lack the externally visible components of previous monuments and exhibit greater investment in the funerary assemblages rather than in the monuments themselves.
3. In spite of comprehensive pillaging in antiquity, Xiongnu tombs still yield a considerable amount of material evidence, with numerous imports attesting to access to long-distance socio-economic networks.
4. The increased homogeneity of Xiongnu tombs compared to khirigsuurs and slab burials, including their placement in clusters or cemeteries, suggests a greater regularity in funerary customs and ideologies.
5. Xiongnu tombs exhibit a greater emphasis and standardization in the inclusion of faunal remains, particularly pastoral resources, in the funerary assemblage.

The emergence of Xiongnu material culture regimes thus represents a considerable disruption of previous mortuary traditions. The disappearance of alternative burial practices, at least those that are visible in the landscape, and increasing standardization in the funerary assemblage imply some degree of control over ideologies on display in mortuary practice and an overall adherence to a particular funerary tradition.

DISCUSSION

Turning to the interpretation of faunal materials to elucidate human-animal relationships during the Bronze-Iron Age transition in Mongolia, I maintain that it is not possible to discuss faunal remains in the absence of other lines of evidence. Mobile pastoralism involves complex knowledge of land management and herd composition. Nomadic pastoralist communities are intricately tied to both the landscape in which they graze their livestock and to the livestock itself. Animals are not only used as a means of subsistence but for secondary resources as well, such as milk, fur, wool, and bone. Therefore, a transition to nomadic pastoralism should encompass a development over time so the animals on which pastoralists depend become increasingly incorporated into other aspects of daily and ritual life. Similarly, some aspects of material culture would be expected to reflect new economic and social constraints resulting from the adoption of more mobile ways of life. This is certainly the case in Mongolia during the Bronze-Iron Age transition.

The presence of livestock in khirigsuurs and slab burials demonstrates access to and a preference for pastoral resources in mortuary ritual. Yet the monuments themselves, especially khirigsuurs, reveal tremendous variability both locally and regionally in how and where these rituals were enacted and how faunal remains were used. The inherently local expression of monumentality by both khirigsuurs and slab burials may indicate material culture as demonstrative of an adjustment to new regimes of land use as a result of grazing practices. However, this local articulation of monumentality also suggests limited range of mobility; if pastoralism was in fact practiced by the builders of these monuments, it was likely conducted on a smaller, more concentrated scale. The presence of horse trappings in slab burials is also indicative that, over time, at least a segment of the population rode horses and that the equipment associated with this activity had become incorporated into mortuary ritual. This is important because while a horse bit in itself is purely a functional item, upon deliberate placement into a burial it assumes a symbolic role as well. Although it is only possible to speculate on the true nature and extent of nomadic pastoralism in Mongolia prior to the formation of the Xiongnu confederacy, it was evidently present. However, the appearance of Xiongnu material culture regimes acts as a watershed that fundamentally alters the archaeological record in Mongolia.

The widespread and substantial increase in livestock included in Xiongnu mortuary assemblages gives the overall impression of an increased reliance on pastoral resources. Similarly, the addition of livestock in more prominent ceremonial features, such as buried stone cists and pits, is consistent with an expected increase in the inclusion of pastoral resources in ritual behavior, particularly as animal husbandry becomes more firmly entrenched across Mongolia. The same can also be said for the recurrent practice of placing crania and hooves of livestock in a niche in the northern section of the burial. Indeed, the evidence, including perforated astragali, all points to the adoption of a repertoire of symbolism and ritual behavior consistent with the widespread practice of pastoralism. The fact that animals included, such as sheep, goat, cattle, and horses, typically require seasonal movement if herded on a larger scale further suggests pastoralism of the nomadic variety. However, additional lines of evidence indicate that Xiongnu mortuary practices also entailed a political dimension.

The emergence of large elite tomb complexes in parts of Mongolia concurrent with a standardization of mortuary practice on the local level that disrupts previous mortuary tradition is evidence of political change on a broader, regional scale. The appearance of Xiongnu material culture and associated mortuary traditions at BGC and elsewhere suggests the formation, whether by adoption or coercion, of a distinct "Xiongnu" identity. This identity was a conscious separation from previous ones expressed symbolically in local landscapes of mortuary stone monuments consisting of khirigsuurs and slab burials. Xiongnu mortuary practices thus emphasize a distinct

Xiongnu identity while discontinuing the ideological expression of local lineages of leadership to incorporate outlying regions into a broader "Xiongnu" political economy. The reinforcement of this identity included the symbolic use of animals in mortuary ritual. The appearance of a niche containing faunal remains in both the elite tomb complexes and in local Xiongnu ring tombs suggests that the ideology directing the use of pastoral resources in funerary ritual emanated from the core of the Xiongnu confederacy. The Ivolga settlement and the dietary evidence emerging from Burkhan Tolgoi should caution that although Xiongnu mortuary practices incorporate symbols and ritual ideologies consistent with nomadic pastoralism, these are not necessarily an accurate reflection of local subsistence strategies. Instead, pastoral resources were an expedient symbolic funerary currency used by the Xiongnu elite in mortuary rituals that emphasized and reinforced political and social unity.

This cursory examination of animal exploitation in Xiongnu mortuary practice leads to a number of observations. Primarily, this overview has illustrated the significant investiture of faunal resources into the Xiongnu funerary rite that was not necessarily indicative of economic activities. The widespread nature of this occurrence suggests an overarching socio-political influence of the Xiongnu uppermost elite on mortuary customs found at both micro- and macro-regional scales. However, this influence and the ideology it sought to convey, namely of belonging to a cohesive Xiongnu pastoralist system, are easily overlooked if faunal resources are treated merely as economic units and if other aspects of material culture are not also taken into account. The predominant use of faunal material of a pastoral variety thus suggests that despite variance in subsistence regimes, pastoral resources carry a particular symbolic weight that can be recognized throughout the extent of the study region.

ACKNOWLEDGMENTS

I would like to thank William Honeychurch and Chunag Amartuvshin, as well as the Institute of Archaeology and the National University in Ulaanbaatar, the Timothy P. Mooney Foundation at UNC Chapel Hill, and all the international and Mongolian students, staff, and volunteers who have contributed to the work at BGC over the years. Any errors contained herein are entirely my own.

REFERENCES

Allard, Francis, and Diimajav Erdenebaatar. 2005. "Khirisgsuurs, Ritual, and Nomadic Pastoralism in the Bronze Age of Mongolia." *Antiquity* 79 (305): 547–63.

Anthony, David. 2007. *The Horse, the Wheel, and Language: How Bronze-Age Riders from the Eurasian Steppes Shaped the Modern World.* Princeton: Princeton University Press.

Berlin, Brent, Dennis E. Breedlove, and Peter H. Raven. 1973. "General Principles of Classification and Nomenclature in Folk Biology." *American Anthropologist* 75 (1): 214–42. http://dx.doi.org/10.1525/aa.1973.75.1.02a00140.

Brosseder, Ursula, Yeruul Erdene, Damndinsuren Tsverendorj, Chunag Amartuvshin, Tsagaan Turbat, and Tugsuu Amgalantugs. 2011. "Twelve AMS-Radiocarbon Dates from Xiongnu Period Sites in Mongolia and the Problem with Chronology." *Studia Archaeologica Instituti Archaeologici Academiae Scientiarum Mongolicae* 31 (4): 53–70.

Cauvin, Jacques. 2000. *The Birth of the Gods and the Origins of Agriculture*. Cambridge: Cambridge University Press.

Chang, Claudia. 2006. "The Grass Is Greener on the Other Side: A Study of Pastoral Mobility on the Eurasian Steppe of Southeastern Kazakhstan." In *Archaeology and Ethnoarchaeology of Mobility*, ed. Frederic Sellet, Russell Greaves, and Pei-Lin Yu, 184–200. Gainesville: University Press of Florida.

Christian, David. 2001. *A History of Russia, Central Asia and Mongolia: Inner Eurasia from Prehistory to the Mongol Empire*. Oxford: Blackwell.

Crabtree, Pamela J. 1989. "Zooarchaeology in Complex Societies: Some Uses of Faunal Analysis for the Study of Trade, Social Status, and Ethnicity." In *Archaeological Methods and Theory*, vol. 2, ed. Michael B. Schiffer, 155–205. Tucson: University of Arizona Press.

Cribb, Roger. 1985. "The Analysis of Ancient Herding Systems: An Application of Computer Simulation in Faunal Studies." In *Beyond Domestication in Prehistoric Europe: Investigations in Subsistence Archaeology and Social Complexity*, ed. Graeme Barker and Clive Gamble, 75–106. New York: Academic.

Cribb, Roger. 1991. *Nomads in Archaeology*. Cambridge: Cambridge University Press. http://dx.doi.org/10.1017/CBO9780511552205.

Davydova, A. V. 1995. *Ivolginskii Arkheologicheskii Kompleks: Ivolginskoe Gorodishche*. St. Petersburg: AziatIKA.

Davydova, A. V. 1996. *Ivolginskii Arkheologicheskii Kompleks: Ivolginskii Mogil'nik*. St. Petersburg: AziatIKA.

Di Cosmo, Nicola. 2002. *Ancient China and Its Enemies: The Rise of Nomadic Power in East Asian History*. Cambridge: Cambridge University Press. http://dx.doi.org/10.1017/CBO9780511511967.

Driver, John C. 1990. "Meat in Due Season: The Timing of Communal Hunts." In *Hunters of the Recent Past*, ed. Leslie B. Davis and Brian Reeves, 11–33. London: Unwin Hyman.

Frachetti, Michael. 2008. *Pastoralist Landscapes and Social Interaction in Bronze Age Eurasia*. Berkeley: University of California Press.

Frye, Richard N. 1996. *The Heritage of Central Asia: From Antiquity to the Turkish Expansion*. Princeton: Markus Wiener.

Grimal, Nicolas. 1992. *A History of Ancient Egypt*. Oxford: Blackwell.

Grousset, René. 1970. *The Empire of the Steppes: A History of Central Asia*. New Brunswick: Rutgers University Press.

Härke, Heinrich. 1997. "The Nature of Burial Data." In *Burial and Society: The Chronological and Social Analysis of Archaeologial Burial Data*, ed. Claus K. Jensen and Karen H. Nielsen, 19–28. Aarhus: Aarhus University Press.

Holtorf, Cornelius. 2007. *Archaeology Is a Brand: The Meaning of Archaeology in Contemporary Popular Culture*. Oxford: Archaeopress.

Honeychurch, William H. 2003. "Inner Asian Warriors and Khans: A Regional Spatial Analysis of Nomadic Political Organization and Interaction." PhD diss., Department of Anthropology, University of Michigan, Ann Arbor.

Johannesson, Erik G., and Michelle L. Machicek. 2008. "From the Saddle to the Grave: Mobility and Mortuary Remains in Iron Age Mongolia." Paper presented at the Graduate Archaeology Association, Oxford University, Conference: Challenging Frontiers: Mobility, Transition, and Change. Oxford, April 4–5.

Konovalov, Prokopii B. 1976. *Khunnu v Zabaikal'e*. Ulan-Ude, Russia: Buriatskoe Knizhnoe izd-vo.

Kuzmina, Elena E. 2008. *The Prehistory of the Silk Road*. Philadelphia: University of Pennsylvania Press.

Lattimore, Owen. 1967. *Inner Asian Frontiers of China*. Boston: Beacon.

Levine, Marsha, Colin Renfrew, and Katie Boyle, eds. 2004. *Prehistoric Steppe Adaptation and the Horse*. Cambridge: McDonald Institute Monograph.

Machicek, Michelle L. 2011. "Reconstructing Diet, Health, and Activity Patterns in Early Nomadic Pastoralist Communities of Inner Asia." PhD diss., Department of Archaeology, University of Sheffield, UK.

Marciniak, Arkadiusz. 2006. *Placing Animals in the Neolithic: Social Zooarchaeology of Prehistoric Farming Communities*. London: University College London Press.

Medin, Douglas L., and Scott Atran. 1999. "Introduction." In *Folkbiology*, ed. Douglas L. Medin and Scott Atran, 1–15. Cambridge: Massachusetts Institute of Technology.

Mellaart, James. 1975. *The Neolithic of the Near East*. New York: Scribner.

Miller, Brian K., Jamsranjav Bayarsaikhan, Tseveendorj Egimaa, and Christian Lee. 2008. "Xiongny Elite Tomb Complexes in the Mongolian Altai: Results of the Mongol-American Hovd Archaeology Project, 2007." *Silk Road* 5 (2): 27–36.

Miniaev, Sergey, and Ludmila Sakharovskaia. 2002. "Soprovoditel'nye Zakhoroheniia 'Tsarskogo' Kompleksa no. 7 v Mogil'nike Tsaram." *Arkheologicheskie Vesti* 9: 86–118.

Miniaev, Sergey, and Ludmila Sakharovskaia. 2006. "Investigation of a Xiongnu Royal Complex in the Tsaraam Valley." *Silk Road* 4 (1): 47–51.

Miniaev, Sergey, and Ludmila Sakharovskaia. 2007. "Investigation of a Xiongnu Royal Complex in the Tsaraam Valley, Part 2: The Inventory of Barrow no. 7 and the Chronology of the Site." *Silk Road* 5 (1): 44–56.

Olsen, Sandra. 2006. "Early Horse Domestication: Weighing the Evidence." In *Horses and Humans: The Evolution of Human-Equine Relationships*, ed. Sandra L. Olsen, Susan Grant, Alice M. Choyke, and László Bartosiewicz, 81–113. BAR International Series 1560. Oxford: British Archaeological Reports.

Parker Pearson, Mike. 2000. *The Archaeology of Death and Burial*. College Station: Texas A&M University Press.

Pearson, Mike, and Michael Shanks. 2001. *Theatre/Archaeology*. London: Routledge.

Rudenko, Sergey. 1969. *Die Kultur der Hsiung-nu und die Hugelgräber von Noin Ula*. Bonn, Germany: Habelt.

Shepard, Paul. 1996. *The Others: How Animals Made Us Human*. Washington, DC: Island.

Smith, Andrew B. 1992. *Pastoralism in Africa*. Athens: Ohio University Press.

Smith, Brian K. 1991. "Classifying Animals and Humans in Ancient India." *Man* 26 (3): 527–48. http://dx.doi.org/10.2307/2803881.

Stenning, Derrick J. 1959. *Savannah Nomads*. London: International African Institute by Oxford University Press.

Tsybiktarov, Aleksandr D. 2003. "Eastern Central Asia at the Dawn of the Bronze Age: Issues in Ethno-cultural History of Mongolia and the Southern Trans-Baikal Region in the Late Third–Early Second Millennium BC." *Archaeology, Ethnology and Anthropology of Eurasia* 3: 107–23.

Wapnish, Paula. 1995. "Towards Establishing a Conceptual Basis for Animal Categories in Archaeology." In *Methods in the Mediterranean: Historical and Archaeological Views on Texts and Archaeology*, ed. David B. Small, 233–73. Leiden: Brill.

Willis, Roy, ed. 1990. *Signifying Animals: Human Meaning in the Natural World*. London: Unwin Hyman. http://dx.doi.org/10.4324/9780203169353.

Wright, Joshua. 2006. "The Adoption of Pastoralism in Northeast Asia: Monumental Transformations in the Egiin Gol Valley, Mongolia." PhD diss., Department of Anthropology, Harvard University, Cambridge, MA.

Yablonsky, Leonid T. 2000. "Scythians and Scythian Triad." In *Kurgans, Ritual Sites, and Settlements: Eurasian Bronze and Iron Age*, ed. Jeannine Davis-Kimball, 3–9. Oxford: British Archaeological Reports.

6

Kalas and Kurgans

Some Considerations on Late Iron Age Pastoralism within the Central Asian Oasis of Chorasmia

Michelle Negus Cleary

There is more evidence for the presence and practice of pastoralism within the ancient oases of Central Asia in antiquity than has hitherto been widely recognized. The old "steppe versus sown" paradigm still overshadows Eurasian scholarship, where the fertile oases have been characterized as farming enclaves fighting off the predations of raiding pastoralists from the surrounding desert-steppe zones (Frachetti et al. 2010:623). The assumption that the steppe was a peripheral zone of exclusively specialized pastoral production and that the oases were centralized areas of intensive agricultural production has been questioned by many scholars who have shown the multiple resource variability of steppe pastoralist systems (Chang 2008; Frachetti 2008; Frachetti et al. 2010:623) and the dominant presence of "steppic" groups within the oases (Rapin 2007; Stride, Rondelli, and Mantellini 2009). Others, such as Bella I. Vaynberg (1979a, 1979b, 1981, 2004), have demonstrated the long-term occupation of western areas of the Chorasmian oasis by a mixed agropastoralist population. The presence of pastoral sites and landscape features, such as *kurgans* (burial mounds), within the oasis and in close proximity to the Chorasmian fortresses, or *kalas*, seems to be clear evidence that the oasis was not solely a zone of sedentary agriculturalists. Furthermore, the evidence suggests that fortified sites were not only used by sedentary communities but also by pastoral groups and that they served multiple purposes and remained important places in the landscape over time.

Chorasmia, also known as "Khorezm" or "Choresmia" in antiquity, is the term for a Late Iron Age polity and archaeological culture that existed from the eighth/

DOI: 10.5876/9781607323433.c006

FIGURE 6.1. Chorasmia and western Central Asia regional map (after Bregel 2003: map 3).

seventh century BC until the fourth century AD (Tolstov 1948b, 1962; Rapoport, Nerazik, and Levina 2000). It was located within the oasis formed by the Amu-Darya River delta, where it fans out before emptying into the Aral Sea (see figure 6.1). This culture can be broadly characterized by distinctive ceramic and artifact assemblages, the appearance and proliferation of monumental mud-brick fortified enclosures and architecture, and irrigation networks. It has been seen as a peripheral area, an outpost of the Achaemenid world, and even as a culturally conservative backwater by some (e.g., Helms 2006:9) or as an influential centralized state by others (e.g., Tolstov 1948b:103, 1953:129, 134, 1960:11; Khozhaniyazov 2006:40).

The archaeology of the Amu-Darya delta area was initiated and dominated by the multidisciplinary Khorezm Expedition (directed by Sergey P. Tolstov until the late 1960s) during the twentieth century. Their Kompleksnaya Ekspeditsiya (Complete Expedition) incorporated specialists from a great diversity of fields, such as geology, archaeology, ethnography, history, anthropology, linguistics, and the like, to completely study a particular geographic area (Frumkin 1962:339). As a result, an enormous amount of archaeological work was carried out in the delta, with a particular focus on the ancient period and its Chorasmian culture.

The simultaneous appearance of monumental fortified sites with large canals around the seventh/sixth century BC and their subsequent proliferation in the "Early Antique" period (known as "Kangiui" in the Chorasmian chronology), the time from the fourth century BC to the first century AD, led Tolstov to interpret Chorasmia as a centralized, urbanized, agrarian state (Tolstov 1948b:103, 1953:129, 134, 1960:11), following the hydraulic state model proposed by Wittfogel (1957;

see Stride, Rondelli, and Mantellini 2009:74–75). This was in accordance with the Soviet Marxist theoretical framework of historical materialism (Frumkin 1962:335–36; Trigger 1989:207), which also had echoes in Western socio-evolutionary archaeological theory. In Tolstov's interpretation of Chorasmia, the fortresses played the dual role of defending the irrigation system as well as enforcing and entrenching the despotic, administrative state required for the construction, implementation, and management of a complex irrigated agricultural economy.

This interpretation of the available archaeological data for pastoralism or pastoral practices within the Chorasmian oasis (Amu-Darya delta) on the whole has tended to promote an urban-rural and sedentary farmer–nomadic herder set of binary dualisms.[1] Fortresses were considered urban centers or fortified villages with evidence of fully sedentary occupation in combination with irrigation agriculture, and open settlements were rural villages, part of a productive hinterland (Nerazik 1976). The fertile oasis zones of the delta were characterized as a sedentary enclave of farmers barricading themselves from the predations of nomadic pastoralists located in territories on their peripheries (Tolstov 1948b:122; Lavrov 1950:20; Rapoport and Trudnovskaya 1958:347, 366; Mambetullaev 1978:87; Negmatov 1994; Rapoport, Nerazik, and Levina 2000:24, 52; Khozhaniyazov 2006:42, 119, 123n304; Yagodin 2007:46–47, 75; Betts et al. 2009:38–39). Consequently, any fortified sites or dwellings found within the oasis zone were interpreted as belonging to a Chorasmian sedentary urban state, and any sites or features found outside or on the geographic periphery of this zone must therefore have been "nomadic."

However, the evidence from western Chorasmia, where there are no irrigation systems and no large-scale farming but many fortresses, does not fit Tolstov's interpretation. For this area, the only possible viable economy for local inhabitants would have been pastoralism. Vaynberg and others have suggested that many of these fortified sites in a "peripheral" area were military outposts of the "centralized" Chorasmian state to control local pastoral populations and their economy (Tolstov 1960:13; Vaynberg 1979b:172, 2004:9, 247; Bizhanov and Khozhaniyazov 2003:36, 49). Although this is a plausible explanation, there are alternative ways of interpreting the archaeological material assembled by the Khorezm Expedition and examining some of the evidence for pastoral practices and pastoral groups within the Chorasmian oasis territory.

Perhaps confusingly, Vaynberg (1981:122, 125–26) herself has offered such alternative interpretations. She suggested that certain western Chorasmian fortified sites were administrative, economic, and ritual centers for local pastoral groups rather than fortified urban settlements or even state military outposts. Vaynberg (1979b, 1981) argued that the Chorasmian population was multicultural, at least in the western part of the delta, with both pastoralists and farmers enmeshed within

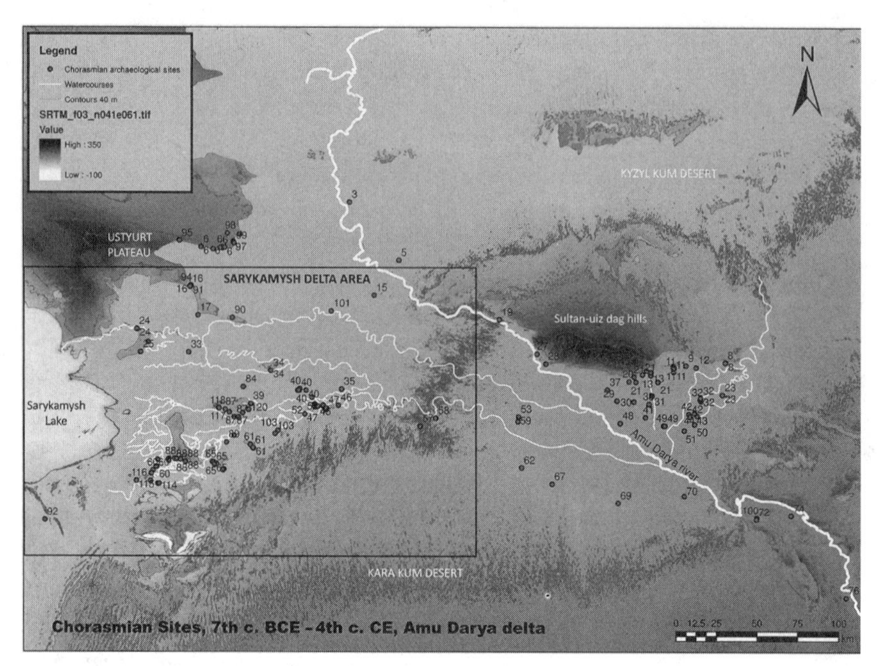

FIGURE 6.2. GIS map of ancient Chorasmia (Amu Darya delta oasis) showing all sites dating between seventh century BC and fourth century AD.

the same socio-political system but sometimes with variable cultural traits, such as mortuary practices. This interpretation was innovative in that it deviated from the set paradigm of oasis = farmers and steppe = nomadic pastoralists. Her work raises questions about the level of integration of pastoral and agricultural communities within the polity of ancient Chorasmia and about power and domination within the oasis. The steppe versus sown paradigm is not well supported by the archaeological evidence here, where pastoral campsites, mortuary sites, settlements, and fortresses coexisted at different points in time, and requires examination.

The assumptions inherent in the binary dualisms of walled site = elite farmers and funerary monument = nomads have dominated research agendas, data collection, and interpretation in Chorasmian and Central Asian archaeology. There has been division within the scholarship between those looking exclusively at "settled" sites versus those looking exclusively at "nomadic" sites or features. This kind of division has been pervasive throughout Central Asian archaeology (Stride, Rondelli, and Mantellini 2009:83). In Chorasmia, for example, Vadim N. Yagodin investigated the kurgan cemeteries immediately surrounding Bol'shoi Aibugir-kala (figures 6.2, 6.3) (Yagodin 1982, 2007:45–46), while Mambetullaev (1978, 1990) investigated

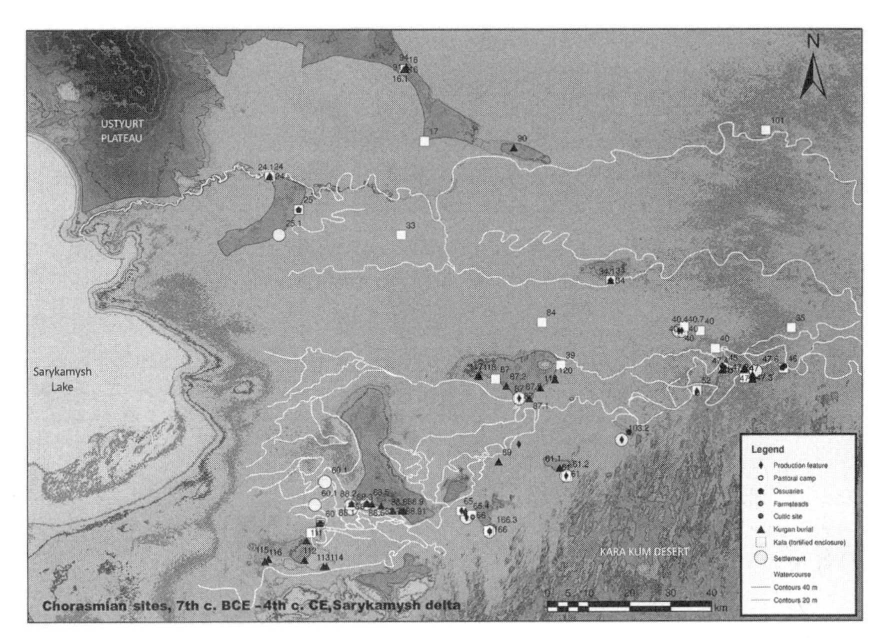

FIGURE 6.3. GIS map of archaeological sites in the Sarykamysh delta area, seventh century BC–fourth century AD.

the fortress of Bol'shoi Aibugir-kala (6.0) itself. This division has made understanding and interpreting the archaeological material difficult and often myopic.

A survey of this archaeological material is presented here to better understand what meanings and roles these kalas had and how they related to the kurgans and other sites in terms of a settlement regime. This may in turn shed some light on the interaction between pastoral and sedentary groups that occurred in the Amu-Darya delta from the seventh century BC to the fourth century AD. Here, I survey the architectural and structural remains of fortified sites, settlements, and burial mounds from western (left bank) Chorasmia (focusing on the Sarykamysh delta area), primarily derived from published Russian archaeological reports. By assembling these "pastoral" and "settled" data together in a GIS and employing remote sensing tools, a better understanding and clearer presentation is gained of the spatial relationships between different elements of the archaeological landscape.

Questioning the steppe-sown paradigm in Chorasmia keys into the work of other scholars. In Central Asia, Biscione (1981), Lecomte (1999), and Stride (Stride, Rondelli, and Mantellini 2009) have suggested that the presence of fortresses in combination with irrigation systems does not necessarily indicate a sedentary, agrarian state but may instead represent a more local, communal organization (Negus Cleary 2007:338). Similarly, Philip (2001:181–82, 201–2, 2003:109) questioned the assumption that Early Bronze Age walled sites were the administrative centers of city-states (the product of sedentary elite farmers) and megalithic funerary monuments were the products of mobile pastoral groups. Baker Brite's recent work on the fortified site of Kara-tepe in fourth–ninth century AD Chorasmia showed that a semi-mobile population relied on a diversity of production strategies necessary for subsistence in an unstable, arid environment (Baker Brite 2011).

Furthermore, analysis of the Chorasmian archaeological landscape also ties into archaeological debates, such as how archaeology should, or might, engage with fortified structures and burial monuments as key elements in the ancient landscape (Chapman 1995; Creighton 2002; Johnson 2002; Allard and Erdenebaatar 2005; Mercer 2006; Platt 2007; Wright 2007; Creighton and Liddiard 2008). Questioning the function and use of Chorasmian "fortresses" and "urban sites" ties into critiques about defining sites and settlements (e.g., Brück 1999) and the identification of pastoral sites (see Koryakova 2000:14).

DEFINITIONS

The term *kala* is used throughout Central Asia to denote a fortified enclosure or fortified dwelling (often within a fortified enclosure) (Szabo and Barfield 1991; Horne 1994:83; Lamberg-Karlovsky 1994) and to describe ancient and medieval fortified sites. Here I use the term *kala* to provide distance from the more deterministic term *fortress*, which has strong functional and political connotations that perhaps do not fit the archaeological evidence for enclosure sites in Central Asia. The term *kala* may help to convey something of the multiple meanings and functions of this architectural form that go beyond "war" or "status" (Johnson 2002; Creighton and Liddiard 2008). Mud-brick walled sites have a long history, going back at least to the Bronze Age Bactrian-Margiana Archaeological Complex (Hiebert 1994; Sarianidi 2002) through to the ancient period in Chorasmia, Margiana, Parthia, Sogdia, and Bactria, as well as in the steppe—for example, the Xiongnu fortified enclosures of the Late Iron Age (Rogers, Ulambayar, and Gallon 2005; Honeychurch and Amartuvshin 2007).

The kala is a traditional, formulaic architectural type and has been adapted to multiple purposes, often varying greatly over time (Hiebert 1992; Lamberg-Karlovsky 1994; Negus Cleary 2007, 2008). Such varied purposes include use as elite residences, temples, ceremonial structures, secure trading spaces or caravanserai, walled gardens, vineyards and orchards, refuges, cattle corrals, fortified settlements, manor houses, shelters, and military garrisons (cf. Tolstov 1958; Nerazik 1966; Horne

1991:49–51; Szabo and Barfield 1991; Lamberg-Karlovsky 1994; Gardin 1995; Abdoullaev 2001:204; Khozhaniyazov 2002; Rogers, Ulambayar, and Gallon 2005; Honeychurch and Amartuvshin 2007). I suggest that the kala was also a response to the often extreme climatic conditions of Central Asia, creating a protected space that mitigated the effects of hot summer and freezing winter winds and sandstorms (Taaffe 1990:21). It is hoped that the examination of the Chorasmian kalas here will further emphasize the versatility of this building type by setting out its role in an apparently pastoral context.

PASTORALISM IN CHORASMIA

"Pastoralism" and "nomadic" are key concepts discussed in this chapter, and it is important to define their meanings as used here. There are many manifestations of pastoralism and nomadism (Khazanov 1984; Cribb 1991:15–20; Koryakova 2000:15; Howell-Meurs 2001:96–97; Salzman 2002). Cribb (1991:16–20), Chang (2008:331–34), Dyson-Hudson and Dyson-Hudson (1980:18), Frachetti (2008:368–69), and Salzman (2002) clearly make the point that nomadic pastoralists employed many variable subsistence strategies for their social and economic existence in response to dynamic environmental, political, economic, and ideological contexts and that their defining characteristic should be their great adaptability. Salzman (2002:256) has written, "We must also keep in mind that 'settled' and 'nomadic', rather than being two types, are better thought of as opposite ends of a continuum with many gradations of stability and mobility."

It is not proposed here that the Chorasmians (or tribes living within the Amu-Darya delta) were highly mobile nomadic pastoralists (as defined by Khazanov 1984:7); rather, they more likely had a diverse agropastoral economy, perhaps some form of "distant-pastures husbandry" (Howell-Meurs 2001:96) or "transhumance" (Cribb 1991:19) in which some members of the population tended livestock over short, seasonal migrations but the majority of the population remained sedentary or at least semi-sedentary. Perhaps a more appropriate term for Chorasmia that has been recently used is *mobility* rather than *nomadism*. Mobility is a more generalized term defined as strategy of movement for procuring resources of any kind, and it is characterized as inclusive of a great variety of manifestations and is not confined to the movement of households (Bernbeck 2008; Frachetti 2008; Baker Brite 2011:7–8). Baker Brite (2011) identifies the population of Kara-tepe during the fourth–eighth centuries AD as semi-mobile and employing diverse subsistence strategies in response to fluctuating environmental conditions.

It is known that the peoples inhabiting the Amu-Darya delta during the Bronze Age and Early Iron Age were mobile (or semi-mobile) communities and part of a steppic cultural tradition, that is, the Tazabagyab, Amirabad, and Kelteminar cultures that were related to the Early and Late Andronovo steppic cultures (Tolstov 1948a:67–68, 1962:47). A portion of the "steppic" or "Saka" pastoral Kuiusai culture of the seventh to fourth centuries BC may have developed into the Chorasmian culture in the central and eastern areas of the oasis under the influence of Achaemenid or other external contacts from the south.[2] Importantly, there are steppe elements in Chorasmian art (Kidd 2011; Kidd, Negus Cleary, and Baker Brite 2012), the use of names associated with steppic groups (Livshits 2003:169), and tribal signs (*tamgas*) that appear on Chorasmian coins and mud bricks (Vaynberg 1977; Gertman 1991; Helms and Yagodin 1997:52). Given Chorasmia's location on the periphery of the Achaemenid world and well within the desert-steppe zone, it has been widely thought to have been associated with nomadic groups such as the Massagetae, Dahae, and Saka (Tarn 1951:81; Brentjes 1996:5; Bregel 2003:maps 3, 4; Livshits 2003:169). As a result of these factors, it is highly probable that pastoral practices were key elements of the social, cultural, and economic life of the ancient Chorasmian oasis.

THE CHORASMIAN CONTEXT

The archaeological material collected by the Khorezm Expedition personnel, including Vaynberg and others from the Institute of History, Archaeology and Ethnology in Nukus (e.g., V. N. Yagodin, G. Khozhaniyazov, M. M. Mambetullaev, Iu. P. Manylov), shows that various types of pastoral sites or landscape features dating to the Late Iron Age–ancient Chorasmian period (seventh/sixth century BC–fourth century AD) were present in the Sarykamysh (or Prisarykamysh) delta. The most obvious and readily attributable to pastoral groups are the mortuary monuments in the form of kurgan and podboi-catacomb–type burials.[3] They are found within and around the Chorasmian oasis (Amu-Darya delta), mostly on high ground in peripheral locations (Vaynberg 1981; Yablonsky 1990:288; Bader and Usupov 1995:24–28; Yagodin 2007), and even within and around Chorasmian fortified enclosures, such as at Mangyr-kala (**34.0**) (Vaynberg 1981:126). Other types of pastoral sites include nonagricultural settlements and pastoral camps (ibid.:122).

In the context of Chorasmian (and western Central Asian) archaeology, there has been fairly good preservation of this ancient landscape as a result of the very arid conditions, the isolation of the area, and the lack of later reoccupation of the sites. Chorasmia lay within the oasis zone formed by the delta of the Amu-Darya (ancient Oxus) River. It was surrounded on the northeast and east by the Kyzyl Kum and on the southwest by the Kara Kum desert, to the northwest by the elevated Ustyurt Plateau, and by the Aral Sea and its marshes to the north (see figure 6.1). Within the delta, small oases were formed by the branches of the Amu-Darya, as the great river fanned out into smaller braided distributaries. In between these oases existed areas of sandy desert, desert-steppe, and expanses of flat clay pans (*takyrs*), *solonchaks* (salty soil deposits usually associated with salt marshes), and in the west, isolated elevated plateaus (see figure 6.2).

Agricultural areas, canals, and associated settlements appear to have been predominately sited off the Amu-Darya's distributaries rather than on the main course of the river itself, as this was highly prone to flooding. Similarly, the Late Iron Age pastoral sites appear to have been located along the banks of these river distributaries. The pastoral sites and features are predominately located in the western half of the Amu-Darya delta, close to the Ustyurt Plateau and other smaller elevations, such as Tuz-gyr, Butentau-gyr, Tarym-kaya, Gyaur-gyr, and Kanga-gyr. They increase in numbers at the southwest and western peripheries of the delta. (However, this could be a result of differential preservation of more ephemeral pastoral sites in the central areas of the delta where later medieval and modern intensive occupation may have destroyed or completely obscured pastoral sites or monuments.) This far western area of the Amu-Darya delta is known as the Prisarykamysh delta, a subdelta related to the large Sarykamysh lake that often forms at the western edge of the Amu-Darya delta (see figure 6.2) when there is a high volume of water flow from the Amu-Darya through its western distributaries into the Sarykamysh and then out to the south by way of the Uzboi channel and ultimately to the Caspian Sea (see Létolle 2000; Létolle et al. 2007; Boomer et al. 2009:83–84).

Importantly, there is very little evidence for irrigation systems in the vicinity of most ancient sites in the Sarykamysh and western Chorasmian area. This is a very arid region where dry farming is impossible and where agriculture was only practiced in small irrigated plots directly off the river channels (the Amu-Darya's western distributaries: Chermen-yab, Daryalyk, Daudan-Darya, Kanga-Darya, Tuny-Darya) (Vaynberg 1979b:171, 173, 1981:123, 2004:9). Archaeological investigations by the Khorezm Expedition personnel have shown that these river channels were flowing in the ancient period until at least the second century BC (Tolstov 1962:93; Andrianov 1969:151; Vaynberg 2004:6), which would have made habitation along their banks possible. Since no irrigation systems were found (except in certain locations generally to the east, such as Kalaly-gyr 1 and 2), it appeared that populations living in this area of the Amu-Darya delta would have relied on pastoralism as the mainstay of their economy (Tolstov 1962:104; Vaynberg 1979b:172–73).

In western Central Asia, the presence of nomadic populations has generally been identified by their burial monuments (Rapin 2007:31). In Chorasmia, defining or identifying sites or landscape features as having been created, inhabited, or used by pastoral peoples, or "nomads," has been largely based on their type, that is, kurgans were nomad monuments, while fortresses were monuments of settled farmers. They have also been defined by their lack of proximity to canal systems, so, for example, Vaynberg (1979a, 1979b, 1981, 2004) has identified pastoral settlements largely on their location in peripheral areas devoid of irrigation. This seems unproblematic for the dispersed, low-density settlements and the kurgan monuments; however, it causes more problems when trying to interpret the fortified enclosures present in areas where there was clearly no irrigation agriculture.

CHORASMIAN CHRONOLOGY AND THE DATING OF SITES

Chorasmian chronology was set by the Khorezm Expedition (Tolstov 1948a, 1948b, 1962) and has more or less remained unchanged. It is based on relative dating of bronze tri-lobal arrowheads and ceramic assemblages (Vorob'eva 1959) and coins for the later periods (Vaynberg 1979a; Helms and Yagodin 1997:45–47; Helms et al. 2001:136–37). Although absolute dating has been established for two eastern (right bank) Chorasmian sites by the Karakalpak-Australian Archaeological Expedition (Helms et al. 2001, 2002; Betts et al. 2009), the work has not yet progressed to a synchronization of the ceramic chronology and the radiocarbon dating. The reliance on this pottery-based chronology is problematic, and any analysis of the sites, including that presented here, is constrained by the limitations of the chronology (see discussion on a related issue in Wright 2008:54). The different periods for the Late Iron Age are set out below. The names "Kangiui" and "Kushan" are problematic when applied to Chorasmia and therefore are included in inverted commas (quotation marks) (see comments in Helms and Yagodin 1997:45–47; Helms 1998, 2006:14, n30; Khozhaniyazov 2006:22, n55).

Archaic	seventh/sixth–fifth/fourth centuries BC
Antique	
Early "Kangiui"	fourth–third centuries BC
Late "Kangiui"	second century BC–first century AD
Early "Kushan"	first–second centuries AD
Late "Kushan"	second–fourth centuries AD

All of the archaeological sites for western Chorasmia have been dated based on ceramics and arrowheads or other diagnostic finds if they exist. Most of these finds have come from secure stratigraphic contexts, but some sites or features are dated only from surface finds. This variability in chronological information is represented for each site on tables 6.1 and 6.2; where the site's dating is reliant on surface finds only, this is noted in the text below. As a result, the issues discussed here in relation to the Chorasmian sites can be only very broad and preliminary in nature.

KUIUSAI CULTURE

Importantly, a separate archaeological culture has been identified within the area of western Chorasmia, known as the Kuiusai culture. This culture is characterized by distinct ceramic forms, kurgan cemeteries, turquoise processing, copper and iron working, a lack of irrigation agriculture, and an apparent lack of fortified sites (Vaynberg 1979a; Rapoport et al. 2000:23). Vogelsang pointed out the connections between Kuiusai culture material and Yaz II/III and the Ancient Dakhistan culture further to the south and southwest of Central Asia (Vogelsang 1992:289). The Kuiusai have been regarded as one of the Saka nomadic groups (Itina 1979; Negmatov 1994:449). Vaynberg also considered it possible that the Chorasmian seventh/sixth century BC period site of Kiuzely-gyr may have been a part of the Kuiusai culture population that had come under Achaemenid influence (Vaynberg 1979b:173). The chronology for the Kuiusai is contiguous with the Chorasmian Archaic and Early "Kangiui" periods, Kuiusai Early period, seventh–fifth centuries BC (Levina 1979:183), and Kuiusai later period, fifth–fourth BC (Itina 1979:5; Vaynberg 1979a:45).

This culture is considered to have been semi-sedentary in nature and fully pastoral in subsistence, with strong steppic ties (Itina 1979; Vaynberg 1979a; Negmatov 1994:449), and yet it existed in conjunction with the Archaic and early "Kangiui" Chorasmian cultures. Chorasmian ceramics were present at most Kuiusai sites, as was the use of ossuary burials in some instances (discussed below). The archaeological material presented by Vaynberg illustrated the close integration between the Kuiusai and Chorasmian cultures, until the fourth century BC when Kuiusai ceramics disappear but Chorasmian ceramic forms continue, many settlements and fortified sites remain in use, and the kurgan burial tradition carries on. There are, however, changes in the kurgan mortuary tradition, for example, the podboi-catacomb–type kurgan burials that appear from the second century BC (e.g., the tombs of Tumek-kichidzhik) and represent for Vaynberg (1979b:173–74) an influx of new pastoral nomadic groups.

SPATIAL ANALYSES OF THE SARYKAMYSH AREA

A database of sites and landscape features was compiled from the published archaeological information by Vaynberg, Tolstov, Bizhanov, Khozhaniyazov, and others

Table 6.1 List of sites from the Sarykamysh delta area arranged in groups that appear to be spatially and/or temporally related. Dating abbreviations:

1 = 7th–4th centuries BC; 1A = 7th–5th centuries BC; 1B = 6th–5th centuries BC; 1C = 5th–4th centuries BC
2 = 4th–2nd centuries BC; 2A = 4th century BC; 2B = 4th–3rd centuries BC; 2C = 3rd–2nd centuries BC; 2D = 3rd–1st centuries BC
3 = 2nd–1st centuries BC
4 = 1st–4th centuries AD; 4A = 1st–2nd centuries AD; 4B = 1st–3rd centuries AD; 4C = 2nd–4th centuries AD; 4D = 3rd–4th centuries AD
5 = 4th–5th centuries AD

ID	*Fortified Enclosure*	*Period*	*ID*	*Kurgans/burials*	*Period*	*ID*	*Inhabited sites*	*Period*	*References (all sites)*
6.0	Aibugir-kala bol'shoi	1C, 2B, 4C	6.1	Aibugiir-kala kurgans	2B?	6.0	Aibugir-kala bol'shoi fortress	1C, 2B, 4C	Mambetullaev 1978:87–88
			95.0	Kulmagambet kurgan cemetery	1C–2B				Yagodin 2007:46
			93.0	Berniyaz kurgan cemetery	varies				
24.0	Butentau-kala 1	1B-1C, 2B, 4C	24.1	Butentau-kala 1 kurgans	?	25.1	Butentau 3 settlement	2B	Bizhanov and Khozhaniyazov 2003:40
25.0	Butentau-kala 2	2B	25.0	Butentau-kala 2 ossuary	4A				Bizhanov and Khozhaniyazov 2003:38–39
			25.1	Butentau-kala 3 graves	?				Bizhanov and Khozhaniyazov 2003:33–34
17.0	Devkesken-kala	2	16.1	Devkesken Wall kurgan	2C?	17.0	Devkesken-kala	2	Khozhaniyazov and Khakiminiyazov 1997

continued on next page

Table 6.1—*continued*

ID	*Fortified Enclosure*	*Period*	*ID*	*Kurgans/burials*	*Period*	*ID*	*Inhabited sites*	*Period*	*References (all sites)*
16.0	Devkesken Wall	1C, 2B	91.0	Sab'inel kurgan cemetery	1C				Khozhaniyazov 2006:52–54, 62–63
			94.0	Kummetel 2 kurgan cemetery	1C				Yagodin 2007:46
			90.0	Chash-tepe kurgan complex	5A				Rapoport and Trudnovskaya 1979:39
66.0	Gyaur-kala 1 Chermen-yab	2, 3	66.0	Gyaur-kala ossuaries	2? 4?	66.1	Gyaur-kala I Chermen-yab farmstead	2B	Vaynberg 1979b:171, 2004:248
65.0	Gyaur-kala 2 Chermen-yab	2?, 4?				65.3	Gyaur-kala Chermen-yab pottery kilns	2?, 4	Nerazik 1976:16–17
						65.4	Gyaur-kala 2 settlement	2?, 4	Iusupov 1979:100;
166.0	Gyaur 3 kala	2				166.3	Gyaur 3 settlement (+ production)	2	Tolstov 1958:20, 22, figure 4
45.0	Kalaly-gyr 2	2	47.1–47.5	Yasy-gyr 1, 2, 3, 5 kurgans	2D	45.1	Kalaly-gyr 2 farmsteads	2?	Rapoport and Lapirov-Skoblo 1963:143n5
46.0	Kalaly-gyr 1	1C, 2	47.4	Yasy-gyr 4 kurgans	1A, 4B	45.2	Kalaly-gyr 2 vineyard/fields	2?	Vaynberg 1979a:7, 25, 1979b:168–69
47.6	Yasy-gyr 2 kala	?	46.0	Kalaly-gyr 1 (ossuaries)	4C	45.3	Kalaly-gyr 2 vineyard 2	2?	Vaynberg 1981: 122, 123–24, 1994:75

continued on next page

Table 6.1—*continued*

ID	*Fortified Enclosure*	*Period*	*ID*	*Kurgans/burials*	*Period*	*ID*	*Inhabited sites*	*Period*	*References (all sites)*
			45.0	Kalaly-gyr 2 (ossuaries)	3?, 4	46.0	Kalaly-gyr 1	1C	Vaynberg 2004:9, 11, figure 2/1
						47.6	Yasy-gyr settlement	?	Vaynberg 1981:122
						52.2	Kiuzely-gyr pastoral camp	?	Bizhanov and Khozhaniyazov 2003:48
60.0	Kanga-kala	2, 3, 4A	88.1	Tarym-kaya 1 kurgans	2A	88.2	Tarym-kaya 1 fortified settlement	1C, 2	Itina 1979:6; Iusupov 1979:100
60.2	Kanga-kala 2	1B	88.4	Tarym-kaya 2 and 3 kurgans	1C, 2	88.3	Tarym-kaya 1 open settlement	2A	Vaynberg 1979a:27, 1979b:171–72
			88.4	Tarym-kaya 2 and 3 ossuaries	1C	88.5	Tarym-kaya 2 open settlement	2	Tolstov 1958:94, 1960:231
			60.0	Kanga-kala burials and ossuaries	4D	60.1	Kanga-kala to Tuz-gyr settlements	1, 2	Bizhanov and Khozhaniyazov 2003:48
			60.3	Kanga-kala kurgans	?	60.4	Kanga-kala 2 settlement	1B	Vaynberg 1981:122
			111.0 - 116.0	Kang-gyr kurgans 1 to 6	?				Shown on the Soviet topographic map K-40–093, 1985

continued on next page

Table 6.1—*continued*

ID	*Fortified Enclosure*	*Period*	*ID*	*Kurgans/burials*	*Period*	*ID*	*Inhabited sites*	*Period*	*References (all sites)*
103.0	Kuiusai-kala	1	103.0	Kuiusai-kala kurgans and ossuaries	1	103.2	Kuiusai 2 open settlement	1A	Vaynberg 1979a:7, 25
34.0	Mangyr-kala	4C?	34.1	Mangyr-kala kurgans	4?	34.0	Mangyr-kala (?)	4C?	Vaynberg 1979b:168, 1981:126, 2004:247
61.0	Shakh-senem	3, 4	61.1	Shakh-senem kurgans	2, 3 / 5?	61.2	Shakh-senem extramural	3, 4	Andrianov 1969:figure 47; Vaynberg 1979b:170, 2004:248
87.5	Tuz-gyr 2 kala	2C	87.1	Tuz-gyr 2 (south) kurgans	1?, 2	87.5	Tuz-gyr 2 settlements	2B, 4A	Vaynberg 1979b:167–68, 175
87.3	Tuz-gyr 1 kala	4B	87.2	Tuz-gyr 3 (southwest) kurgans	4B	87.0	Tuz-gyr 2 pottery production sites	4A	Nerazik 1976:17
39.0	Akcha-gelin	2D							Tolstov 1958:26

TABLE 6.2 List of sites from the Sarykamysh delta area, including ceramic types. The sites are listed in alphabetical order, and the site ID number can be used to locate them on figures 6.2 and 6.3. A = Archaic Chorasmian pottery; Ch = Chorasmian pottery ("Kangiui" or "Kushan"); Ku = Kuiusai ceramics.

Site ID	*Site Name*	*Site type*	*Ceramics type*	*Comment*	*Reference*
16	Devkesken wall	Defensive installation	Ch	Archaic to Early Kangiui	Khozhaniyazov 2006:63
16.1	Devkesken wall kurgan	Kurgan/nomad burial	?		
17	Devkesken-kala	Large fortified site	Ch	Kangiui	Khozhaniyazov 2006:63, 366
24	Butentau-kala I	Fortified site	Ki + Ch	Kiuzely-gyr–Kangiui types	Bizhanov and Khozhaniyazov 2003:34
24.1	Butentau-kala I kurgans	Kurgan/nomad burial	?		
25	Butentau-kala II	Fortified site	?		
33	Kurgan-kala	Large fortified site	?		
34	Mangyr-kala	Large fortified site	Ch	Kushan-Afrighid	Tolstov 1958:22
34.1	Mangyr-kala kurgans	Kurgan/nomad burial	?	no pottery found in excavated kurgans	Vaynberg 1979b:168
35	Kunya-uaz	Large fortified site	Ch	Kangiui-Kushan	Nerazik 1958:367–96
35.1	Kunya-uaz settlement	Settlement	Ch	Kangiui-Kushan	Nerazik 1976:19
39	Akcha-gelin	Fortified site and settlement	Ch	Kangiui	Tolstov 1958:26
40	Turpak-kala	Fortified site	Ch	Kangiui-Kushan	Tolstov 1958:26, 28; Nerazik 1976:20
40.2	Turpak-kala 2	Fortress	?		
40.3	Turpak-kala 3	Fortress	Ch	Kangiui 4th–3rd centuries BC	Nerazik 1976:21
40.4	Turpak-kala 2 oasis settlement and kilns	Settlement	Ch	Late Kushan	Nerazik 1976:19

continued on next page

Table 6.2—*continued*

Site ID	*Site Name*	*Site type*	*Ceramics type*	*Comment*	*Reference*
40.5	Turpak-kala 4	Fortified site	Ch	Early Antique (Kangiui)	Nerazik 1976:19
40.7	Turpak-kala 1 oasis settlement and kilns	Settlement	A	Archaic	Nerazik 1976:20
45	Kalaly-gyr 2	Cultic center	Ch	Early Kangiui	Vainberg 1994:75
45.1	Kalaly-gyr 2 farms	Farmstead	?	assumed to be Kangiui	Vaynberg 2004:19, figure 2/1, 2/5
45.2	Kalaly-gyr 2 vineyard/fields	Farmstead	?	assumed to be Kangiui	Vaynberg 2004:19, figure 2/1, 2/5
45.3	Kalaly-gyr 2 vineyard 2	Farmstead	?	assumed to be Kangiui	Vaynberg 2004:19, figure 2/1, 2/5
46	Kalaly-gyr 1	Large fortified site	Ch	Early Kangiui, then Kushan	Rapoport and Lapirov-Skoblo 1963
47	Yasy-gyr 2 kala	Fortified site	?	?	Bizhanov and Khozhaniyazov 2003:48
47.1 – 47.5	Yasy-gyr 3, 4, and 5 kurgans	Kurgan/nomad burial	Ch + Ku	Early graves only K; later graves C	Vaynberg 1981:124
47.6	Yasy-gyr settlement	Settlement	?		Vaynberg 1981:122
52	Kiuzely-gyr	Urban center	Ki (A)		Rapoport, Nerazik, and Levina 2000:26–29
52.2	Kiuzely-gyr pastoral camp	Pastoral camp	?		Vaynberg 1981:122
57	Kiunerli-kala	Fortified site	Ch	Kangiui	Tolstov 1948a:102
57.1	Kiunerli settlement	Settlement	Ch	Kangiui 4th–3rd centuries BC	Nerazik 1976:18
60	Kanga-kala	Fortified site	Ch + A + Var	"Varvar" Uzboi type ceramics	Tolstov 1958: 72, 1962:230

continued on next page

Table 6.2—*continued*

Site ID	Site Name	Site type	Ceramics type	Comment	Reference
60.1	Kanga-gyr to Tyz-gyr settlements	Settlement	Ch + Ku	Kangiui and Kuiusai	Vaynberg 1981:122
60.2	Kanga-kala 2	Fortified site	Ku		Vaynberg 1979b:173
60.3	Kanga-kala kurgans	Kurgan/nomad burial	?		
60.4	Kanga-kala 2 settlements	Settlement	Ku		Vaynberg 1979b:173
61	Shakh-senem	Fortified site	Ch	4th–3rd centuries BC	Rapoport 1958:404
61.1	Shakh-senem kurgans	Kurgan/nomad burial	?	1st–3rd and 4th–5th centuries AD	Durdyev, Babakov, and Iusupov 1978; Vaynberg 1979b:170
61.2	Shakh-senem settlement	Settlement	Ch	Kangiui-Kushan	Andrianov 1969:figure 47
65	Gyaur-kala II Chermen-yab	Fortified site	Ch	?	Nerazik 1976:17
65.1	Gyaur-kala Chermen-yab pottery kilns	Production site	Ch	Kangiui-Kushan	Nerazik 1976:17
65.4	Gyaur-kala II Chermen-yab settlement	Settlement	Ch	Kangiui-Kushan	Nerazik 1976:17
66	Gyaur-kala I Chermen-yab	Fortified site	Ch	Kangiui	Nerazik 1976:16
66.1	Gyaur-kala I Chermen-yab farmstead	Farmstead	Ch	Kangiui 4th–3rd centuries BC	Nerazik 1976:16
66.7	Gyaur 4	Kurgan/nomad burial	?	4th–1st centuries BC	Vaynberg 2004:248

continued on next page

Table 6.2—*continued*

Site ID	Site Name	Site type	Ceramics type	Comment	Reference
87	Tyz-gyr II pottery production sites	Production site	Ch	Kushan	Nerazik 1976:17
87.2	Tuz-gyr III (Southwest) kurgans	Kurgan/nomad burial	K + C + others	4th–3rd centuries BC + 1st–3rd centuries AD	Trudnovskaya 1979:105
87.3	Tuz-gyr 1 kala	Fortified site	others	1st–3rd centuries AD	Vaynberg 1979b:175
87.4	Tuz-gyr 2 kala	Fortified site	Ch	4th–3rd centuries BC	Nerazik 1976:17; Vaynberg 1979b:175
87.5	Tuz-gyr 2 settlements	Settlement	Ch	4th–3rd centuries BC + 1st–3rd centuries AD	Nerazik 1976:17; Vaynberg 1979b:175
87.6	Tuz-gyr northern kurgan group	Kurgan/nomad burial	?		
88.1	Tarym-kaya I kurgans	Kurgan/nomad burial	Ch + Ku	4th–3rd or 3rd–2nd centuries BC	Vaynberg 1979b:172
88.2	Tarym-kaya I fortified settlement	Settlement	Ch + Ku	5th–2nd centuries BC	Vaynberg 1979b:171
88.3	Tarym-kaya I open settlement	Settlement	Ch + Ku	5th–2nd centuries BC	Vaynberg 1979a:27, 1979b:171
88.4	Tarym-kaya II and III kurgans	Kurgan/nomad burial	Ch + Ku	5th–4th/4th–3rd centuries BC	Itina 1979:6; Iusupov 1979:100
88.5	Tarym-kaya II settlement	Settlement	Ch + Ku	4th–3rd centuries BC	Vaynberg 1979b:171
89	Tumek-kichidzhikh	Kurgan/nomad burial	Ch + Ku	Early graves only K; later graves C	Vaynberg 1981:124
90	Chash-tepe	Kurgan/nomad burial	Ch	3rd–4th centuries AD	Rapoport and Trudnovskaya 1979:162

continued on next page

Table 6.2—*continued*

Site ID	*Site Name*	*Site type*	*Ceramics type*	*Comment*	*Reference*
91	Sab'inel	Kurgan/nomad burial	?	5th–4th centuries BC	Yagodin 2007:48
94	Kummetel-2 kurgan cemetery	Kurgan/nomad burial	?	5th–3rd centuries BC	Yagodin 2007:46
103	Kuiusai-kala	Fortified site	Ch	Late Kangiui	Tolstov 1958:70
103.2	Kuiusai 2 settlement	Settlement	Ku	Early Kuiusai, 7th–6th centuries BC	Vaynberg 1979a:7.
111–116	Kanga-gyr kurgans 1 to 6	Kurgan/nomad burial	?		Shown on the Soviet topographic map K40–093, 1985
166.2	Gyaur 3 circular building Chermen-yab	Cultic center	Ch	Early Kangiui	Vaynberg 2004:247
166.3	Gyaur 3 settlement Chermen-yab	Settlement	Ch	Early Kangiui	Vaynberg 2004:247

to gain a clearer picture of the Sarykamysh area's archaeology. This facilitated mapping of the sites in a GIS and allowed further analyses.

The GIS map (see figure 6.2) displays the spatial distribution of all mappable sites in the Sarykamysh delta (59 of 62) that date from the seventh century BC to the third century AD. These include many of the fortified sites, burial monuments, settlements, and one pastoral camp discussed by Vaynberg (1979a, 1979b, 1981, 2004) and others (Tolstov 1948a, 1948b, 1958, 1960, 1962, 1970; Nerazik 1976; Mambetullaev 1978; Khozhaniyazov 1982, 2006; Yagodin 2007). The GIS map (figure 6.2) illustrates how close many of the kurgan sites in particular are to the fortresses and that they were clearly within what has been regarded as ancient Chorasmian territory (Khozhaniyazov 2006:73). As mentioned, dating is problematic because of its reliance on the relative Chorasmian periodization; because many sites and features have not been exhaustively investigated, there is great variability in the datable material. Mapping the sites and features has not been a simple task, since most of the maps in the archaeological publications are very large-scale, if there are maps at all. I have used maps and descriptions in Russian publications, Soviet military topographic maps, as well as remote sensing—primarily through the Google Earth satellite imagery platform—to map each site or feature.

It is apparent that sites are spatially and often temporally and culturally linked in a dispersed, low-density settlement pattern. Some of these linkages between sites were highlighted by Vaynberg (1979a:25–27, 1979b:171, 1981:123–24), in particular if they shared similar stratigraphy and cultural material (see the examples of Tuz-gyr, Yasy-gyrs, and Kalaly-gyrs 1 and 2 below). Table 6.1 presents a preliminary attempt to group together the kalas, burial sites, and settlements that appear to be related in the form of a sort of dispersed or non-nucleated settlement pattern, for example, where a fortified enclosure may have been linked with populations also using nearby burial sites and settlements. Sites are grouped together based on spatial proximity, shared land forms, and common chronological material culture. There are limitations to these analyses because of the differential investigation of many sites (i.e., some were intensively investigated and excavated, others only had surface inspections) and the relatively broad chronological divisions of the Chorasmian periodization (see earlier comments on chronology). The suggested groupings of sites assembled in table 6.1 require further confirmation through more detailed investigations of their material culture, a project for future research. Yet their spatial proximity and the rough indications of temporal proximity suggest patterns in the occupation of the Late Iron Age landscape that are worth exploring further.

In this interpretation, the settlements and pastoral camps are seen as evidence of habitation, of everyday domestic production and reproduction activities taking place (see Brück 1999:55–62). Kalas are viewed as evidence of monumental, communal endeavor, of tenure and the marking and protection of territory (Koryakova 2000; Allard and Erdenebaatar 2005). Potentially, they are also representative of administrative control and power regimes within the oasis (Negus Cleary 2007, 2008; Kidd, Negus Cleary, and Baker Brite 2012). The mortuary sites are viewed as places of ceremony and ritual, as well as memory and marking, and the kurgans as monuments of tenure and territory. Wright (2007) makes the point that Xiongnu burial monuments that incorporated kurgans were places of group ritual and performance and of markers of memory and territory for communities rather than spaces reserved for elites (for a discussion of tenure versus territoriality in mortuary monuments of pastoral groups, see Chapman 1995). They may also have been markers of movement (migration routes, trade routes, travel) (cf. Koryakova 2000; Allard and Erdenebaatar 2005:558; Wright 2007). Importantly, sometimes these categories overlap, that is, they appear at the same site—for example, the use of Akcha-gelin fortress as a settlement as well as a fortified enclosure and the use of the fortified walls of Kalaly-gyr 2 and Kalaly-gyr 1 as necropoleis for ossuaries.

The fortified enclosures appear to have been located almost exclusively on high ground, which is quite different from the fortified sites of eastern (right bank) Chorasmia, where the majority of fortresses are located on the plain (see figure 6.2). However, this difference may be a result of the fact that there is far less high ground and no plateaus in eastern Chorasmia, and, where possible, most kalas are located on higher ground and consolidated sand dunes. Settlements are unfortified, of low density, and dispersed, with the exception of Akcha-gelin (**39.0**) (see table 6.1). They are often found close to kalas. Kurgans are all located on high ground, near cliff edges where they can be viewed, and most often close to kalas (e.g., Mangyr-kala [**34.0**], Bol'shoi Aibugir-kala [**6.0**], Kanga-kala 1 [**60.0**], Butentau-kala 1 [**24.0**]). The kalas themselves also appear to be foci of mortuary activity, apart from the kurgan cemeteries located in their immediate vicinity, with several of them used as ossuary necropoleis (e.g., Butentau-kala 2 [**25.0**], Kalaly-gyr 1 [**46.0**] and 2 [**45.0**], and Kuiusai-kala [**103.0**]). Importantly, this mortuary use occurs after the site has been abandoned (discussed later). Both settlements and kalas were located close to natural river courses and did not rely on complex irrigation systems, with the exception of Shakhsenem (**61.0**), which was reliant on a supply canal (Andrianov 1969:figure 47), and Kalaly-gyr 2 (**45.0**), with its small-scale canals to facilitate some agriculture (Vaynberg 2004:figure 2/1, 2/2, 2/5). Given this scant evidence for farming and irrigation, the majority of Late Iron Age communities must have relied on pastoralism as the mainstay of their economy. The kalas do appear to have been centers for dispersed, low-density settlements, but the question remains, what kind of centers?

Interestingly, the kalas are more often located in the peripheral areas of the delta than in centralized locations (see figures 6.2 and 6.3). This may be in part a result of differential preservation of sites or a preference for high ground that is more available on the outskirts of the Amu-Darya delta. It may also be because the enclosures were useful for cattle corrals, and more grazing land was available in the peripheral areas. It may have also been for defense against enemy incursions, as posited by Khozhaniyazov (2006). A third possibility, as at Bol'shoi Aibugir-kala, is that the location facilitated trade with more mobile nomadic groups (Mambetullaev 1978:87–88). Alternatively it may have been for all these reasons in combination. Certainly, all the kalas were located close to water resources and often had open settlements or production sites nearby.

For a more detailed examination of the landscapes surrounding several of these locales with kalas, kurgans, and other features, it is useful to "zoom in" to a smaller scale. The locales around Kalaly-gyr 1 and 2, Kanga-kala to Tarym-kaya, and Mangyr-kala have been chosen as examples here because they are representative of different manifestations of the settlement pattern and in part because Kalaly-gyr 1 and 2 and Kanga-kala are signature sites where a broader picture of their landscape context may be useful. These are some of several examples of such spatial and temporal connectedness that could be given (see GIS map figure 6.3; tables 6.1 and 6.2).

KALALY-GYR GROUP

Between the now dry Southern Daudan and Chermen-yab river channels are flat alluvial plains punctuated by several low plateaus known as Kalaly-gyr and the Yasy-gyrs (see figure 6.4). The plateaus divide up this flat landscape, effectively forming shallow "valleys" in between the Yasy-gyr and Kalaly-gyr plateaus. Kalaly-gyr 1 (**46.0**) is a large (63 ha) fortified enclosure site situated on the plain in the east. It lies near the Southern Daudan river channel, and canals carried water from this river to the site, dating to the sixth–fifth centuries BC (Tolstov 1962:94; Andrianov 1969:154–55). The fortification walls were constructed of large, square sun-dried mud bricks, with double walls and an "archers gallery" in between, externally projecting towers and loopholes—all clearly in the ancient Chorasmian architectural tradition (see plan in figure 6.5a). The only structure found within the walls is a large "palace" complex, luxuriously fitted out with many halls and courts. The site was occupied intermittently. The first period relates to the site's construction at the end of the fifth century/early fourth century BC (Rapoport and Lapirov-Skoblo 1963:n5); the second period involves a partial occupation of the palace only during the fourth through second centuries BC (ibid.:143); during a third period the site was used as an ossuary necropolis, from the second to the beginning of the fourth centuries AD (Tolstov 1958:161; Rapoport and Lapirov-Skoblo 1963:143), and for production of Chorasmian "Kushan" ceramics (Tolstov 1958:164–65). There is also

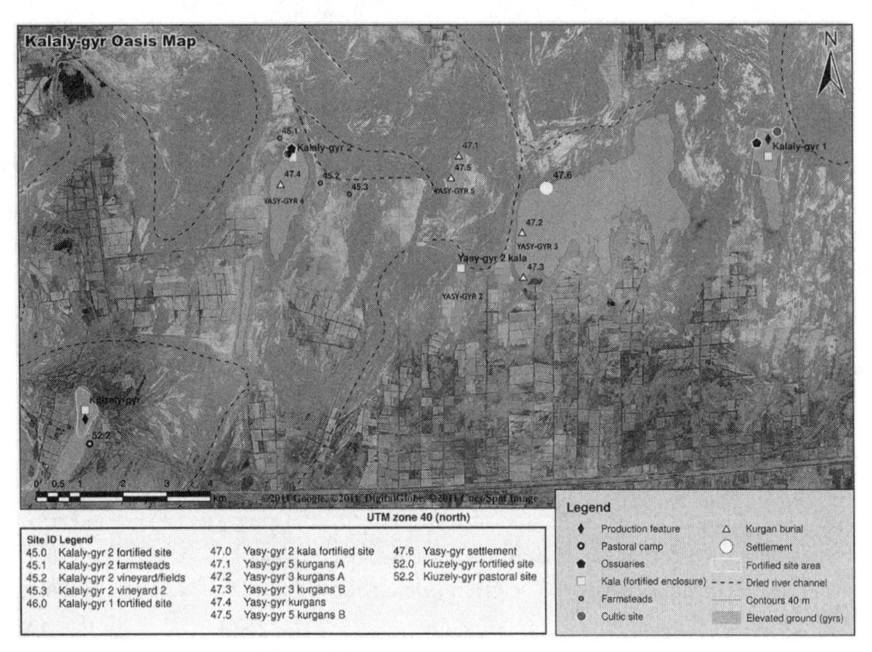

FIGURE 6.4. GIS map of the Kalaly-gyr-area oasis.

evidence of mortuary practices carried out here with the preparation of the dead in *dakhmas,* or places where bodies were exposed for excarnation by wild animals and the elements (ibid.). Layers of manure were excavated that had been periodically burned out and overlain with more manure from later occupation (Rapoport and Lapirov-Skoblo 1963:150–51), suggesting a pastoral use of the site after the second century BC. The use of the site for pottery production when there was no intramural contemporary permanent habitation raises questions about where these potters may have lived.

The second fortified site, located 10.5 km to the west of Kalaly-gyr 1, is Kalaly-gyr 2 (45.0). Kalaly-gyr 2 is located on high ground at the northern end of the western Yasy-gyr plateau, overlooking the plain and the Southern Daudan river channel to the north. This fortress was approximately 3 ha in area and had a large, monumental barbican-type gateway and an unusual plan that followed the topography of the cliff edge along its northwestern edge, but it had two straight sides forming a right angle along the southern and eastern sides (see plan in figure 6.5b). The construction is typical of Chorasmian fortresses, using mud bricks, double galleried walls, and towers. The site is dated from the mid-fourth to the early second century BC (Andrianov 1969:154–55; Vainberg [Vaynberg] 1994:75). Vainberg

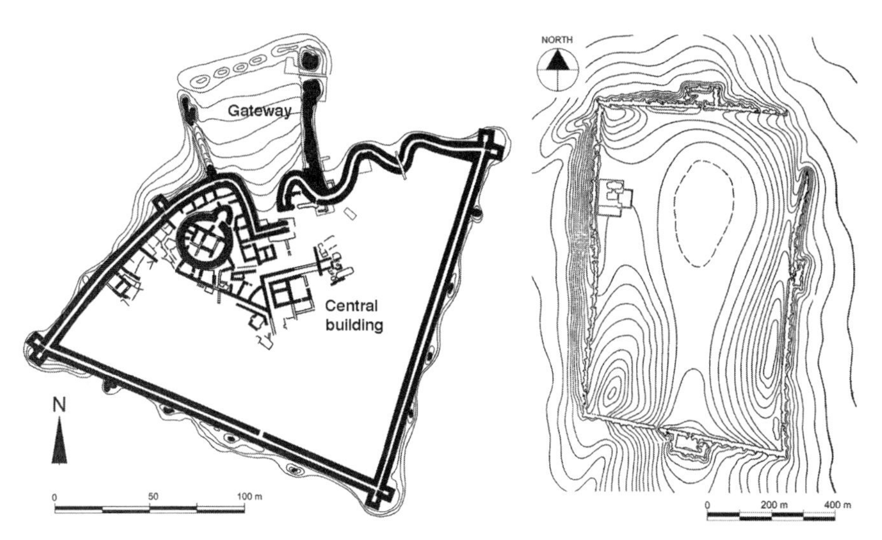

FIGURE 6.5. (a) Kalaly-gyr 2 site plan. (b) Kalaly-gyr 1 site plan (after Rapoport and Lapirov-Skoblo 1963: fig.1).

(1994:75, 77; Vaynberg 2004:9, 17) regarded it as a cult center for the surrounding pastoral populations and not a permanent settlement. The site was accessible from the entire Sarykamysh delta by way of the Southern Daudan river bed, which is why Vaynberg (Vainberg 1994:77) considered it to have been an important regional center. Interestingly, it was not occupied at the same time as Kalaly-gyr 1, and, similarly, it had buildings that have been interpreted as elite or cultic in nature but not domestic (ibid:75; Vaynberg 2004).

Another fortified site was Yasy-gyr 2 kala (**47.0**), located on a low promontory overlooking a river channel and plain between Yasy-gyr 2 and Yasy-gyr 5. Dating for this site is unknown. This kala is a different type than the geometric Kalaly-gyr kalas; it was formed using an earthwork to defend the end of the promontory and employed the naturally steep sides of the promontory as barriers on the remaining sides. The preserved height of the earthwork rampart was 10 m, and approximate dimensions of the site were 100 × 120 × 90 m. This is very similar construction and morphology to Butentau-kala 1 and Kanga-kala 2, and this type of kala has been interpreted as having been constructed by pastoral groups (discussed below) (Bizhanov and Khozhaniyazov 2003:48; Vaynberg 2004:figure 2/1).

Several habitation sites and features were identified from surface surveys conducted by Vaynberg's expedition. Temporally (based on surface finds) and spatially associated with the Kalaly-gyr 2 fortress (**45.0**) were some fields (**45.2**) and several vineyards (**45.3**) and therefore a small farming settlement (**45.1**) serviced by irrigation canals coming off the Southern Daudan river channel (Vaynberg 2004:19, figure 2/1, 2/2, 2/5; see figure 6.4). The settlement (**47.6**) located on the northern edge of the Yasy-gyr elevation was unfortified and, according to Vaynberg (1981:122), located outside the area of irrigation and was therefore a pastoral settlement. In terms of other domestic or settlement sites, there was the pastoral camp (**52.2**) near Kiuzely-gyr (**52.0**). Although Vaynberg (ibid.) discusses both the Yasy-gyr settlement and the Kiuzely-gyr camp in context with other "Antique" period (i.e., seventh/sixth century BC–fourth century AD) sites in Chorasmia, I am not aware of any published information about their dating.

On the edges of the plateaus facing the "valley" (or bounded) plain to the west of the Kalaly-gyr 2 plateau are groups of kurgan burials known as Yasy-gyrs 1–5 (**47.1–47.5**) (see figure 6.4). These appear to have been multi-period cemeteries with a date range from the seventh/sixth century BC to the first/second century AD (Vaynberg 1979b:168–69, 1981:123–24). There is a variety of burial traditions represented, including inhumation directly on the ground surface with mound raised above (Kurgans 2 and 4, Yasy-gyr 4) and podboi-catacomb burials (e.g., Kurgans 7, 14, and 15; Yasy-gyr 4), a steppic cultural tradition. One kurgan burial in the catacomb dating to around the first/second century AD appears to have incorporated Mazdean burial ritual, with only the cleaned bones interred (Tolstov 1970:400).[4]

This Mazdean burial tradition with the cleaned bones interred usually in ossuaries is considered synonymous with sedentary Chorasmians (cf. Vaynberg 1981:122, 2004:11), and for the Kalaly-gyr area such ossuary burials are found only in a later period, well after the fortified sites have been abandoned—for example, in the second–fourth centuries AD for Kalaly-gyr 1 (Rapoport and Lapirov-Skoblo 1963:143n5) and sometime between the second/first century BC and the early centuries AD for Kalaly-gyr 2 (Vaynberg 1981:124, 2004:4). The fact that the ossuaries were found buried in the structures of these kalas clearly displays the changed use of the site and seems to suggest that these ossuaries were not the burials of a local sedentary community, although they could be those from a settlement further afield. Evidence also suggested that the abandoned pottery kilns on the towers of Kalaly-gyr 1 were reused as dakhmas during this later period (Tolstov 1958:164–65). It appears that the kurgan burials may have been the only localized burials that could be contemporary with the occupied life of the kala or the agricultural settlement associated with Kalaly-gyr 2, yet no clear material evidence links the kalas and the kurgans here. Perhaps the occupants of both kalas buried their dead further afield, or perhaps there were very few permanent inhabitants of these sites, resulting in an apparent absence of cemeteries. However, the burial tradition for Chorasmia is problematic for the seventh–fifth/fourth centuries BC, as there are no known examples of ossuary burials and only two known examples of fourth century BC ossuaries (Bizhanov and Khozhaniyazov 2003:49).

TARYM-KAYA–KANGA-KALA GROUP

Between the plateaus of Tarym-kaya and Kanga-gyr are a series of sites and landscape features (see figures 6.6 and 6.7) that are classified as belonging to the Kuiusai culture but located within Chorasmia, and the presence of Chorasmian ceramics indicates inclusion within the Chorasmian cultural sphere.

The plateau of Tarym-kaya is a notable landscape feature, as its southwestern part is the highest elevation in the Sarykamysh delta and has been used for orientation by people moving through the area (Vaynberg 1979b:171). The southwestern promontory of the Tarym-kaya plateau has three kurgan cemeteries, Tarym-kaya 1, 2, and 3. In addition, three contemporaneous settlements at Tarym-kaya are associated with the kurgans nearby (see figure 6.7).

Tarym-kaya 1 kurgan group (**88.1**) is located near the edge of the southwestern cape of Tarym-kaya (see figures 6.6 and 6.7). It comprises about fifty-two kurgan mounds of varying sizes (Vaynberg 1979a:table XIIIv, g), sited near the edge of the plateau. Five kurgans were excavated by the Khorezm Expedition. Kurgan 18 was a typical Kuiusai culture burial, with Kuiusai ceramics (Vaynberg 1979b:171). Kurgans 3, 4, 7, and 9 were podboi burials (pit-type tombs with a side niche in which the body was interred) with artifacts including handmade early "Kangiui" Chorasmian pottery, glass fragments, bone facings for composite bows, iron artifacts including a bronze buckle with an iron uvula, an iron knife, and beads. The dating of these artifacts is uncertain, though they are most likely from the fourth century BC (ibid.:172).

A small fortified settlement, known as Tarym-kaya 1 fortified settlement (**88.2**), was located on the very edge of the cliffs, approximately 350 m southwest of the most southerly Tarym-kaya 1 kurgans (see figure 6.7). The site was fortified by means of a wall constructed to control access to the settlement from the plateau, while the other sides were protected by the steep cliff edges (see plan in figure 6.6). The fortified area was approximately 0.25 ha but would have been slightly larger, as erosion has destroyed parts of the cliffs. The eastern half of the site was filled with buildings that excavations revealed were houses with spaces for cooking, food preparation, and storage. The western half was larger and had no structural remains, but there were abundant traces of manure (Vaynberg 1979a:25). This strongly suggests that the settlement was used by a pastoral community where cattle could be corralled in one half of the protected area and people inhabited the other half. This was a very small settlement, perhaps a pastoral station for several families. According

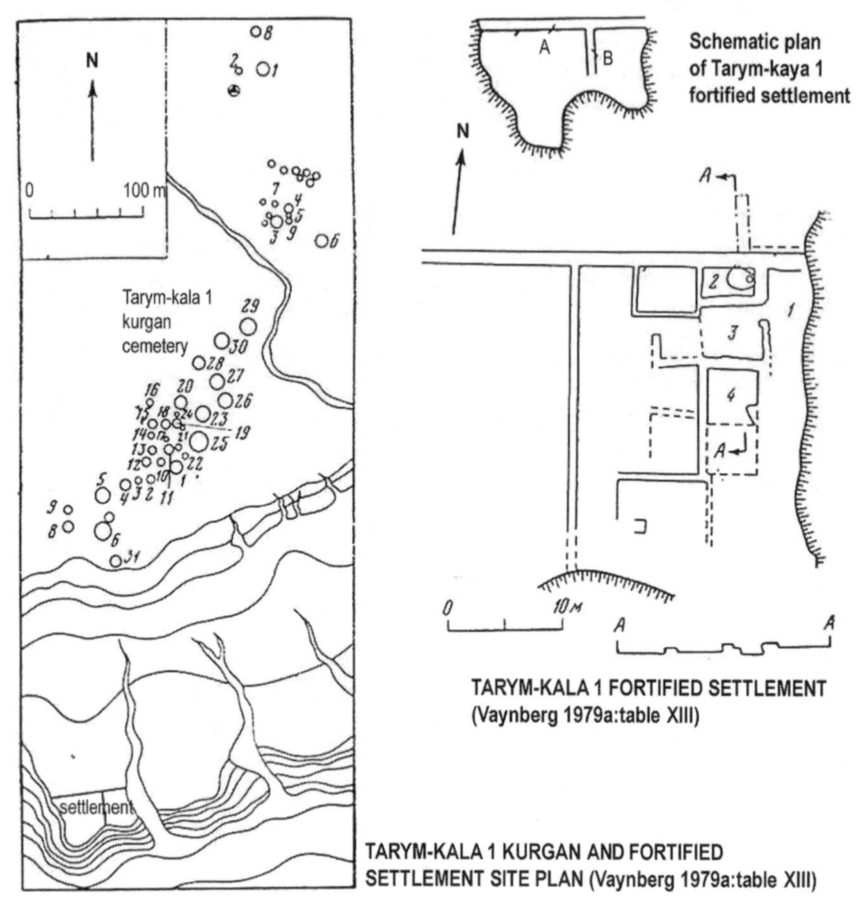

Figure 6.6. Tarym-kaya 1 fortified settlement and kurgan cemetery.

to Vaynberg, this settlement was associated with the kurgan cemetery. The ceramics were predominantly of Kuiusai type, but there were also many Chorasmian ceramics dating to the Late "Archaic" and Early "Kangiui" periods (ibid.:26). More than ten bronze tri-lobal (so-called Scythian) arrowheads were found across the intramural areas of the settlement, and Vaynberg noted that they were of the same types as those found on other Chorasmian sites and so-called Sarmatian sites. These arrowheads also confirm the early dating of fifth–fourth and fourth–second centuries BC (ibid.:26–27).

A second settlement was found at the base of the plateau, known as the Tarym-kaya 1 open settlement (**88.3**). It comprised a continuous strip of low-density settlement along the banks of the river channel, where small farmsteads were preserved

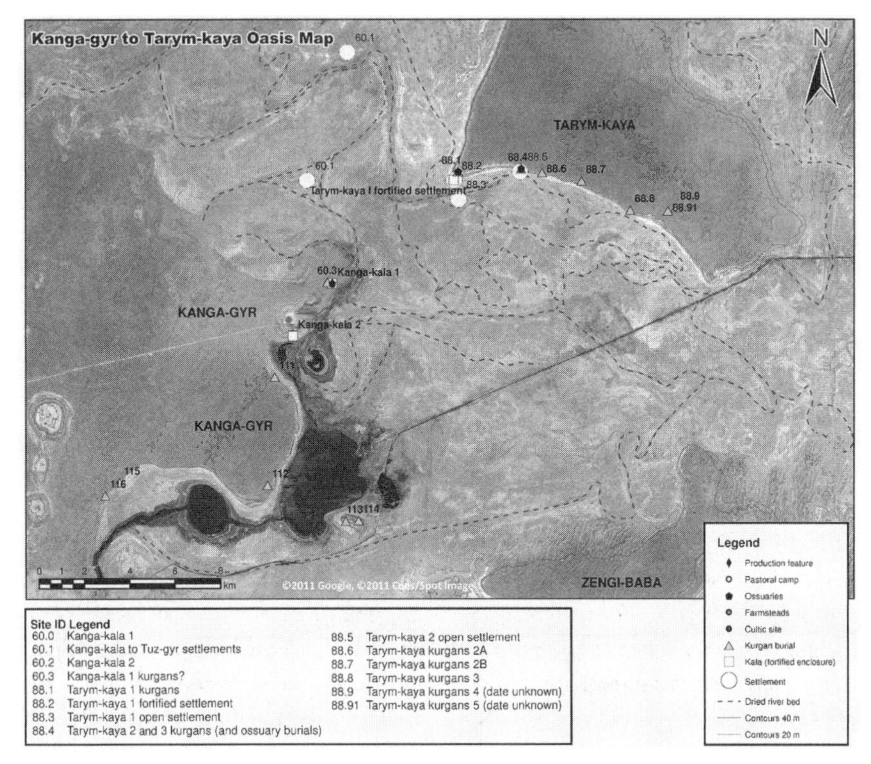

FIGURE 6.7. Tarym-kaya to Kanga-gyr GIS landscape map.

as building remains and occupation layers spaced out 50–100 m from each other in proximity to the riverbed (see figure 6.7). There were no traces of an irrigation system (Andrianov 1969:160; Vaynberg 1979a:6, 27, 1979b:171; Bizhanov and Khozhaniyazov 2003:49). The archaeological material (surface finds) was the same as that from Tarym-kala 1 fortified settlement (**88.2**) at the top of the cliffs, the main difference being that Chorasmian ceramics formed the largest group of finds with Kuiusai type pottery forming a much smaller proportion (Vaynberg 1979a:27).

Tayrm-kaya 2 was a very similar undefended, low-density settlement (**88.5**) that also lay at the foot of the Tarym-kaya, further to the southeast. This settlement was located on the western bank of the river channel, close to the cliffs and near the kurgan tombs of Tarym-kaya 2. There were no traces of irrigation. The material culture (surface finds) was contemporary with that of the kurgans above and with the Tarym-kaya 1 open settlement (Vaynberg 1979b:171).

The Tarym-kaya 2 and 3 kurgan cemeteries (**88.4**, see figure 6.7) contain kurgan tombs of the fifth–fourth centuries BC (Iusupov 1979:100). These kurgans are of great interest because they appear to show a blending of burial traditions—the use of the steppic kurgan burial mounds, with not only primary interments of bodies but also secondary burials of cleaned bones in ceramic ossuaries in a more Mazdaean burial rite (Itina 1979:6).[5] Most of the ceramics used for these ossuaries are typical "Kangiui" period Chorasmian types (Iusupov 1979).

A series of low-density, dispersed, unfortified settlements noted on the plains between the elevations of Kanga-gyr to Tarym-kaya and Tuz-gyr (**60.1**, see figure 6.7) (Vaynberg 1979b) were very similar to the Tarym-kaya 1 and 2 open settlements. These sites were not excavated, but there were significant surface finds, including domestic areas with large numbers of *khumi* (storage jars) dug into the ground. Ceramics were both Kuiusai and Chorasmian, with a significantly higher percentage of Chorasmian pottery at these sites. These open settlements were located adjacent to the riverbeds—there were no traces of an irrigation system—and they were apparently pastoral settlements (Vaynberg 1979a:27).

The first fortified site in the area was Kanga-kala 2 (**60.2**, see figure 6.7), which dates from the sixth–fifth century BC, based on surface finds (Bizhanov and Khozhaniyazov 2003:48). It was formed by a single low, wide rampart with a ditch in front, which created a protected area at the end of a triangular promontory on the plateau of Kanga-gyr (similar to Butentau-kala 1). At the base of the fortress of Kanga-kala 2 was found a Kuiusai settlement (**60.3**), dating to the same period as the kala. Kuiusai type pottery was found in both the kala and the associated settlement (Vaynberg 1979b:173).

After the abandonment of Kanga-kala 2 and its settlement, a second kala of Kanga-kala 1 (**46.0**, see figure 6.7) was constructed close by. It occupied a strategic and visually dynamic position on a cape of the plateau that juts out toward the east and overlooks the entire valley plain that lies between Kanga-gyr and the Tarym-kaya plateau. The fortress was constructed in apparently typical Chorasmian architecture of double mud-brick walls, with a 2.4-m-wide "archers' gallery" down the center (see plan in figure 6.8). Rectangular towers projected from the walls, and a defensive ditch encircled the fortress on at least one side (Tolstov 1958:71). However, the loopholes had much shorter shafts, and Tolstov (1962:229) insinuates this was because the walls were built by "barbarians" (i.e., nomads) who were replicating the Chorasmian military architecture. Kanga-kala 1 was constructed no later than the fifth century BC. Based on ceramic finds, the fortress was occupied for a relatively long period (by Chorasmian standards), probably intermittently from the fifth century BC until the fourth century AD (Tolstov 1958:72, 1962:230). The fortress was of medium size, with an area of approximately 4.7 ha. Traces of intramural structures could be seen on the surface, including a large complex of rooms inside the western wall and in the center of the enclosure. There was abundant Chorasmian pottery

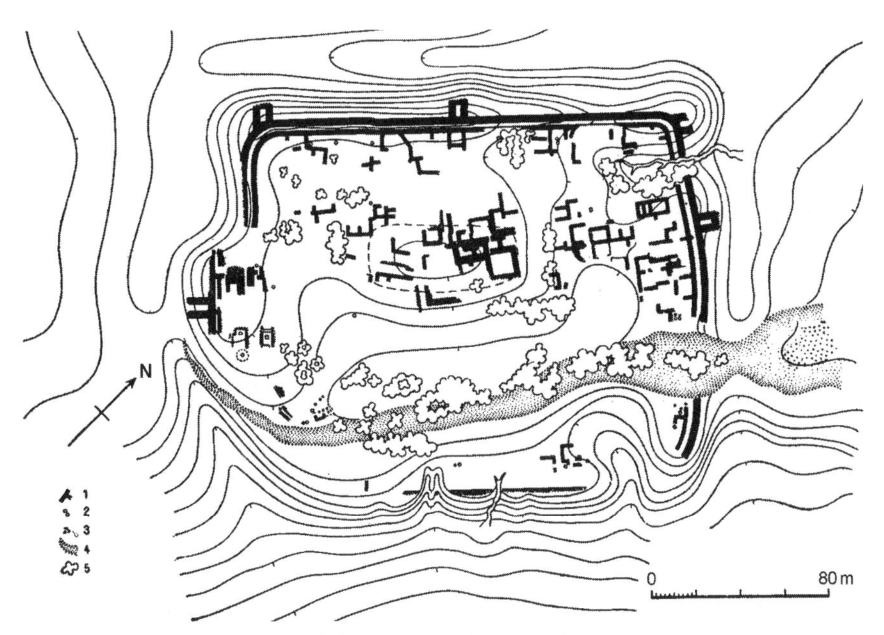

FIGURE 6.8. Kanga-kala plan (after Tolstov 1962: fig.138).

from the "Kangiui" and "Kushan" periods and a pottery kiln found on the southern wall. The location of the kiln on the fortification wall strongly suggests that it postdates the main occupation of the kala, which ended sometime in the fourth century AD (Tolstov 1962:230). Other finds included Scythian bronze tri-lobal arrowheads, ceramic ossuaries from the fortification wall, a clay sculpture of a horse, sets of blue paste beads, and a single silver coin of Chorasmian manufacture (Tolstov 1958:71–72). It is assumed that the fortress was destroyed by a military attack in the fourth century AD, as a large breech was found in the eastern wall, and the upper strata show evidence of a great fire and destruction (Tolstov 1960:13, 1962:230).

The fortress of Kanga-kala 1 (**46.0**) must have served as a center for the local pastoral communities, at least up until the fourth century BC when the local Kuiusai-Chorasmian settlements on the plain were no longer in use (Vaynberg 1979a, 1979b). The plan of the site does not support it having had a sizable population, given that there were structures against the fortification walls and a large central building (Tolstov 1958:71) (see plan in figure 6.8). It has also been interpreted as a military outpost of the political state of Chorasmia whose function was economic and political control of the local pastoral communities of the Sarykamysh area (Vaynberg 1979b:172, 2004:9, 247). Certainly, the material culture from the fortress is represented in the Russian literature as only Chorasmian. Yet there are indications that the fortress was not solely military or administrative in nature. There are small domestic quarters located near the western wall and gateway (Tolstov 1962:229–31) and, importantly, the so-called *naos* building excavated in 1953: a rectangular room, 4.4 × 4.8 m, with a rectangular platform hearth made of bricks in the center around which were sixteen human skulls and piles of associated long bones. These burials appear to be cleaned bones similar to Mazdaean mortuary practice, yet they were not interred in ossuaries as was usual for Chorasmian burials (Tolstov 1958:92–94, 1962:230–32). These burials have been dated to the third–fourth centuries AD based on associated finds, such as an iron blade, a glazed vessel, and a carved stone cosmetic box (Tolstov 1958:94, 1962:231). There are also some ossuaries found associated with the fortification walls (Tolstov 1958:72). These finds indicate a ritual component to the use of the site, at least in the later periods. The personal items mentioned, such as the beads and the horse sculpture, also indicate a more domestic and possible elite occupation of the fortress. These domestic and ritual components possibly rule against the primary use of the site as a martial installation of the state to control the outer pastoral territories; and, of course, its function may have changed over time (or had multiple uses simultaneously).

The satellite imagery available for the Kanga-gyr area also shows circular features (**60.3**) consistent with kurgan burial mounds in the immediate vicinity of the fortress. Several groups of them appear in linear arrangements that are also consistent with kurgans. One group is aligned with the western wall of the fortress (see figure 6.9). I have found no discussion of these features in the Russian literature, and certainly their dating is unknown. Yet their presence immediately outside the fortress is analogous to the kurgans and fortresses of Mangyr-kala and Bol'shoi Aibugir-kala,[6] and kurgans are documented within the fortified area of Butentau-kala 1 (ibid.:81). Potentially, there are other kurgan cemeteries associated with fortified Chorasmian sites that have not been mentioned in the archaeological literature, as discussed earlier.

MANGYR-KALA AND KURGANS

On the southwestern promontory of the plateau of Mangyr is the fortified enclosure site of Mangyr-kala (**34.0**) (also known locally as Khalap-kala). This site overlooks the plain and the course of the northern Daudan river channel below. The kala is composed of two adjoining fortified enclosures having a combined area of around 27 ha, and they are of irregular shape informed by the topography of the cliff promontory (see site plan in figure 6.10). The fortification walls are of typical ancient Chorasmian construction and use Chorasmian type mud bricks. These primary fortified enclosures are adjoined by two larger secondary enclosures that extend out

FIGURE 6.9. Digital Globe satellite image showing Kanga-kala 1 fortified site and kurgans as pale circles to the immediate west and northwest of the fortress.

to the northwest and northeast. These enclosures are formed by embankments constructed of local crushed stone and are much lower and smaller in height and width (Vaynberg 1979b:168).

There were no apparent preserved structures within any of the enclosures. Chorasmian ceramics were found on the surface in the fortress and the enclosures, with the most numerous those types known as Late "Kushan" period (second–fourth centuries AD), with a second group Chorasmian "Afrighid" ceramics (fourth–eighth centuries AD) and a third not very numerous group of medieval pottery (Tolstov 1958:22). Only a single excavation trench has been placed over the fortification walls at this site (Durdyev, Babakov, and Iusupov 1978), and no subsurface testing of the intramural area has been done.

Within the secondary enclosures are many kurgans (**34.1**). In an example of the division that prevailed in the scholarship between "sedentary" and "nomadic" material, the published representations of the fortress of Mangyr-kala ignored completely the huge kurgan cemetery within and around this fortified site (Tolstov 1958:20–22, figure 4; Durdyev, Babakov, and Iusupov 1978; Khozhaniyazov 2006:74, 77, 83, figure 41). Vaynberg's (1979b:168, figure 1, 1981:126) work was the exception, as she included both the kurgans and enclosures and discussed their relationship to one another.

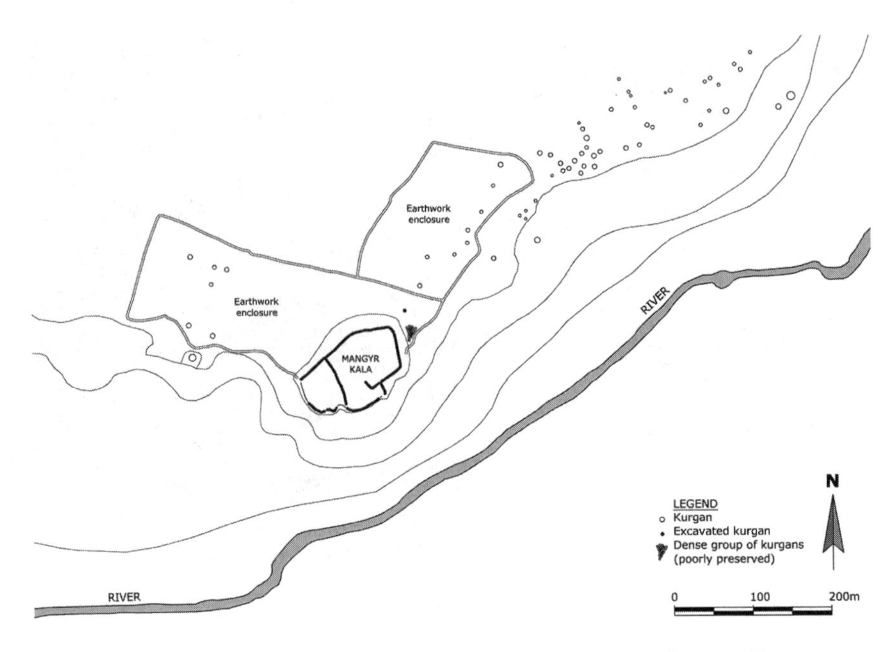

FIGURE 6.10. Mangyr-kala and kurgans site plan, with information from satellite imagery (after Vaynberg 1979b: fig.1).

According to Vaynberg's (1979b:figure 1) plan, no kurgans were found within the mud-brick fortress proper, but in the immediately adjoining northwestern enclosure there are at least seven kurgans and a very dense concentration of smaller kurgans in the southeast corner, where the mud-brick fortifications and the stone embankment meet. There are eight large kurgans in the northeastern enclosure and at least forty-eight more extending out along the cliff top immediately to the east of the enclosures.

The position of both the fortress and the kurgans was clearly selected to view the lowland plain below and, in turn, to *be* viewed from this plain. Only one kurgan was excavated and no datable evidence was found (ibid.:168), but Vaynberg stated that the clear spatial and structural relationships between the kurgans and the walls and embankments make it possible to assume temporal proximity between all features in the complex. However, the question remains as to whether the kurgans and secondary enclosures are later additions to the site. Vaynberg (ibid.) also noted the comparative similarities between the Mangyr-kala kurgan cemetery and the third–fourth centuries AD Chash-tepe kurgan necropolis (**90.0**) (Rapoport and Trudnovskaya 1979:165), which correlates with the Mangyr-kala pottery finds. However, secondary usage of such a highly visible location in the landscape by other tribal groups after Chorasmian occupation of the kala may have ceased is also a likely possibility that perhaps had political connotations in terms of the new group staking its claim over a strategically vital place.

There appear to have been no traces of settlement or production activities in the vicinity of Mangyr-kala. (The closest settlement evidence appears to have been an "Archaic" and early "Kangiui" settlement 10 km to the east [Andrianov 1969:156].)

The fortress of Mangyr-kala itself has numerous surface artifacts but no discernible structures to suggest its use as a place of permanent habitation, although the rocky ground may not preserve traces of ephemeral structures. There have not been extensive excavations at the site, but aerial reconnaissance and photography was taken in 1948, as well as three extensive surveys of the site and analysis of the surface ceramics (1950, 1968, 1977), the excavation of a kurgan (1968), and an excavation trench across the northwest corner of the primary enclosure walls (1977) (Tolstov 1958:20, 22, figure 4; Durdyev, Babakov, and Iusupov 1978; Vaynberg 1979b:168, figure 1).

It is interesting that this fortress, which dates to the end of the ancient period, was located in an area with no irrigation system but with a natural watercourse and is surrounded by kurgans both internally and externally. The visual monumentality of the fortress would have been enhanced by the positioning of these large kurgans along the cliff edge extending out from the fortress itself. It is possible that the outer annex enclosures may have been built for some purpose connected with the kurgans. This may indicate that the fortress was used at some point by pastoral "nomadic" peoples rather than a sedentary farming community. Although the ceramic sherds found within the site are Chorasmian rather than Kuiusai or other overtly steppic types, this does not preclude a pastoral population, since nomadic groups often used the ceramics of settled peoples (Koryakova 2000:15).

DISCUSSION

The previous examples show that fortified sites were part of a landscape occupation pattern that included low-density settlements located on river channels and mortuary monuments in the form of kurgan cemeteries located on nearby promontories or associated with kalas, as illustrated in the GIS maps. The fortified enclosures of the ancient Sarykamysh delta area were clearly not the permanent fortified settlements of sedentary farmers, given the scant evidence for irrigation agriculture. Many of the kalas had large areas of open space within them that were devoid of structures (Negus Cleary 2007, 2008), including Kiuzely-gyr (Tolstov 1960:12), Kalaly-gyr 1 (Tolstov 1948a:80, 1948b:94; Rapoport, Nerazik, and Levina 2000:31; Khozhaniyazov 2006:72), Ayaz-kala 1 (Manylov and Khozhaniyazov 1981), and Ayaz-kala 3 (Bolelov 1998:117). There may have been ephemeral dwellings, such as tents, reed housing, and wattle-and-daub–style houses, used within these enclosures that have not survived in the archaeological record, and there is some evidence for the use of this housing in the Sarykamysh area (e.g., at Kiuzely-gyr [Rapoport, Nerazik, and Levina 2000], Kalaly-gyr 2 [Vaynberg 2004:6], Gyaur 3 settlement [Vaynberg 2004:243], Butentau-kala 2 [Bizhanov and Khozhaniyazov 2003:36–37], and Butentau 3 settlement [ibid.:40, figure 6B]). There is also the possibility of the use of composite housing (combination of permanent buildings and tented/ephemeral structures), as has been the case in the Near East (Cribb 1991:154) and Xiongnu fortified sites (Rogers, Ulambayar, and Gallon 2005:803–12).

Shakh-senem is an exception, having been the fortified center for its extramural agricultural settlement (Rapoport 1958:404; Andrianov 1969:figure 47), and the later dating of the kurgans in its vicinity supports this (Vaynberg 1979b:170). Akchagelin also has medium-density structures assumed to have been housing within its enclosure, as well as Chorasmian pottery (Tolstov 1958:26), although there is not the convincing extramural evidence of agriculture at this site as there is for Shakh-senem. Despite some irrigated fields near Kalaly-gyr 2 (discussed earlier), they are not extensive and there is little evidence for settlement in the area (Vaynberg 2004). The unusual fourth–third centuries BC enclosure site of Gyaur-kala 1 Chermen-yab had the remains of ditches inside the extensive secondary enclosure that were interpreted as drainage for orchards or gardens, yet there is little evidence that the intramural spaces of this site were for permanent habitation. There are, however, indications of an extensive, low-density settlement outside the enclosures, near the water channels (Nerazik 1976:16).

For many areas of the Sarykamysh delta, settlements were instead like that of Gyaur-kala 1 Chermen-yab, unfortified and located on the un-irrigated plain near river channels. Vaynberg demonstrated that there were extensive open settlements in areas devoid of irrigation, such as the Kanga-gyr–Tuz-gyr settlements, Tarym-kaya 1 and 2 open settlements, Gyaur 3 settlement (Vaynberg 2004:243–47), Yasy-gyr 2 settlement (Vaynberg 1981:122), and Tuz-gyr 2 settlement (Nerazik 1976:17; Vaynberg 1979b:175)—clearly marking them as communities heavily reliant on pastoral production. The contemporary kalas in the vicinity of many of these settlements, plus the presence of kurgan cemeteries, strongly indicate that Vaynberg's linkage of the kalas with these pastoral communities was correct.

Hiebert (1992:111) has described the settlement regime of the Bronze Age Bactrian-Margiana Complex as a non-nucleated settlement pattern typical of the pre-modern oases of Central Asia. This was a pattern of dispersed and separate building complexes rather than of towns and cities. The kala formed the prominent building type in this dispersed, low-density pattern (ibid.; Lamberg-Karlovsky 1994:400–405; Negus Cleary 2007:349–50). Other scholars have also identified similar patterns where non-nucleated rural settlements clustered around a fortified site and were located near river channels or irrigation systems (Leriche 1977:307–8; Biscione 1981:207–8; Vogelsang 1992:273). Nomadic/pastoral settlements have been characterized as low-density, of medium size (i.e., < 100 ha), and dispersed (Fletcher 1986:71; Cribb 1991:133). All of these characteristics appear to fit the settlement and occupation evidence for western Chorasmia.

Vogelsang (1992), Stride and colleagues (2009), and Rapin (2007) have also raised the possibility that fortified enclosure sites during specific periods may not have been constructed or occupied by sedentary farming communities but may instead have been the centers (perhaps itinerantly occupied) of "nomadic" (or perhaps "steppic" is more apt) elites. While the pastoral settlements of the Sarykamysh delta may not have been for elites, the kalas possibly were. Chang proposes a reconstruction for Iron Age steppe agropastoral settlements of the Talgar in Kazakhstan, where the local social organization may have been egalitarian for the farmers and herders but hierarchical at a regional level where the elite "maintained and cultivated the image of nomadism" that was tied to the construction of monumental kurgan tombs (Chang 2008:329). In western Chorasmia, there are both kurgan tombs and monumental enclosures, often in close landscape and temporal contexts. The seventeenth–twentieth centuries AD Karakalpaks, who also inhabited the Amu-Darya delta, built similar mud walled enclosures, often with a large building inside that was the residence of a local governor. These enclosures were used by the local population (including the governor) as refuges from Turkmen raiders (Khozhaniyazov 2002:147, 155) and perhaps serve as a good analogy for the ancient Chorasmian kalas.

The question remains as to the nature of the relationship between the kalas and the kurgans and pastoral settlements. Vaynberg included many fortified sites such as Kalaly-gyr 2, Kanga-kala, Butentau-kalas 1 and 2, and Mangyr-kala as important centers for pastoral groups (Vaynberg 1979b:169, 1981:122–28, 2004:9, 247), but her characterization of the nature of these sites differed. She interpreted Kanga-kala 1, Kalaly-gyr 2, Mangyr-kala, and Shakh-senem as military outposts of the Chorasmian state, located to economically control the local pastoral population (Vaynberg 2004:9, 247). Bizhanov and Khozhaniyazov agree, but they regard all fortified sites as military installations of an assumed Chorasmian state and propose that their purpose was to protect the borders of the Chorasmian state from nomadic incursions (Bizhanov and Khozhaniyazov 2003:36, 46). Vaynberg (1979b:169, 174) singled out Butentau-kalas 1 and 2 and Kanga-kala 2, among others, as having been built and operated/inhabited by the local pastoral populations and considered them evidence of the pastoralists' mastery of this outlying region of Chorasmia. Vaynberg's interpretation of fortresses as centers for local pastoral tribes was unusual, although it follows in the vein of Tolstov's early interpretation of the enormous fortified sites of Kiuzely-gyr and Kalaly-gyr 1 as large cattle enclosures of the Chorasmians, whose early economy was dominated by pastoralism (Tolstov 1948a:80; Rapoport, Nerazik, and Levina 2000:25, 28). However, there is obviously a chronological disparity here because although Kanga-kala 2 is earlier than the supposed military outpost of Kanga-kala 1, Butentau-kala 2 dates from the fourth century BC. If it was built and used by pastoral groups, this does not fit with a military takeover of the area by the Chorasmian state in the fifth/fourth century BC. Also, this view of domination and control does not entirely fit with Vaynberg's interpretation of Kalaly-gyr 2 as a cultic center of the Sarykamysh pastoral population. The picture is very complex and not as simple as one group or another being dominant. Certainly, the relationships among the kalas, kurgans, and pastoral and agricultural settlements fluctuated over the ancient period.

The interpretation of Bol'shoi Aibugir-kala and Kuiusai-kala is perhaps more indicative of how we should be viewing these sites. Yagodin (2007:46) and Mambetullaev (1978:87–88) interpreted Bol'shoi Aibugir-kala as a center for trade between Chorasmians and nomadic groups of the northern Ustyurt Plateau, where part of the population of the site was composed of steppic nomads (or non-mobile members of the nomadic tribe) and part by settled Chorasmians. The material culture of the site was viewed in terms of this mixed population. Vaynberg (1981:124) similarly suggested that the remains of ossuaries and kurgans found within Kuiusai-kala were representative of the mixed nature of the Chorasmian population of the left bank. Her interpretation of Kalaly-gyr 2 was also non-military and presented the site as a religious cult center for all local pastoral communities within the Sarykamysh delta during the fourth–second centuries BC (Vainberg 1994:77; Vaynberg 2004:9).

Western (left bank) Chorasmia, in particular the Sarykamysh area, is regarded as markedly different in character to eastern (right bank) Chorasmia. The Chorasmian pottery present at many of the kala sites in this peripheral area is of an inferior grade (mainly because of poorer preparation of the clay and lower-standard firing techniques), for example, at Bol'shoi Aibugir-kala (Mambetullaev 1978:86), Akchungul' (Khozhaniyazov 1982:84–85), Butentau-kala 1, Gyaur-kala Chermen-yab 1 and 2, and Kanga-kala (Vorob'eva 1959:219). It appears that this pottery was locally made, as evidenced by pottery kilns at the fortified sites of Kiuzely-gyr (Vorob'eva 1959:199–200), Butentau-kala 1 (Bizhanov and Khozhaniyazov 2003:33), Bol'shoi Aibugir-kala (Mambetullaev 1978), Tuz-gyr, Kanga-kala 1, and Kalaly-gyr 1 (Vorob'eva 1959:199, 210, figure 53) and at the settlement sites of Gyaur 3 settlement (Vaynberg 2004:243) and Tuz-gyr 2 settlement (Nerazik 1976:17). Vaynberg (1979a:27, 1981:122) also noted that substandard mud bricks were used at Kanga-kala 1 and at the open settlements on the Kanga-gyr–Tuz-gyr plain.

Vorob'eva considered this an indication that these sites were outside (or outlying) the central area of the Chorasmian state (Vorob'eva 1959:219) and so agreed with Mambetuallaev that this was a result of the influence of local nomad population mixing with the settled Chorasmians (ibid.; Mambetullaev 1978:86). The evidence for poorer-quality Chorasmian ceramics and building materials from the far western areas of Chorasmia, in conjunction with evidence of localized pottery production, does suggest that a different socio-economic situation was present in this area when compared with eastern Chorasmia.

Importantly, the construction technology of the kalas and their architectural style are essentially the same as in eastern/central Chorasmia (with variations in quality), with the exception of the Butentau-kalas 1 and 2, Yasy-gyr 2 kala, Kanga-kala 2, and the annex enclosures of Mangyr-kala. The different building style and formal composition of these kalas—earthworks raised to form a vertical barrier with a ditch in front, cutting off the edge of the plateau as a protected zone—is probably the main reason for Vaynberg's interpretation of them as the products of pastoral groups. However, the Butentau-kalas have only Chorasmian pottery present, no steppic or Kuiusai ceramics (Tolstov 1958:82; Bizhanov and Khozhaniyazov 2003:34–35, 38–39). Bizhanov and Khozhaniyazov (ibid.:33, 38, 46) include them as installations of the Chorasmian state yet admit that their construction is not Chorasmian in style. The annex enclosures of Mangyr-kala are also earthworks rather than mud-brick, or *pakhsa,* construction, and they contain kurgans (Vaynberg 1979b:168). Tolstov mentioned that Kanga-kala 1, constructed around the fifth century BC, appears to have been built by "barbarians" who reworked the typical Khorezmian form for their own use (Tolstov 1962:229), which raises the possibility that even the later kalas may have been built by the local pastoral communities in emulation of eastern/central Chorasmian examples rather than by military cohorts of a centralized Chorasmian state.

The attempt to emulate a particular building style is a conscious decision that assumes a culturally accepted architectural language considered "appropriate" (see Rogers, Ulambayar, and Gallon 2005:812). As highly visible and highly symbolic built forms, like other monumental architecture, fortified enclosures embody important social and political meanings and values and have great symbolic significance (Lawrence and Low 1990; Trigger 1990; Moore 1996; Smith 1999; Creighton 2002; Neustupný 2006; Creighton and Liddiard 2008). The formulaic nature of fortress architecture is replicated to communicate to a wide public audience so their meaning is very clear. Importantly, that meaning is not simply one of state authority but can also be about community, status, territory, land tenure, and ritual (Rowlands 1972; Moore 1996; Smith 1999; Johnson 2002; Philip 2003:112–14; Smith 2003; Jakubiak 2006:141; Creighton and Liddiard 2008).

The fact that there is a great variety of plan forms for the kalas is another objection to viewing them as the product of a centralized military state, as pointed out by Betts (2006:149). For example, Mangyr-kala, Butentau-kala, Kanga-kala 2, and Bol'shoi Aibugir-kala had topographically determined plans; rectangular plan forms can be seen at Kanga-kala 1, Kalaly-gyr 1, and Kuiusai-kala; and there is the unusual attempt at both a triangular and topographically determined plan for Kalaly-gyr 2. Despite the varied site morphologies, the architectural language is very martial in style, with arrow-shaped loopholes, crenellated battlements, projecting towers, and bastioned gateways (Khozhaniyazov 2006). These were prominent and imposing pieces of monumental architecture and highly visible markers in the landscape that involved the coordinated effort of many people for their construction. They were also strategically sited for defensive purposes, as well as for surveillance. Defense cannot be discounted as the primary motivation for their construction, but they need not have been state installations. Instead, they may have served as centers for more localized communities with multiple functions, such as refuges, corrals, elite residences, secure trading locations, and territorial markers (Negus Cleary 2007, 2008; Kidd, Negus Cleary, and Baker Brite 2012).

One result of this survey of the Sarykamysh delta kalas and their related features has been to highlight the changing use of these sites over time, in particular as foci for mortuary activities. Some fortified sites have kurgan burials immediately outside their walls[7] and ossuary interments within their walls, as at Kanga-kala 1 (discussed earlier), also Mangyr-kala (Durdyev, Babakov, and Iusupov 1978; Vaynberg 1979b:168) and Kuiusai-kala (Vaynberg 1981:124). There are kurgan cemeteries located on high ground in the vicinity of Kalaly-gyr 1 and 2 (i.e., the Yasy-gyr kurgans) and Bol'shoi Aibugir-kala (Yagodin 2007:46), while these fortresses also have ossuary necropoleis within them but dating to later periods. Most interesting, the ossuary necropoleis only appear in or around fortified sites well after they have been abandoned (i.e., there are no habitation layers that are of the same time period as the ceramic ossuaries), as mentioned previously for Kalaly-gyrs 1 and 2, Kanga-kala 1, and Butentau-kala 2 (Bizhanov and Khozhaniyazov 2003:39). Butentau-kala 1 has several kurgan burial mounds within the fortified enclosure, although these may date to the medieval period (Tolstov 1958:81). Within Kanga-kala 1 there was also the unusual "naos" building, with excarnated bones arranged in groups around a central hearth (discussed earlier), and an almost identical example excavated at Kunya-uaz (Nerazik 1958:380–82) that has been interpreted as the burials of invading Khionites (White Huns/Hephtalites), who invaded the northeastern borders of Sassanid Iran around this time, fourth century AD (Tolstov 1962:231).

This evidence seems to suggest that many fortified sites in this area underwent a change of use, perhaps from initial functions as cattle corrals, elite residences, administrative centers, trading posts, and refuges to itinerant pastoral camps and spaces for necessary but unpleasant activities such as pottery and metallurgical production, excarnation of the dead, and interment of their remains. It is interesting that these important but noxious activities took place in abandoned, monumental, and highly visible sites. The kalas may have also been attractive for burials (both kurgans and ossuaries) because of their location on the highest and most prominent landscape features available, and there seems to have been a marked preference for cemeteries on high ground in Chorasmia, which remains the case today. Their status as highly significant sites, the monuments of a past culture, may have also been an important factor in their use as cemeteries.

The reuse of abandoned sites for important production or mortuary activities, and possibly periodically as refuges and cattle corrals, also illustrates the decentralized and dispersed nature of the occupation of the ancient landscape, where different activities took place in different locales spread out around the delta. Settlements similarly were dispersed and low in density, sited close to water and a fortified site. The dispersal of the core activities and infrastructure suggests that the ancient population may have been very mobile.

There is also evidence for periodic reuse of the kurgan cemeteries over the ancient period. This was noted for Berniyaz (Yagodin 2007:46), Charyshly groups 1 and 2 (Iusupov 1986:44–49), Kaskazhol (Yagodin 2007:45), Tumek-kichidzhik (Vaynberg 1979b:171), and Yasy-gyr 4 (ibid.:169). Several of these cemeteries have evidence of changes and variability in kurgan burial practices, which is again suggestive of diversity in local population (ibid.).

CONCLUSION

It is apparent from the archaeological evidence amassed by Vaynberg, Tolstov, and others from the Sarykamysh delta of western Chorasmia that the fortified sites here were used, and many probably built, by pastoral groups. There are clear linkages between the fortified sites and the ancient pastoral settlement pattern, at least during specific periods. The kalas most certainly performed a defensive function, both practically and symbolically, and were clear statements of elite power, presence, and affluence within the oasis territory (Betts 2006:153–55; Betts et al. 2009:43–44, 53; Kidd, Negus Cleary, and Baker Brite 2012). Yet there was a variety of functions, sizes, shapes, and locations for these structures, which changed over time. This situation appears to have been different from that of eastern (right bank) Chorasmia, where there is little evidence for nonagricultural communities. The exact sociopolitical role of the western kalas within the Chorasmian polity requires further investigation and consideration.

During the earlier period of the sixth–second centuries BC, some kalas were built and used as humble refuges for people or cattle and others as grander enclosures, with elite or cultic building complexes and imposing military-style architecture. Most kalas of this earlier period do not appear to have been settlements themselves but probably contained residences, perhaps for elites, and were certainly linked into a non-nucleated settlement pattern. Importantly, many kalas appear to have been built and used by both pastoral and farming communities rather than having been the product of sedentary communities alone. While some may have been military fortresses of a state polity, the evidence is conflicting and perhaps suggests that although some kalas were founded as fortresses, they were subsequently reused in different ways.

During the later period, second century BC–fourth century AD, many kalas in the Sarykamysh area underwent a change of use. There were fluctuations in the local population, with evidence of different mobile groups in the area, but the presence of Chorasmian material culture continued during this time. The kalas were no longer occupied as residences, settlements, cultic centers, or military installations but saw use as production and mortuary sites and possibly also as cattle corrals and refuges. This clearly shows the continued importance of such sites as markers of land tenure and as significant places within the landscape.

The fluctuating yet enduring use of places in this landscape, in particular the kalas, indicates the importance of multiple resource subsistence strategies and pastoral adaptation. While the archaeological evidence seems to suggest that pastoralism dominated in western Chorasmia, some communities employed irrigation agriculture, probably in combination with pastoralism. These mobile, or pastoral, communities employed diverse modes of production as required in this arid area reliant on distributaries of the Amu-Darya.

The underlying message here is one of multiplicity. Vaynberg (1979b:175–77, 1981:123–24) has pointed out the multicultural nature of the population of western Chorasmia. The Chorasmian (and Kuiusai) kalas were of a single archetypal form but were used and reused in many different ways and perhaps by different groups over time. The fact that this multifunctional built form was used by pastoral groups is perhaps perfectly appropriate, given the well-acknowledged resourcefulness and great adaptability of these peoples.

ACKNOWLEDGMENTS

I wish to thank Dr. William Anderson and Damjan Krsmanovic for their pertinent and insightful comments on this manuscript; Dr. Fiona Kidd for invaluable discussions about all things Chorasmian; Dr. Gairatdin Khozhaniyazov for his generosity and unique knowledge of the kalas; Shamil Amirov for his help with research; and Professor Vadim N. Yagodin and Associate Professor Alison Betts for their support and encouragement. This chapter owes everything to the legacy of Vaynberg's varied, prolific, and insightful body of work. Any errors in this chapter are my own.

NOTES

1. The binary dualism is inherent in Western philosophy, where one term is privileged and considered more important than the other, for example, light/dark, heaven/earth, mind/body, and the "other" term is defined only by what it is not (cf. Irigaray 1985; Grosz 1994:4–6), for example, nomads defined as all things not associated with sedentary civilization, such as uncivilized, unstable, mobile, barbarian. Dualisms such as nomad/sedentary, agricultural/pastoral, and desert/sown "perpetrate gross distortions of our ability to understand the relationship between the two" (Cribb 1991:16). Instead, each of these dimensions should be viewed as a continuum, with variations and gradations between one and the other (ibid.).

2. The origins of the Chorasmian culture are somewhat contentious, with different diffusionist and migrationist theories expounded (cf. Henning 1951:42–43; Gnoli 1980:91–127; Vogelsang 1992:16; Helms 1998:86–87). However, several notable Chorasmian scholars suggest that the relatively sudden appearance of monumental architecture, more developed irrigation canals, and some southern ceramic forms around the seventh–sixth centuries BC was the combined result of contact with the Achaemenid Empire influencing the local Andronovo/Amirabad and Kuiusai cultures (Vaynberg 1979b:172; Rapoport 1994:161; Yablonsky 1995:251; Khozhaniyazov 2006:36).

3. Koryakova has listed the commonly accepted archaeological markers of nomadic pastoralism as a kurgan burial ground, relative absence of permanent settlements or the presence of campsites, absence of or very limited farming, wheeled transport, and the presence of the bones of animals such as sheep, goats, cattle, camels, or horses. She cautioned that these so-called markers have rarely been successful in identifying nomadic groups archaeologically and suggested the importance of rethinking our theoretical approach to the study of nomadic material culture (Koryakova 2000:14).

4. There are other excavated examples of this blending of Mazdean mortuary ritual with the steppic kurgan burial mounds in Chorasmia at Tarym-kala II and III dating to the fifth–fourth centuries BC (Iusupov 1979:100). This has been interpreted as clear evidence of the blending of Kuiusai and Chorasmian cultural traditions in western Chorasmia (Itina 1979:6; Iusupov 1979:100).

5. According to Grenet (Boyce and Grenet 1991:193–94), these are the earliest ossuary interments known, and although the lack of ritual and iconographic details does not allow

certainty about these people's religious beliefs, the Tarym-kaya ossuaries do provide some limited support for the existence of nominally Zoroastrian communities in the fifth–fourth centuries BC in the north of Central Asia.

6. Also, the name of the fortress known as Kurgan-kala (in the vicinity of Mangyr-kala) is highly suggestive of another fortified enclosure site with associated kurgans. Kurgan-kala is briefly mentioned in Khozhaniyazov (2006:30, 31, 75, 77).

7. The Devkesken fortified wall has a large kurgan at the western end located in a small enclosure attached to the wall at the cliff edge (Khozhaniyazov 2006:62–63).

REFERENCES

Abdoullaev, Kamoludin A. 2001. "La Localisation de la Capitale des Yuëh-chih." In *La Bactriane au Carrefour des Routes et des Civilisations de l'Asie Centrale: Actes du Colloque de Termez 1997*, ed. Pierre Leriche and Vincent Fourniau, 197–214. Paris: Maisonneuve and Larose, IFEAC.

Allard, Francis, and Diimaajav Erdenebaatar. 2005. "Khirigsuurs, Ritual and Mobility in the Bronze Age of Mongolia." *Antiquity* 79: 547–63.

Andrianov, Boris V. 1969. *Drevnie Orositel'nye Sistemy Priaral'ya* [Ancient Irrigation System of the Aral Area]. Moscow: Nauka.

Bader, Andrey, and Khemra Usupov. 1995. "Gold Earrings from North-West Turkmenistan." In *In the Land of the Gryphons: Papers on Central Asian Archaeology in Antiquity*, ed. Antonio Invernizzi, 23–38. Firenze: Casa Editrice.

Baker Brite, Elizabeth. 2011. "The Archaeology of the Aral Sea Crisis: Environmental Change and Human Adaptation in the Khorezm Region of Uzbekistan ca. AD 300–800." PhD diss., Department of Anthropology, University of California, Los Angeles.

Bernbeck, Reinhard. 2008. "An Archaeology of Multisited Communities." In *The Archaeology of Mobility: Old World and New World Nomadism*, ed. Hans Barnard and Willeke Wendrich, 43–77. Los Angeles: Cotsen Institute of Archaeology, University of California.

Betts, Allison V. G. 2006. "Chorasmians, Central Asian Nomads and Chorasmian Military Architecture." In *The Military Architecture of Ancient Chorasmia (6th century B.C.–4th century A.D.)*, ed. Gairatdin Khozhaniyazov, 131–55. Persika 7. Paris: De Boccard.

Betts, Allison V. G., Vadim N. Yagodin, Svend W. Helms, Gairatdin Khozhaniyazov, S. Amirov, and Michelle Negus Cleary. 2009. "The Karakalpak-Australian Excavations in Ancient Chorasmia, 2001–2005: Interim Report on the Fortifications of Kazakl'i-yatkan and Regional Survey." *Iran* 47: 33–55.

Biscione, Raffaele. 1981. "Centre and Periphery in Later Protohistoric Turan: The Settlement Pattern." In *Proceedings of the South Asian Archaeology, 1979: Papers from the Fifth International Conference of the Association of South Asian Archaeologists in Western Europe*, ed. Herbert Härtel, 203–13. Berlin: Museum für Indische Kunst der Staalichen Museen Preussischer Kulturbesitz.

Bizhanov, E. B., and Gairatdin Khozhaniyazov. 2003. "Arkheologicheskii Compleks Butentau" [The Archaeological Complex of Butentau]. *Arkheologia Priaral'ya* 6: 32–59.

Bolelov, S. B. 1998. "Krepost' Ayaz-kala 3 v Pravoberezhnom Khorezme" [The Ayaz-kala 3 Fortress in Right-Bank Khorezm]. In *Priaral'ie v Drevnosti i Srednevekov'e: K 60-letiu Khorezmskoi Arkheologo-Ètnograficheskoi Èkspeditsii*, ed. E. E. Nerazik and L. S. Efimova, 116–35. Moscow: Vostochnaya literatura RAN.

Boomer, I., B. Wünnemann, A. W. Mackay, P. Austin, P. Sorrel, C. Reinhardt, D. Keyser, F. Guichard, and M. Fontugne. 2009. "Advances in Understanding the Late Holocene History of the Aral Sea Region." *Quaternary International* 194 (1-2): 79–90. http://dx.doi.org/10.1016/j.quaint.2008.03.007.

Boyce, Mary, and Frantz Grenet. 1991. *A History of Zoroastrianism*, vol. 3. Leiden: Brill.

Bregel, Yuri. 2003. *An Historical Atlas of Central Asia*. Leiden: Brill.

Brentjes, Burchard. 1996. *Arms of the Sakas and Other Tribes of the Central Asian Steppes*. Varanasi, India: Rishi.

Brück, Joanna. 1999. "What's in a Settlement? Domestic Practice and Residential Mobility in Early Bronze Age Southern England." In *Making Places in the Prehistoric World: Themes in Settlement Archaeology*, ed. Joanna Brück and Melissa Goodman, 52–75. London: University College London Press.

Chang, Claudia. 2008. "Mobility and Sedentism of the Iron Age: Agropastoralists of Southeast Kazkhstan." In *The Archaeology of Mobility: Old World and New World Nomadism*, ed. Hans Barnard and Willeke Wendrich, 329–42. Los Angeles: Cotsen Institute of Archaeology, University of California.

Chapman, Robert. 1995. "Ten Years After—Megaliths, Mortuary Practices, and the Territorial Model." In *Regional Approaches to Mortuary Analysis*, ed. Lane A. Beck, 29–51. New York: Plenum. http://dx.doi.org/10.1007/978-1-4899-1310-4_2.

Creighton, Oliver H. 2002. *Castles and Landscapes*. London: Continuum.

Creighton, Oliver H., and Robert Liddiard. 2008. "Fighting Yesterday's Battle: Beyond War or Status in Castle Studies." *Medieval Archaeology* 52 (1): 161–69. http://dx.doi.org/10.1179/174581708x335477.

Cribb, Roger. 1991. *Nomads in Archaeology*. Cambridge: Cambridge University Press. http://dx.doi.org/10.1017/CBO9780511552205.

Durdyev, D., O. Babakov, and K. Iusupov. 1978. "Raboty v Severnoi Turkmenii" [Works in Northern Turkmenistan]. *Arkheologicheskie Otkrytiya* 1977 goda: 546.

Dyson-Hudson, Rada, and Neville Dyson-Hudson. 1980. "Nomadic Pastoralism." *Annual Review of Anthropology* 9 (1): 15–61. http://dx.doi.org/10.1146/annurev.an.09.100180.000311.

Fletcher, Roland. 1986. "Settlement Archaeology: World-Wide Comparisons." *World Archaeology* 18 (1): 59–83. http://dx.doi.org/10.1080/00438243.1986.9979989.

Frachetti, Michael D. 2008. "Variability and Dynamic Landscapes of Mobile Pastoralism in Ethnography and Prehistory." In *The Archaeology of Mobility: Nomads in the Old and in the New World*, ed. Hans Barnard and Willeke Wendrich, 366–96. Los Angeles: Cotsen Institute of Archaeology, University of California.

Frachetti, Michael D., Norbert Benecke, Alexei N. Mar'yashev, and Paula N. Doumani. 2010. "Eurasian Pastoralists and Their Shifting Regional Interactions at the Steppe Margin: Settlement History at Mukri, Kazakhstan." *World Archaeology* 42 (4): 622–46. http://dx.doi.org/10.1080/00438240903371270.

Frumkin, Grégoire. 1962. "Archaeology in Soviet Central Asia and Its Ideological Background." *Central Asian Review* 10 (4): 334–42.

Gardin, Jean-Claude. 1995. "Fortified Sites of Eastern Bactria (Afghanistan) in Pre-Hellenistic Times." In *In the Land of the Gryphons: Papers on Central Asian Archaeology in Antiquity*, ed. Antonio Invernizzi, 83–103. Firenze: Casa Editrice.

Gertman, A. N. 1991. "Syrtsovyi Kirpich Kaparas i Elkharas [Mud Bricks from Kaparas and Elkharas]." In *Drevnosti Iuzhnogo Khorezma*, ed. Marianna A. Itina, 277–86. Trudy Khorezmskoi Arkheologo-Etnograficheskoi Ekspeditsii 16. Moscow: Nauka.

Gnoli, Gherardo. 1980. *Zoroaster's Time and Homeland: A Study on the Origins of Mazdeism and Related Problems*. Naples: Istituto Universitario Orientale.

Grosz, Elizabeth. 1994. *Volatile Bodies: Towards a Corporeal Feminism*. St. Leonards, New South Wales, Australia: Allen and Unwin.

Helms, Svend W. 1998. "Ancient Chorasmia: The Northern Edge of Central Asia from the 6th Century BC to the Mid-4th Century AD." In *Worlds of the Silk Roads: Ancient and Modern: Proceedings from the Second Conference of the Australasian Society for Inner Asian Studies (A.S.I.A.S), Macquarie University, Sydney, Australia, September 21–22, 1996*, ed. David Christian and Craig Benjamin, 77–96. Silk Road Studies 4. Brepols, Turnhout: Ancient History Documentary Research Centre, Macquarie University.

Helms, Svend W. 2006. "Preface." In *The Military Architecture of Ancient Chorasmia (6th Century BC–4th Century AD)*, ed. Gairatdin Khozhaniyazov, trans. Svend W. Helms, 8–16. Persika 7. Paris: De Boccard.

Helms, Svend W., and Vadim N. Yagodin. 1997. "Excavations at Kazakl'i-yatkan in the Tashk'irman Oasis of Ancient Chorasmia: A Preliminary Report." *Iran* 35: 43–65.

Helms, Svend W., Vadim N. Yagodin, Allison V. G. Betts, Gairatdin Khozhaniyazov, and Fiona Kidd. 2001. "Five Seasons of Excavations in the Tash-K'irman Oasis of Ancient Chorasmia, 1996–2000: An Interim Report." *Iran* 39: 119–44.

Helms, Svend W., Vadim N. Yagodin, Allison V. G. Betts, Gairatdin Khozhaniyazov, and Michelle Negus. 2002. "The Karakalpak-Australian Excavations in Ancient Chorasmia:

The Northern Frontier of the 'Civilised' Ancient World." *Ancient Near Eastern Studies* 39: 3–44. http://dx.doi.org/10.2143/ANES.39.0.501773.

Henning, Walter B. 1951. *Zoroaster—Politician or Witch-Doctor? Ratanbai Katrak Lectures 1949*. Oxford: Oxford University Press.

Hiebert, Fredrik T. 1992. "The Oasis and City of Merv (Turkmenistan)." *Archeologie Islamique* 3: 111–27.

Hiebert, Fredrik T. 1994. *Origins of the Bronze Age Oasis Civilization in Central Asia*. Cambridge, MA: Peabody Museum of Archaeology and Ethnology, Harvard University.

Honeychurch, William, and Chunag Amartuvshin. 2007. "Hinterlands, Urban Centres, and Mobile Settings: The 'New' Old World Archaeology from the Eurasian Steppe." *Asian Perspective* 46 (1): 36–64. http://dx.doi.org/10.1353/asi.2007.0005.

Horne, Lee. 1991. "Reading Village Plans: Architecture and Social Change in Northeastern Iran." *Expedition* 33 (1): 44–52.

Horne, Lee. 1994. *Village Spaces: Settlement and Society in Northeastern Iran*. Smithsonian Series in Archaeological Inquiry. Washington, DC: Smithsonian Institution Press.

Howell-Meurs, Sarah. 2001. *Early Bronze and Iron Age Animal Exploitation in Northeastern Anatolia: The Faunal Remains from Sos Höyük and Büyüktepe Höyük*. Bar International Series 945. Oxford: Archaeopress.

Irigaray, Luce. 1985. *Speculum of the Other Woman*. Ithaca, NY: Cornell University Press.

Itina, Marianna A. 1979. "Ot Redaktora" [From the Editor]. In *Trudy Khorezmskoi Arkheologo-Etnograficheskoi Ekspeditsii, XI: Kochevniki na Granitsakh Khorezma*, ed. Marianna A. Itina, 5–6. Moscow: Nauka.

Iusupov, K. 1979. "Kurgany Mogil'nikov Tarym-kaya II i III" [Burial Mounds of Tarym-kaya II and III]. In *Trudy Khorezmskoi Arkheologo-Etnograficheskoi Ekspeditsii, XI: Kochevniki na Granitsakh Khorezma*, ed. Marianna A. Itina, 94–100. Moscow: Nauka.

Iusupov, K. 1986. *Drevnosti Uzboya* [The Ancient Uzboi]. Ashkhabad: Akademiia Nauk Turkmenskoi SSR.

Jakubiak, Krzysztof. 2006. "The Origin and Development of Military Architecture in the Province of Parthava in the Arsacid Period." *Iranica Antiqua* 41: 127–50. http://dx.doi.org/10.2143/IA.41.0.2004764.

Johnson, Matthew. 2002. *Behind the Castle Gate: From Medieval to Renaissance*. London: Routledge.

Khazanov, Anatoly M. 1984. *Nomads and the Outside World*. Trans. J. Crookenden. New York: Cambridge University Press.

Khozhaniyazov, Gairatdin. 1982. "Akchungul'—Nov'ii Pamyatnik Epokhi Antichnosti Severo-Zapadnogo Khorezma" [Akchungul'—New Antique Monument of Northwestern Khorezm]. *Arkheologia Priaral'ya* 1: 81–86.

Khozhaniyazov, Gairatdin. 2002. "Les Fortifications Karakalpakes aux XVIIe–XIXe Siecles." *Cahiers d'Asie Centrale* 10 : 139–66.

Khozhaniyazov, Gairatdin. 2006. *The Military Architecture of Ancient Chorasmia (6th Century B.C.–4th Century A.D.)*. Trans. S. W. Helms. Persika 7. Paris: De Boccard.

Khozhaniyazov, Gairatdin, and Z. Khakiminiyazov. 1997. *Gorodishche Devkesken—Vazir*. Uzbekistan: Bilim.

Kidd, Fiona J. 2011. "Complex Connections: Figurative Evidence from Akchakhan-kala and the Problematic Question of Relations between Khorezm and Parthia." *Topoi Orient-Occident* 17 (1): 229–76.

Kidd, Fiona, Michelle Negus Cleary, and Elizabeth Baker Brite. 2012. "Public vs. Private: Perspectives on the Communication of Power in Ancient Chorasmia." In *The Archaeology of Power and Politics in Central Asia: Regimes and Revolutions*, ed. Charles W. Hartley, G. Bike Yazıcıoğlu, and Adam T. Smith, 91–121. Cambridge: Cambridge University Press.

Koryakova, Ludmila. 2000. "Some Notes about the Material Culture of Eurasian Nomads." In *Kurgans, Ritual Sites, and Settlements: Eurasian Bronze and Iron Age*, ed. Jeannine Davis-Kimball, Eileen M. Murphy, Ludmila Koryakova, and Leonid T. Yablonksy, 13–18. BAR Centre for the Study of Eurasian Nomads [cited July 27, 2004]. http://www.csen.org/BAR%20Book/BAR.%20Part%2001.TofC.html, accessed January 22, 2011.

Lamberg-Karlovsky, C. C. 1994. "The Bronze Age *Khanates* of Central Asia." *Antiquity* 68 (259): 398–405.

Lavrov, V. A. 1950. *Gradostroitel'naya Kul'tura Srednei Azii* [The Urban Planning Culture of Central Asia]. Moscow: State Publishing House, Architecture and Urban Planning.

Lawrence, Denise L., and Setha M. Low. 1990. "The Built Environment and Spatial Form." *Annual Review of Anthropology* 19 (1): 453–505. http://dx.doi.org/10.1146/annurev.an.19.100190.002321.

Lecomte, Olivier. 1999. "Vehrkana and Dehistan: Late Farming Communities of South-West Turkmenistan from the Iron Age to the Islamic Period." *Parthica: Incontri di Culture nel Mondo Antico* 1: 135–70.

Leriche, Pierre. 1977. "Problèmes de la Guerre en Iran et en Asie Centrale dans l'Empire Perse et a l'Époque Hellénistique." In *Le Plateau Iranien et l'Asie Centrale des Origines à la Conquête Islamique: Leurs Relations à la Lumière des Documents Archéologiques: [Actes du Colloque], Paris, 22–24 Mars 1976*, ed. J. Deshayes, 297–312. Paris: Éditions du Centre National de la Recherche Scientifique.

Létolle, René. 2000. "Histoire de l'Ouzboi, cours Fossile de l'Amou Darya: Synthèse et Éléments Nouveaux." *Studia Iranica* 29: 195–240.

Létolle, René, Philip Micklin, Nikolay Aladin, and Igor Plotnikov. 2007. "Uzboy and the Aral Regressions: A Hydrological Approach." *Quaternary International* 173–74: 125–36. http://dx.doi.org/10.1016/j.quaint.2007.03.003.

Levina, L. M. 1979. "Poseleniya VII–V vv. do n.e. i 'Shlakovye' Kyrgany Iuzhnykh Raionov Syrdar'inskoi del'ty" [Settlements of the Seventh–Fifth Centuries B.C. and "Slag" Kurgans of the Southern Region of the Syr-Dayra Delta]. In *Trudy Khorezmskoi Arkheologo-Etnograficheskoi Ekspeditsii, XI: Kochevniki na Granitsakh Khorezma*, ed. Marianna A. Itina, 178–90. Moscow: Nauka.

Livshits, Vladimir A. 2003. "Three Silver Bowls from the Isakovka Burial-Ground no.1 with Khwarezmian and Parthian Inscriptions." *Ancient Civilizations from Scythia to Siberia* 9 (1-2): 147–72. http://dx.doi.org/10.1163/157005703322114874.

Mambetullaev, Myrzamurat. 1978. "Gorodishche bol'shaya Aibugiir-kala (Raskopi 1976 Goda)" [The Ancient Settlement of Large Aibugiir-kala (Excavations of 1976)]. *Vestnik Karakalpakskogo Filiala Akademika Nauk UzSSR* 4: 80–88.

Mambetullaev, Myrzamurat. 1990. "Gorodishche bol'shaya Aibugiir-kala (Raskopki 1976–1977 i 1981 gg.)" [The Ancient Settlement of Greater Aibugiir-kala (Excavations from 1976–77 and 1981)]. *Arkheologia Priaral'ya* 4: 91–131.

Mambetullaev, Myrzamurat, and Gairatdin Khozhaniyazov. 1978. "Raskopki Gorodishcha Bol'shaya Aibugir-kala" [Excavations of the Ancient Settlement of Large Aibugir-kala]. *Arkheologicheskie otkr'itiya* 1977 goda: 529.

Manylov, I. P., and Gairatdin Khozhaniyazov. 1981. "Gorodishcha Ayazkala 1 i Burlykala (k Izucheniiu Fortifikatsii Drevnego Khorezma)" [Ancient Settlements of Ayazkala 1 and Burlykala (A Study of the Fortifications of Ancient Khorezm)]. In *Arkheologicheskie Issledovaniya v Karakalpakii*, ed. I. Kos'imbetov and V. N. Yagodin, 32–47. Tashkent, Uzbekistan: Fan.

Mercer, David. 2006. "The Trouble with Paradigms: A Historigraphical Study on the Development of Ideas in the Discipline of Castle Studies." *Archaeological Dialogues* 13 (1): 93–109. http://dx.doi.org/10.1017/S1380203806001838.

Moore, Jerry D. 1996. *Architecture and Power in the Ancient Andes: The Archaeology of Public Buildings*. Cambridge: Cambridge University Press. http://dx.doi.org/10.1017/CBO9780511521201.

Negmatov, Numan N. 1994. "States in North-Western Central Asia." In *History of Civilizations of Central Asia*, vol. 2: *The Development of Sedentary and Nomadic Civilizations: 700 B.C. to A.D. 250*, ed. Janos Harmatta, B. N. Puri, and G. F. Etemadi, 441–56. History of Civilizations of Central Asia, 4 vols. Paris: UNESCO.

Negus Cleary, Michelle. 2007. "The Ancient Oasis Landscape of Chorasmia: The Role of the Kala in Central Asian Settlement Patterns." In *Social Orders and Social Landscapes: Proceedings of the 2005 University of Chicago Conference on Eurasian Archaeology*, ed.

Laura Popova, Charles Hartley, and Adam T. Smith, 334–58. Newcastle: Cambridge Scholars Press.

Negus Cleary, Michelle. 2008. "Walls in the Desert: The Phenomenon of Central Asian Urbanism in Ancient Chorasmia." In *Art, Architecture and Religion on the Silk Roads*, ed. Ken Parry, 51–78. Brepols, Turnhout: Ancient History Documentary Research Centre, Macquarie University.

Nerazik, E. E. 1958. "Arkheologicheskoe Obsledovanie Gorodishche Kunya-uaz v 1952 g" [Archaeological Investigation of the Ancient Settlement of Kunya-uaz in 1952]. In *Trudy Khorezmskoi Arkheologo-Etnograficheskoi Ekspeditsii*, vol. 2, ed. Sergey P. Tolstov and T. A. Zhdanko, 367–96. Moscow: Nauka.

Nerazik, E. E. 1966. *Sel'skie Poseleniya Afrigidskogo Khorezma: Po Materialam Berkut-Kalinskogo Oazisa* [Rural Settlements of Afrighid Khorezm: Based on the Berkut-Kala Oasis]. Moscow: Nauka.

Nerazik, E. E. 1976. *Sel'skoe Zhilishche v Khorezme (I–XIV v.v.): Iz Istorii Zhilishcha i Sem'i: Arkheologo-Etnograficheskie Ocherki* [Rural Dwellings in Khorezm (1st–14th Centuries): The History of the Home and Family: Archaeological-Ethnographic Essays]. Trudy Khorezmskoi Arkheologo-Etnograficheskoi Ekspeditsii 9. Moscow: Nauka.

Neustupný, Evžen. 2006. "Enclosures and Fortifications in Central Europe." In *Enclosing the Past: Inside and Outside in Prehistory*, ed. Anthony Harding, Susanne Sievers, and Natalie Venclová, 1–4. Sheffield Archaeological Monographs 15. Sheffield: J. R. Collis.

Philip, Graham. 2001. "The Early Bronze I–III Ages." In *The Archaeology of Jordan*, ed. Burton MacDonald, Russell B. Adams, and Piotr Bienkowski, 163–232. Sheffield: Sheffield Academic Press.

Philip, Graham. 2003. "Early Bronze Age of the Southern Levant." *Journal of Mediterranean Archaeology* 16 (1): 103–32. http://dx.doi.org/10.1558/jmea.v16i1.103.

Platt, Colin. 2007. "Revisionism in Castle Studies: A Caution." *Medieval Archaeology* 51 (1): 83–102. http://dx.doi.org/10.1179/174581707x224679.

Rapin, Claude. 2007. "Nomads and the Shaping of Central Asia: From the Early Iron Age to the Kushan Period." In *After Alexander: Central Asia before Islam*, ed. Joe Cribb and Georgina Herrmann, 29–72. Oxford: British Academy, Oxford University Press. http://dx.doi.org/10.5871/bacad/9780197263846.003.0003.

Rapoport, I. A. 1958. "Raskopki Gorodishcha Shakh-Senem v 1952 Godu" [Excavations at the Ancient Settlement of Shakh-Senem in 1952]. In *Trudy Khorezmskoi Arkheologo-Etnograficheskoi Ekspeditsii 2: Arkheologicheskie i Etnograficheskie Rabot'I Khorezmskoi Ekspeditsii 1949–1953*, vol. 2, ed. Sergey P. Tolstov and T. A. Zhdanko, 397–420. Moscow: Nauka.

Rapoport, I. A. 1994. "The Palaces at Topraq-Qal'a." In *The Archaeology and Art of Central Asia: Studies from the Former Soviet Union*, ed. Boris A. Litvinskii and Carol

A. Bromberg, 161–85. New Series, vol. 8. Bloomfield Hills, MI: Bulletin of the Asia Institute.

Rapoport, I. A., and M. S. Lapirov-Skoblo. 1963. "Raskopki Dvortsovogo Zdaniya na Gorodishche Kalaly-gyr 1 v 1958 g" [Excavations of the Palace Building at the Ancient Settlement of Kalaly-gyr 1 in 1958]. In *Materialy Khorezmskoi Ekspeditsii 6: Polevye Issledovaniya Khorezmskoi Ekspeditsii v 1958–1961 gg*, ed. Sergey P. Tolstov, 141–56. Moscow: Nauka.

Rapoport, I. A., E. E. Nerazik, and L. M. Levina. 2000. *V nizov'yach Oksa i Yaksarta: Obrazy Drevnego Priaral'ya* [In the Lower Oxus and Yaksartes: Views of the Ancient Aral Area]. Moscow: Indrik.

Rapoport, I. A., and S. A. Trudnovskaya. 1958. "Gorodishche Gyaur-kala." In *Trudy Khorezmskoi Arkheologo-Etnograficheskoi Ekspeditsii 2: Arkheologicheskie i Etnograficheskie Rabot'I Khorezmskoi Ekspeditsii 1949–1953*, ed. Sergey P. Tolstov and T. A. Zhdanko, 347–66. Moscow: Nauka.

Rapoport, I. A., and S. A. Trudnovskaya. 1979. "Kurgany na ancient Vozvyshennosti Chash-tepe" [Kurgans on the Hill of Chash-Tepe]. In *Trudy Khorezmskoi Arkheologo-Etnograficheskoi Ekspeditsii XI: Kochevniki na Granitsakh Khorezma*, ed. M. A. Itina, 151–66. Moscow: Nauka.

Rogers, J. Daniel, Erdenebat Ulambayar, and Matthew Gallon. 2005. "Urban Centres and the Emergence of Empires in Eastern Inner Asia." *Antiquity* 79: 801–18.

Rowlands, Michael J. 1972. "Defence: A Factor in the Organization of Settlements." In *Man, Settlement and Urbanism*, ed. Peter J. Ucko, Ruth Tringham, and G. W. Dimbleby, 447–62. London: Duckworth.

Salzman, Philip C. 2002. "Pastoral Nomads: Some General Observations Based on Research in Iran." *Journal of Anthropological Research* 58 (2): 245–64.

Sarianidi, Viktor I. 2002. "The Fortification and Palace of Northern Gonur." *Iran* 40: 75–87.

Smith, Adam T. 1999. "The Making of an Urartian Landscape in Southern Transcaucasia: A Study of Political Architectonics." *American Journal of Archaeology* 103 (1): 45–71. http://dx.doi.org/10.2307/506577.

Smith, Monica L. 2003. "Early Walled Cities of the Indian Subcontinent as 'Small Worlds.'" In *The Social Construction of Ancient Cities*, ed. Monica L. Smith, 269–89. Washington, DC: Smithsonian Books.

Soviet topographic map K-40-093. 1985.

Stride, Sebastian, Bernardo Rondelli, and Simone Mantellini. 2009. "Canals versus Horses: Political Power in the Oasis of Samarkand." *World Archaeology* 41 (1): 73–87. http://dx.doi.org/10.1080/00438240802655302.

Szabo, Albert, and Thomas J. Barfield. 1991. *Afghanistan: An Atlas of Indigenous Architecture*. Austin: University of Texas Press.

Taaffe, Robert N. 1990. "The Geographic Setting." In *The Cambridge History of Early Inner Asia*, vol. 1, ed. Denis Sinor, 19–40. Cambridge: Cambridge University Press. http://dx.doi.org/10.1017/CHOL9780521243049.003.

Tarn, William W. 1951. *The Greeks in Bactria and India*, 2nd ed. Cambridge: Cambridge University Press.

Tolstov, Sergey P. 1948a. *Drevnii Khorezma: Op'it Istoriko-Arkheologicheskogo Issledovaniya* [Ancient Khorezm: Experience in Historical-Archaeological Research]. Moscow: MGU.

Tolstov, Sergey P. 1948b. *Po Sledam Drevne-Khorezmiiskoi Tsivilizatsii* [In the Footsteps of Ancient Khorezmian Civilization]. Moscow: Nauka.

Tolstov, Sergey P. 1953. *Auf den Spuren der Altchoresmischen Kultur*. Berlin: Verlag Kultur und Fortschritt.

Tolstov, Sergey P. 1958. "Rabot'i Khorezmskoi Arkheologo-Etnograficheskoi Ekspeditsii AN SSSR v 1949–1953 gg" [Works of the Khorezmian Archaeological-Ethnographic Expedition of the USSR in 1949–53]. In *Trudy Khorezmskoi Arkheologo-Etnograficheskoi Ekspeditsii II: Arkheologicheskie i Etnograficheskie Rabot'I Khorezmskoi Ekspeditsii 1949–1953*, ed. Sergey P. Tolstov and T. A. Zhdanko, 7–258. Moscow: Nauka.

Tolstov, Sergey P. 1960. "Results of the Work of the Khoresmian Archaeological and Ethnographic Expedition of the USSR Academy of Sciences 1951–1956." *Journal of the Asiatic Society of Bombay* 34–35 (Old Series): 1–24.

Tolstov, Sergey P. 1962. *Po Drevnim Deltam Oksa i Yaksarta* [In the Ancient Delta of the Oxus and Yaksartes]. Moscow: Nauka.

Tolstov, Sergey P. 1970. "Raboty Khorezmskoi Ekspeditsii" [Work of the Khorezmian Expedition]. *Arkheologicheskie Otkrytiya* 1969 goda: 399–400.

Trigger, Bruce. 1989. *A History of Archaeological Thought*. Cambridge: Cambridge University Press.

Trigger, Bruce. 1990. "Monumental Architecture: A Thermodynamic Explanation of Symbolic Behaviour." *World Archaeology* 22 (2): 119–32. http://dx.doi.org/10.1080/00438243.1990.9980135.

Trudnovskaya, S. A. 1979. "Ranne Pogrebeniya Iugo-zapadnoi Kurgannoi Gruppy Mogil'nika Tuz-gyr." In *Trudy Khorezmskoi Arkheologo-Etnograficheskoi Ekspeditsii XI: Kochevniki na Granitsakh Khorezma*, ed. Marianna A. Itina. 101–110. Moscow: Nauk.

Vainberg, Bėlla I. 1994. "The Kalali-Gir 2 Ritual Center in Ancient Khwarazm." In *The Archaeology and Art of Central Asia: Studies from the Former Soviet Union*, ed. Boris A. Litvinskii and Carol A. Bromberg, 67–80. New Series, vol. 8. Bloomfield Hills, MI: Bulletin of the Asia Institute. [NB: author's name transliterated Vainberg in this publication.]

Vaynberg, Bėlla I. 1977. *Monety Drevnego Khorezma*. Moscow: Nauka.

Vaynberg, Bėlla I. 1979a. "Pamyatniki Kuyusayskoy Kul'turi" [Monuments of the Kuiusai Culture]. In *Trudy Khorezmskoi Arkheologo-Etnograficheskoi Ekspeditsii XI: Kochevniki na Granitsakh Khorezma*, ed. Marianna A. Itina, 1–76. Moscow: Nauka.

Vaynberg, Bėlla I. 1979b. "Kurgannye Mogil'niki Severnoi Turkmenii" (Prisarykamyshckaya Del'ta Amudar'i)" ["Burial Mounds of Northern Turkmenistan"]. In *Trudy Khorezmskoi Arkheologo-Etnograficheskoi Ekspeditsii XI: Kochevniki na Granitsakh Khorezma*, ed. Marianna A. Itina, 167–77. Moscow: Nauka.

Vaynberg, Bėlla I. 1981. "Skotovodcheskie Plemena v Drevnem Xorezme" [Pastoral Tribes in Ancient Khorezm]. In *Kul'tura i Iskysstvo Drevnego Khoremz*, ed. Marianna A. Itina, I. A. Rapoport, N. S. Sycheva, and Bėlla I. Vaynberg, 121–30. Moscow: Nauka.

Vaynberg, Bėlla I. 2004. *Kalaly-gyr 2: Kultovyi Ysentr v Drevnem Khorezme IV–II vv. do n.e.* [Kalaly-gyr 2: Cultic Center of Ancient Khorezm 4th–2nd centuries BC]. Moscow: Vostochnaya Literatura.

Vogelsang, W. J. 1992. *The Rise and Organisation of the Achaemenid Empire: The Eastern Iranian Evidence*. Leiden: Brill.

Vorob'eva, M. G. 1959. "Keramika Khorezma Antichnogo Perioda" [Khorezmian Ceramics of the Ancient Period]. In *Trudy Khorezmskoi Arkheologo-Etnograficheskoi Ekspeditsii IV: Keramika Khorezma*, ed. Sergey P. Tolstov and M. G. Vorob'eva, 63–220. Moscow: Nauka.

Wittfogel, Karl A. 1957. *Oriental Despotism: A Comparative Study of Total Power*. New Haven: Yale University Press.

Wright, Joshua. 2007. "The Organizational Principles of Khirigsuur Monuments in the Lower Egiin Gol Valley, Mongolia." *Journal of Anthropological Archaeology* 26 (3): 350–65. http://dx.doi.org/10.1016/j.jaa.2007.04.001.

Wright, Joshua. 2008. "Non-Graphic Information Systems and Diachronic Transformations in Margiana." In *The Bronze Age and Early Iron Age in the Margiana Lowlands*, ed. I. S. Masimov, Maurizio Tosi, Sandro Salvatori, and Barbara Cerasetti, 47–56. The Archaeological Map of the Murghan Delta Studies and Reports, vol. 2. BAR International Series 1806. Oxford: Archaeopress.

Yablonsky, Leonid T. 1990. "Burial Place of a Massagetan Warrior." *Antiquity* 64: 288–96.

Yablonsky, Leonid T. 1995. "Some Ethnogenetical Hypotheses." In *Nomads of the Eurasian Steppes in the Early Iron Age*, ed. Jeannine Davis-Kimball, Vladimir A. Bashilov, and Leonid T. Yablonsky, 241–52. Berkeley, CA: Zinat.

Yagodin, Vadim N. 1982. "Arkheologicheskoe Izuchenie Kurgannykh Mogil'nikov Kaskazhol i Berniiaz na Ustiurte" [Archaeological Studies of Burial Mounds Kaskazhol and Berniyaz on the Ustyurt]. *Arkheologiya Priaral'ya* 1: 39–81.

Yagodin, Vadim N. 2007. "The Duana Archaeological Complex." In *Ancient Nomads of the Aralo-Caspian Region*, ed. Vadim N. Yagodin, Allison V. G. Bett, and S. Blau, 11–78. Dudley, MA: Peeters.

7

FulBe Pastoralists and the Neo-Patrimonial State in the Chad Basin

MARK MORITZ

The dominant image of the African state in the "pastoralist literature" is that of a state in opposition to nomadic society, sponsor of large-scale technocratic development schemes based on misconceptions of and incompatible with the mobility of pastoral systems that consequently push pastoralists further to the margins (Salih 1990; Klute 1996; Lenhart and Casimir 2001; and other papers in special issues of *Nomadic Peoples* on the topic of nomads and the state; see also Diallo 1999; Niamir-Fuller 1999a; Azarya 2001). The African state is portrayed as a modern bureaucratic state with agentive and hegemonic powers that is in irreconcilable conflict with pastoralists. This image of the African state in the pastoralist literature is quite different from the one that emerges in the political science literature (e.g., Bayart 1993; Chabal and Daloz 1999; van de Walle 2001; Young 2004).

I do not argue here that African states are merely weak and have little impact on nomadic pastoralists. Directly and indirectly, states have drastically altered the lives and livelihoods of nomadic pastoralists in the Chad Basin over the last centuries, for example, by reducing grazing lands and providing vaccinations and veterinary care. I am arguing that studies of African pastoral societies should pay more attention to how the state actually "works" on the ground (see also Chabal and Daloz 1999). This means that the informal politics of the state's elite and bureaucrats, which feature prominently in the political science literature on the African state, have to be considered more explicitly and systematically in the analyses of pastoralists' relationships with the state.

DOI: 10.5876/9781607323433.c007

Analyses in the pastoralist literature on the state have focused on laws and policies of an ideal bureaucratic state rather than on bureaucrats' actions in a neo-patrimonial state. This focus misrepresents the impact of the African state on pastoral societies because it privileges official laws and policies, which are seldom effectively or completely implemented, and because it fails to consider the "real business" of informal politics. African states do not conform to western models of a bureaucratic state, and studies of pastoralists and the state should consider that in their analyses of the state's impact on African pastoral societies. This also means rethinking the dichotomy between what Meir (1988) has labeled the centripetal forces of the state and the centrifugal forces of nomadic pastoralists, in which "states" seek the encapsulation of nomadic pastoralists, while the latter seek to maintain their autonomy (Fratkin 1997).[1] I argue that in the context of the neo-patrimonial state in the Chad Basin, nomadic pastoralists actually seek fuller integration in the state, while state agents prefer the opposite.

In this chapter I examine pastoralists' relationships with the neo-patrimonial state in the Far North Province of Cameroon and discuss how they affect the lives and livelihoods of nomadic FulBe pastoralists. I focus in particular on the role of the state in pastoral development, insecurity, and access to grazing lands. I situate the relationship between state and pastoralists within the historical and geographical context of the Chad Basin, using literature on FulBe pastoralists and the state in northern Cameroon (Mohammadou 1976; Abubakar 1977; Azarya 1978; Mohammadou 1988; Njeuma 1989; Seignobos and Iyebi-Mandjek 2000) and ethnographic data from my own fieldwork in 1994, 1996, 1999, and 2000–2001 with nomadic FulBe pastoralists of the Mare'en sub-ethnic group.

NEO-PATRIMONIAL STATES

African states are today generally referred to as neo-patrimonial states in the political science literature (e.g., Chabal and Daloz 1999; van de Walle 2001). In the neo-patrimonial state the state is an empty facade, as the real business of politics is done informally through clientelistic networks. Through informal politics, politicians, bureaucrats, and elites instrumentalize the apparent disorder to use the state's public resources for personal enrichment and support for their clients. The African state is labeled *neo*-patrimonial because patrimonial practices coexist with the modern bureaucracy of Weber's legal-rational state. In fact, patrimonial practices can only exist because there is a modern bureaucracy with budgets and laws (van de Walle 2001). Studying pastoralists' relationships with the neo-patrimonial state thus requires analysis of both the official bureaucracy and the informal politics of the clientelistic networks that operate within the state. It is critical to avoid treating the abstraction of "the state" as an autonomous agent and instead focus on the individuals who make up the state—for example, the political elite, bureaucrats, military, and police—and how they implement, manipulate, and instrumentalize official policies and laws.

Nomadic FulBe pastoralists' contacts with the Cameroonian state are limited to encounters with lower-level bureaucrats of the agricultural or animal husbandry services, traditional authorities, custom officials, policemen, and district chiefs (*souspréfets*). In most of their dealings with nomadic pastoralists, these agents of the state engage in "informal politics," using the formal bureaucratic system and its official laws, policies, and budgets—for example, vaccinations, permits for transhumance, transhumance tax, poll tax, identity cards, customs—to seek bribes and prebends from pastoralists. Understanding nomadic pastoralists' relationships with the neo-patrimonial state requires an analytical approach that focuses on actual events and the pastoralists' everyday encounters with these representatives of the state. My analysis therefore focuses on the everyday realities as experienced by pastoralists (see also Chabal and Daloz 1999).[2]

THE CHAD BASIN

Examining nomadic pastoralists' relationships with "the state" does not mean that analysis should be limited to one nation-state. African states are colonial inventions, with more or less arbitrary boundaries drawn by European colonial powers in the nineteenth century (see figure 7.1). Although borders have had an impact on the lives of pastoralists, it is not productive to limit the focus to their relations with only one of these post-colonial states. The transhumance of FulBe Mare'en pastoralists in the Far North takes them frequently outside Cameroon, and their kin and kith live scattered across Niger, Nigeria, Chad, Cameroon, and the Central African Republic. Consequently, they come in contact with representatives of different neo-patrimonial states in the Lake Chad Basin. Moreover, events across the border have a direct impact on the lives and livelihoods of nomadic pastoralists in the Far North (e.g., civil war in Chad, religious unrest in Nigeria). It is thus important to examine pastoralists' relations with the state within a larger regional framework, that of the Chad Basin (Krings and Platte 2004; Roitman 2004).[3] In many ways, it is better to think of the Chad Basin not as neighboring states but as the locus of several partially overlapping ecological, cultural, economic, and political zones crossed by political, economic, and criminal transnational networks (Roitman 2004). An example of such a transnational network is the transit of cattle following ancient trade routes from Sudan and Chad through Cameroon to livestock markets and consumers in Nigeria. The volume of this cattle trade is such that one would not be able to grasp the pastoral economy and livestock markets in the Far North if one does not consider this east-west trade flow of cattle (Moritz 2003). Similarly, one cannot understand the current insecurity of cattle raids and road bandits in the Far North if one does not consider the political instability elsewhere in the Chad Basin (Issa 2004).

FulBe Pastoralists in the Far North of Cameroon

FulBe pastoralists in the Chad Basin are part of the largest ethnic pastoralist group in Africa. There are about 20 million speakers of Fulfulde, who can be found throughout West Africa from Senegal in the west to Sudan in the east.[4] FulBe pastoralists have been present in the Chad Basin and the Far North since the eighth century, but the majority of the FulBe came in several waves between the sixteenth and nineteenth centuries in search of pastures for their cattle (Mohammadou 1976; Mohammadou 1988; Seignobos and Iyebi-Mandjek 2000). The Far North offers excellent grazing opportunities for nomadic pastoralists; the Logone floodplain in particular constitutes one of the most important dry-season rangelands in the Chad Basin (Seignobos and Iyebi-Mandjek 2000; Scholte et al. 2006). Pastoralists from Cameroon, Nigeria, and Niger trek each November to the Logone floodplain to exploit the excellent quantity and quality of the rangelands accessible when the water retreats. In the 1970s there were approximately 950,000 cattle in the Far North Province, most of them owned by FulBe and Arab agropastoralists. In recent decades, cattle numbers have declined to about 600,000, of which about 200,000 go on transhumance to the Logone floodplain (Moritz 2003). Sedentary agropastoralists own the majority of the cattle in the Far North, although it remains unclear what percentage, in part because they entrust cattle to nomadic pastoralists (ibid.). Most nomadic pastoralists in the Far North came during the droughts of the early 1970s and 1980s. The sub-ethnic group with the longest residence history, the FulBe Mare'en, came about sixty years ago from Borno, Nigeria. Others came more recently from southeast Niger and Chad.

The FulBe LesDe

When FulBe pastoralists came to the Far North from the periphery of the neighboring Borno Empire, the Hausa states and Baguirmi between the sixteenth and nineteenth centuries were subject to the rule of local chiefs with whom they had to negotiate access to grazing lands (Mohammadou 1976; Abubakar 1977; Seignobos and Iyebi-Mandjek 2000). The conditions under which FulBe pastoralists had gained access to pastures were not always favorable, and often they were exploited

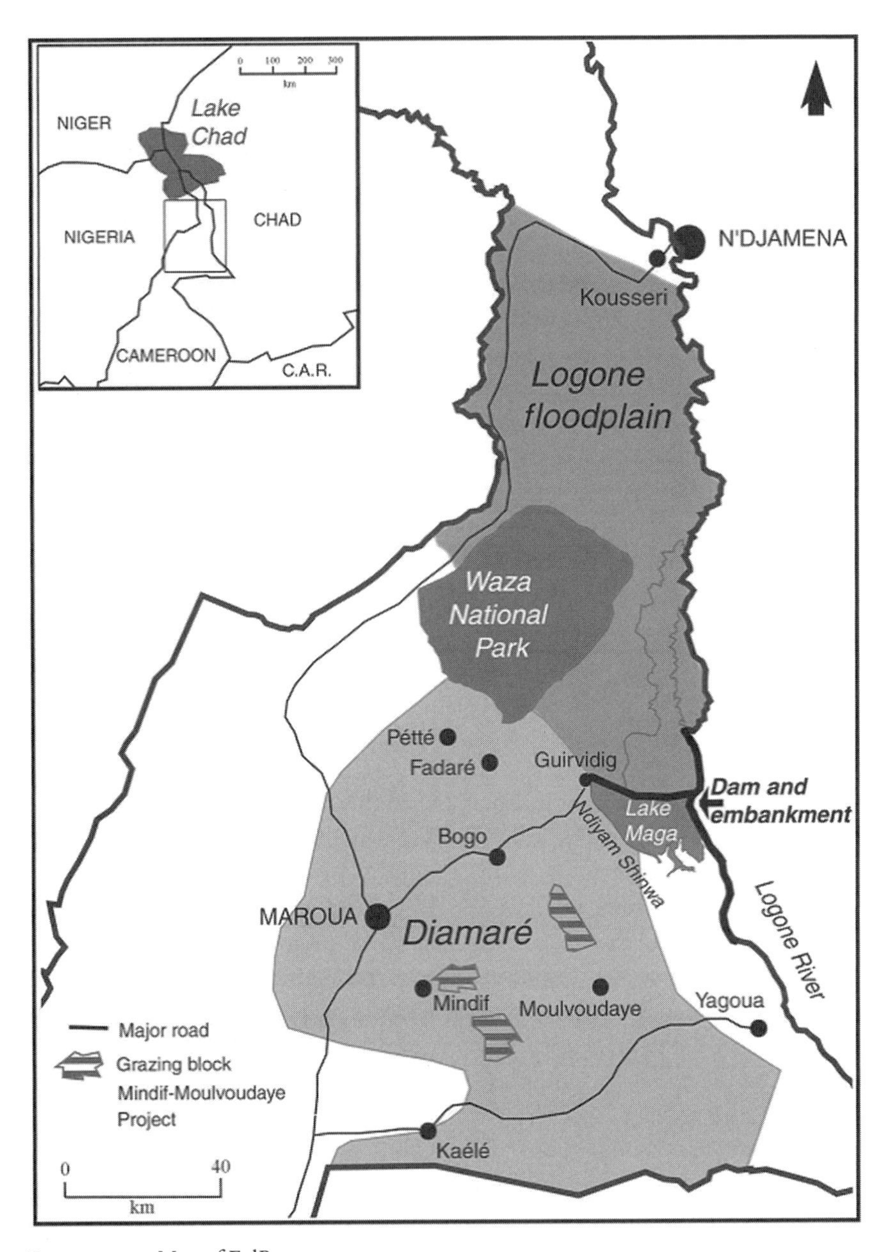

FIGURE 7.1. Map of FulBe area.

by the same rulers who had offered them protection in return for tribute and grazing fees (van Raay 1971). In response, FulBe pastoralists either fled or rebelled against what they considered intolerable conditions (Kirk-Greene 1958; Azarya 1978). Some of these rebellions escalated into wars toward the end of the eighteenth century (Seignobos and Iyebi-Mandjek 2000). Later FulBe wars in the Far North were waged under the cover of a larger FulBe holy war, or jihad, that was started in Sokoto in 1804 by sheikh Uthmān dan Fodio.

When the jihad spread east to the Far North of Cameroon, it resulted in the establishment of eight FulBe *lesDe*.[5] The FulBe lesDe were highly centralized. At the top of the hierarchy was the *laamiiDo* (plural *laamiiBe*), who governed the territory of his lamidat via his secondary and tertiary chiefs, respectively, *lawan'en* and *jawruBe* (singular *lawan*, *jawro*) (see also Azarya 1978; Njeuma 1989:12). The political system of these lesDe can be aptly described as patrimonial (Lacroix 1953; Weber 1964 [1947]; Njeuma 1989). The administration of the lesdi was under direct control of the laamiiDo kin; secondary and tertiary chiefs were generally patrilineal kin, while most council members and soldiers were slaves. The laamiiBe appointed trusted slaves and close kin as lawan'en at the borders of the lesdi to control attacks from Tupuri, Musgum, and Mundang. In the lesdi there was no distinction between public and private property; all land was patrimony of the laamiiDo.

Although the FulBe lesDe were established by pastoral clans, they were not nomadic states (Khazanov 1994); a clear divide developed between the ruling elite and the FulBe who remained nomadic. The ruling elite and many of their followers settled and relied on slaves to cultivate the land and take their herds on transhumance. For FulBe pastoralists, the lesDe provided relatively safe and secure access to rangelands (Moritz, Scholte, and Kari 2002). However, at the borders of emirates there was a constant war between FulBe and non-subjugated populations of Mundang, Giziga, Tupuri, and Musgum. Trade caravans and FulBe herds and villages were at permanent risk of raids from these groups, which made the border areas unsuitable for pastoralists (Beauvilain 1989; Issa and Adama 2000). The majority of the FulBe pastoralists stayed within the limits of the lesDe and did not venture into the no-man's lands or the Logone floodplain because of the risk of cattle raids. Ironically, these so-called no-man's lands (Seignobos and Iyebi-Mandjek 2000), which were also the former slave-raiding areas of the empires of Baguirmi and Borno, are today important transhumance zones because of their historically low population densities.

Nomadic FulBe were incorporated in the FulBe lesDe as separate quarters or villages under the leadership of an *arDo* (nomadic leader, literally, "the one in front" or "the first"), with similar rights and duties as sedentary agricultural and agropastoral populations. The laamiiDo adjudicated conflicts within and between nomadic groups. The nomads, in turn, acknowledged the authority of the laamiiDo and paid tribute and grazing tax (*huDo ceede*, literally grass money). These were collected by the laamiiDo's agents, the *sarkin saanu*, members of the laamiiDo's council in charge of pastoral affairs, and his messengers to the nomads (*ciimaajo*).

Elsewhere, my colleagues and I have discussed the incorporation of nomadic pastoralists in the lesDe as a "contract" between nomads and laamiiBe in which the former paid tax and tribute in exchange for protection of grazing rights and personal safety (Moritz, Scholte, and Kari 2002). We argued that this "contract" had come under pressure when laamiiBe lost power to the state and could no longer uphold their side of the contract, leaving nomadic pastoralists without a sedentary ally. Here I would add that this "contract" was essentially a patron-client arrangement in which nomadic pastoralists were integrated in the lesDe through the patrimonial, clientelistic network of the laamiiDo (see also Njeuma 1989). The taxes and tributes were personal income of the laamiiDo, whose personal commitment secured nomads' access to grazing lands and their personal safety. This commitment was not always strong; a number of laamiiBe cooperated with bandits who raided nomadic pastoralists (Issa 1998) supposedly under his protection. The integration of nomadic pastoralists in the clientelistic network of the laamiiDo fit well with the political system of nomadic FulBe pastoralists, since nomadic leaders were also patrons whose followers depended on their ability to successfully broker access to grazing land.

It is generally to FulBe pastoralists' advantage to change affiliation and "follow" an arDo who has established ties with the "outside world" (Dupire 1970; Burnham 1979:351–52).[6] In their contacts with the state, FulBe Mare'en in the Far North rely on their *arDuBe* (plural of arDo). These arDuBe have no power over their followers but build up a network of clients, the core of which consists of patrilineal (and often matrilineal) kin, by acting as political brokers with the outside world.[7] The legitimacy of Mare'en leaders depends on their ability to successfully maintain contacts with the traditional and government authorities and secure access to grazing lands and personal safety. Successful leaders have large followings; the followings of unsuccessful ones consist only of (close) kin.[8]

THE COLONIAL STATE

The German colonization of Central Africa from 1893 to 1903 met with resistance from the FulBe lesDe, notably that of Maroua, which were ultimately defeated in 1902 by the overwhelming firepower of German machine guns (Dominik 1908). The Germans presented a numerically small administrative and military force and incorporated the laamiiBe in their colonial system of indirect rule. Indirect rule consolidated the power of the FulBe laamiiBe, as populations with acephalous political organizations that previously had not been subjugated by the FulBe were put under the authority of the laamiiBe; for example, when the FulBe laamiiBe levied tribute from these populations, the Germans squashed the resulting revolts (Iyebi-Mandjek and Seignobos 2000; Seignobos and Iyebi-Mandjek 2000).[9] During The First World War, the Germans in the Far North were defeated by the French, who continued to use the laamiiBe in their policy of indirect rule, which from 1917 onward was reformulated in *la politique indigene*. Under this policy, populations with acephalous political organizations, such as the Mundang, Giziga, Tupuri, Musgum, and Masa, were assigned their own political structure independent of the FulBe laamiiBe, although the latter retained considerable power.

It is important to keep in mind that colonial states in Africa were not ideal bureaucratic states with hegemonic power. Colonial rule was dictatorial with patrimonial tendencies. This was particularly true for administrators in the more remote districts, the so-called *rois de la brousse* (king of the bush), who treated their districts as personal fiefdoms and whose rule was marked by a pragmatism and favoritism and an almost arbitrary use of violence. While colonialization drastically changed the economy and political system in the Far North, nomadic FulBe's contact with the Europeans was limited, and the laamiiBe continued to be nomads' primary contacts with the "state." However, the relative "peace" brought by colonialization allowed FulBe pastoralists to move to new grazing lands, and from 1930 onward they ventured farther into the Logone floodplain and the Masa territory of the middle Logone, even though raids continued at the borders of the lesDe (Seignobos and Iyebi-Mandjek 2000).

THE REPUBLIC OF CAMEROON

Cameroon became an independent republic in January 1960, with Ahmadu Ahidjo, a Pullo (singular of FulBe) from the north, as its first president. During his tenure, Ahidjo attempted to unite northern Cameroon as a political force versus the culturally and politically fragmented south by emphasizing a common religion: Islam. In practice, this meant that mostly FulBe and Muslims were appointed to positions of power. Ahidjo also changed the political system to a one-party system and established a highly bureaucratic and authoritarian regime in which the real business of politics was conducted informally and most important decisions were made by the president and his entourage. In 1982 Ahidjo unexpectedly stepped down and Paul Biya succeeded him as president and party leader of the UNC (Union Nationale Camerounaise), which later became the RDPC (Rassemblement Democratique des Peuples Camerounais). Biya ended Ahidjo's policies promoting "northern" power and assigned southerners to administrative positions such as governor, prefet, and sous-prefet. This reversal of policy seriously challenged the dominance of the Muslims and FulBe in the north, including that of the laamiiBe. Today, traditional authorities are officially incorporated in administration of the Cameroonian state, but they are subordinate to the district chiefs, who are the highest local authority.

In the last decade, Cameroon has gone through a process of decentralization and democratization. At the district level, this process involved the introduction of multi-party municipal elections and the devolution of power and collection of some taxes from the sous-prefet to the mayor and municipal council. At the national level, this resulted in the introduction of multi-party elections and greater political freedom, although the Biya regime continues to maintain tight control of the media and close scrutiny of opposition parties.

At the local level, in the Diamare the most important players are the sous-prefet, the laamiiDo, the mayor, and the leaders of political parties—in particular Biya's ruling party, the CPDM (Cameroon People's Democratic Movement), and the UNDP (Union Nationale pour la Democratie et le Progress), an opposition party with much popular support in the Far North. In practice, these elites cooperate and forge alliances to exploit the public resources of the state for personal gain and that of their clientelistic networks, in what have been called "hegemonic exchanges" or "reciprocal assimilation of elites" (Rothchild 1985; Bayart 1993). The hegemonic exchanges of the authorities are best illustrated by the tax collection from nomadic FulBe pastoralists. Traditionally, the laamiiBe collected tax and tribute from nomadic pastoralists, which was their personal income. After independence and especially following the change in the presidency in 1982, the laamiiBe lost power to the state and its officials and had to share nomadic tax revenues with the sous-prefet. Even though the grazing tax has now been transformed into an official transhumance and livestock tax, sous-prefets continued to rely on the laamiiBe personal ties with their client nomads. In practice, this meant that the laamiiBe's agents continued to be in charge of tax collection in the camps. With the decentralization and redistribution of power to the municipalities, mayors have been included in the "hegemonic exchanges" and now also share in nomadic tax revenues.

The fact that multiple authorities are receiving a share of the nomadic tax revenues does not mean that nomadic pastoralists now have multiple patrons on whom they can rely for support. Sous-prefets, for example, do not feel committed to nomadic pastoralists, in part because they are abruptly promoted, demoted, or transferred to different parts of the country. The rapid turnover of state officials affects their legitimacy (and that of the state) in the eyes of nomadic pastoralists. The legitimacy of the authorities depends in large part on the "redistribution" in forms of services or security in return for the extraction of fees and taxes. However, each time a state official changes posts and a new official is appointed, nomadic pastoralists have to reinvest in a new patrimonial network. State officials are thus not reliable patrons for nomadic pastoralists. This is the reason that laamiiBe, who are elected and appointed for life, continue to be more reliable patrons for pastoral nomads, despite the fact that they have lost considerable power.

PASTORAL DEVELOPMENT

The African state has been labeled as weak and without a developmental agenda (van de Walle 2001). Nevertheless, the 1970s and 1980s were eras of large-scale and far-reaching pastoral development programs (Fratkin 1997). Most of these technocratic programs, which aimed at economic development and improving range management, were financed and implemented by international development agencies such as the World Bank, the US Agency for International Development (USAID), and the European Union. Political scientists have pointed out that ruling elites in Africa are quick to embrace new development paradigms and programs proposed by bilateral and multilateral development agencies, not because of the development agenda but because they use development aid as a financial resource (e.g., Chabal and Daloz 1999; van de Walle 2001). As a result, development aid and structural adjustment, intended to reform and develop the economy of African nations, have generally had the opposite effect, as political elites and bureaucrats "instrumentalize" development aid for investment in their own neo-patrimonial politics and clientelistic networks. Since the ruling elite's interest lies primarily in control of development aid, it is doubtful whether African states would have engaged in pastoral development programs were it not for the financial input of the donor countries and nongovernment organizations.

When we examine pastoral development within the analytical framework of the neo-patrimonial state, the emphasis shifts from an analysis of states' official development programs and ideologies—including erroneous assumptions about the inefficiency of extensive pastoralism and pastoralists' aversion toward the market economy and policies to overcome these "deficiencies" through settlement, taxation, and privatization—to an analysis that "follows the money" and examines who has to gain from these development projects. Such an analytical approach is critical for understanding the evolution of pastoral development projects.

An example of instrumentalization of a pastoral development project by political elites is the Mindif-Moulvoudaye Project, which was funded by USAID and the Cameroonian government. The Mindif-Moulvoudaye Project was developed in response to droughts of the early 1970s during which a loss of more than 50,000 cattle was reported (Seignobos and Iyebi-Mandjek 2000). The project's goal was to reform the "anarchic" pastoral system held responsible for overgrazing by implementing a number of measures, including division of land in grazing blocks, introduction of a rotational grazing system, construction of water catchment basins, clearing of firebreaks, and the cultivation of crops for use as fodder.[10] The project was originally planned for the area of Pette and Fadare, but the laamiiDo of Mindif played an important role in getting the project to his lesdi, as it represented an important source of income not only for those people for whom the project was originally intended but also for the governmental and traditional authorities in whose territory the project would be located. The project brought numerous jobs, infrastructure, equipment, and subsidies; and the elites could use their influence to secure these resources for their clients.

Nomadic pastoralists were invited to participate in the project, but they declined to participate in the rotational grazing system and were consequently excluded from other project activities (Cleboski 1985). The laamiiDo of Mindif failed to prevent the project from excluding his nomadic clients from the grazing blocks. Many nomads left the region altogether, as the laamiiDo had probably feared, reducing his income from grazing dues, the loss of which was partially compensated by "revenues" from the project. In 1985 the Americans withdrew personnel and financial support from the Mindif-Moulvoudaye Project, allegedly because of corruption on the part of Cameroonian counterparts, although the project was also ill-conceived and practically all activities had failed. The project continued to exist marginally for another five years but finally folded due to lack of funds in 1990. Soon after that, nomadic pastoralists returned to the area.

In 1994 the French Ministry of Development revived the project using a more participatory and decentralized approach (Reiss 2000). Local agropastoralists themselves were to decide how the pastures should be managed, although nomadic pastoralists were still not welcome. The French project empowered agropastoralists to exclude nomads from pastures to which the latter previously had access. In fact, the activities of both development projects encouraged local populations to view the bush no longer as open access but rather as their exclusive territory. However, the new project failed to keep nomadic pastoralists out of the project area, in part because the laamiiDo, local administrators, and politicians did not support the nomad's exclusion as the elite had little to gain from this relatively small development project.

Even though large-scale development projects have significantly affected nomadic pastoralists' access to grazing lands, "the state" behind these projects is not always visible. Let me illustrate this paradox of the invisible but omnipresent state with a story. In March 2001 I gave Abdu, the twenty-two-year-old son of my host, a ride from the weekly livestock market of Mazera, located in the Logone floodplain, to his father's camp further south in Ndiyam Shinwa. Ndiyam Shinwa refers to the reservoir of Lake Maga created by a dam and embankment of the Logone River for the irrigated rice cultivation of SEMRY II (Societe d'Expansion et de Modernization de la Riziculture de Yagoua II).[11] The dam and embankment drastically changed grazing lands when they were built in 1979; they reduced flooding downstream and flooded areas upstream, where approximately 45,000 hectares of grazing lands were permanently lost to the reservoir and the rice paddies (Loth 2004; Scholte et al. 2006). But the reservoir also created the new grazing lands of Ndiyam Shinwa at the shores of Lake Maga, which pastoralists exploit by following the retreat of the water (as the lake shrinks in the dry season). Abdu had gone to Mazera to see his friends and attend marriage celebrations. He had walked from his father's camp in Ndiyam Shinwa, about 30 km, following the transhumance routes into the floodplain. On our way back from Mazera, we followed a different route by car along the embankment to Pouss, and from there we took the road along the dam to Guirvidig. Midway, in the town of Maga, center of the SEMRY II project, the road crosses the main irrigation canal Mayo Vrick and gives an impressive view of Lake Maga, which at that point is so wide that you cannot see the shores on either side. When we slowed down on top of the dam to take in the view, Abdu was in awe; he had never seen the lake. I was surprised at first; Abdu camped at the shore of this same lake and watered his animals in the lake. How could this be the first time he was seeing the lake? It took me a minute to realize that the shore's reeds blocked the view of the lake and that I myself had never seen the lake from his father's camp at the lake's shores. It was paradoxical that this project, which is so close and immense and had a far-reaching impact on their access to grazing lands and transhumance patterns, had been invisible to Abdu.[12] The paradox of the pastoralist-state relationships in the greater Chad Basin is that while states have drastically altered the lives and livelihoods of nomadic pastoralists over the last centuries, the state and its projects are also often conspicuously absent in their everyday lives. This is another version of the paradox of the weak but at the same time hegemonic state (Young 1994) or the lame leviathan (Callaghy 1984).

Insecurity

Insecurity seems inherent to the neo-patrimonial state, in part because the state is "weak" and unable to maintain security in the periphery of the bush but primarily because the state is responsible for creating and exploiting much of the insecurity. This process, in which politicians and government officials engage in illegal and illegitimate activities, has been referred to as the "criminalization of the state" (Bayart, Ellis, and Hibou 1999). The criminalization ranges from involvement of national politicians in international drugs trade to the involvement of local police in holdups and robberies. The neo-patrimonial state is often the greatest perpetrator of violence in Africa (Chabal and Daloz 1999). This is likely also the case in the Far North of Cameroon, where, for example, gendarme, police, custom officials, and other state agents on duty at roadblocks resort to the threat of violence (and the use thereof) to extort civilians. Nomadic pastoralists without papers, but with cattle and cash, are favorite targets of the gendarme.

Nomadic pastoralists are also one of the favorite targets of bandits (or *coupeurs de route* as they are called in northern Cameroon), as their main criminal activity is holding up cars and buses that travel between the main cities of Ngaoundere, Garoua, Maroua, and Kousseri and local livestock markets.[13] The coupeurs are professional bandits from different ethnic groups and nationalities who operate in the rural areas near the borders (Issa 2004). Many are former soldiers or mercenaries who fought in the Chadian civil war and kept their firearms after demobilization of the armed forces. Some bandits are rumored to be off-duty Cameroonian police and gendarme (ibid.).[14] These transnational groups operate throughout the Chad Basin—robbing in one country, hiding in another. Some of the groups are responsible for carjacking pickup trucks in the cities destined for Sudan and Niger (or taken apart for the local spare parts market), and many enjoy the cooperation of traditional and governmental authorities and thus operate with almost total impunity. In 1998, after an expatriate was shot in a carjacking, the government sent a special unit of the security forces, commonly referred to as the *anti-gang*, to the Far North because of growing insecurity resulting from banditry. The anti-gang operated outside the law and summarily executed suspected bandits, ordinary criminals, and lower-ranking traditional authorities who protected them (see also Amnesty International 1998). The result was selective impunity, since the wealthy and powerful were left alone or allowed to get away. Although the anti-gang now has a permanent base in the Far North, insecurity continues, as was evidenced by an increase in the number of armed robberies near livestock markets and holdups of nomadic camps in 2001.

Nomadic pastoralists are the group most affected by the armed robberies, as they live relatively isolated in the bush. Bandits come to the camps and extort pastoralists under the threat of violence. They frequently announce their arrival in advance to ensure that nomadic pastoralists convert cattle into cash. Mare'en who lived in smaller camps further away in the bush near the border with Chad were at the greatest risk of being robbed. FulBe Mare'en did not report these robberies to the authorities, in part because of the threat of retaliation, which was very real, but also because their previous experiences with the police had been disappointing (Moritz 1995; Scholte, Kari, and Moritz 1996). In the past, FulBe Mare'en have been prosecuted and imprisoned for killing cattle thieves in defense of their herds. But when FulBe Mare'en were killed, police demanded *carburant* (literally "gas," a euphemism for bribe) to conduct investigations and pursue the thieves, as they explained that they did not have the means to do their work. It is equally important to note that the police were able to exploit this weakness of the state for personal gain. When cattle thieves were caught, they were quickly released after payment of bribes. They were not the only authorities to exploit pastoralists; the patrons of nomadic pastoralists, the laamiiBe and lawan'en, have been suspected of collaborating with bandits and cattle thieves by hiding them and their loot.

Access to Grazing Lands

Pastoralists' access to grazing lands in the Chad Basin also has to be examined within the framework of neo-patrimonial politics of the state's agents. Again, a focus on official policies and formal policies, however detrimental to pastoralists, does not accurately reflect how pastoralists' access to grazing lands is affected on the ground.

Grazing lands in the Far North are best described as "annual grazing areas" for mobile pastoralists (Niamir-Fuller 1999b), that is, areas used by one or more ethnic groups in which land is not held in common and no action is undertaken against intruders (Casimir 1992).[15] However, this does not mean that access to grazing lands is regulated by the principle of first-come-first served, as Casimir (ibid.) suggests. Nomadic pastoralists in the Far North have negotiated access to grazing lands through higher-level institutions, in what we have called the "nomadic contract" (Moritz, Scholte, and Kari 2002). In practice, no pastoralists are denied access, as long as they have paid their dues to the authorities.[16] LaamiiBe were generally good in protecting the grazing rights of nomadic pastoralists, as they had a clear interest in ensuring that nomads returned to their territories because of the taxes they paid.[17] However, over the last two decades, especially with the introduction of the multi-party democracy and decentralization in the 1990s, this arrangement between nomadic FulBe and the laamiiBe has come under pressure, as the latter have gradually lost power not only to the state and its agents but also to their own subordinates, the lawan'en and jawruBe (ibid.). This has led to greater ambiguity in the tenure situation for nomadic pastoralists in the Diamare since multiple authorities now claim to "own" the land.

In the past, all lands were owned by the laamiiDo, and there was no distinction between public and private lands in the lamidats. Under colonial rule, all the so-called vacant and ownerless lands were considered public lands and administered by the colonial administration, even though the laamiiBe remained de facto "owners" of the land. This colonial policy was reaffirmed in the 1974 land reform act, which officially abolished customary tenure systems and introduced individual, state, and national lands (Fisiy 1992; van den Berg 1997). The allocation of national lands, the "vacant and ownerless" lands, officially became the prerogative of the district chief, but again, effectively little changed in the way land was allocated (see also van den Berg 1997). In recent decades, as the laamiiBe's power diminished, district chiefs, the official "owners" of the land, are increasingly asserting their authority over national lands. However, the two legal systems continue to coexist, and this has resulted in a situation of apparent institutional ambiguity in land tenure systems, which is exploited by both traditional and government authorities.

I have argued elsewhere that land, like other state resources, has to be considered public goods of the state that can be used by political elites and bureaucrats for personal gain (Moritz 2006). One way authorities make national lands productive in the Far North is by exploiting competing interests over natural resources to create, mediate, and perpetuate conflicts over land. Herder-farmer conflicts in particular have proved relatively easy for authorities to create and exploit through informal politicking (ibid.). The authorities' "politics of permanent conflict" are not always transparent (ibid.). Outwardly, the authorities appear to adhere to the official judiciary process of the bureaucratic state; they follow a protocol and refer to official laws and policies in their decision-making. Informally, however, they make deals with each other that ensure that herder-farmer conflicts continue and are effectively never resolved. We found, for example, that district chiefs and laamiiBe adjudicate but take no action to enforce their decisions and that, consequently, conflicts continue and continue to be "milked" by authorities (ibid.; Moritz, Scholte, and Kari 2002).

From the point of view of government and traditional authorities, there is no ambiguity in who has the authority over national lands, as they cooperate and share the spoils of conflict mediation in a reciprocal accommodation of elites. Authorities in the neo-patrimonial state derive their power and income partly from arbitrage between different groups or networks; thus, it is in their interest to create or perpetuate conflicts between these groups (see also Berry 1993). To a certain extent, the leaders of the FulBe Mare'en also participate in the hegemonial exchange among elites. As representatives of nomadic pastoralists, they take cuts from the taxes and tributes that they collect from their followers and transfer to the laamiiBe as part of the nomadic contract.

Herders and farmers coping with the informal politics of the authorities are increasingly frustrated about the authorities' "appetite," in part because the payment of rents no longer guarantees a favorable outcome. The most likely outcome in conflicts over grazing lands and campsites is a status quo, and this puts the parties already in control of the land at an advantage. The "politics of permanent conflict" thus reaffirm the existing West African pattern that farmers' usufruct rights are more secure than those of herders (Moritz 2006). But they have also increased the costs for farmers, as insecure land tenure requires constant "investments" in patrimonial networks. Nomadic pastoralists' access to grazing land is thus threatened not by implementation of the state's official laws but instead by the informal politics of the state officials on whom nomadic pastoralists rely for access to grazing lands.

CONCLUSION

Although most researchers working with pastoral societies in Africa have been confronted with the informal politics of bureaucrats and the elite and how they affect the lives and livelihoods of pastoralists, they have not yet incorporated this systematically and explicitly in their analytical models. Corruption features frequently in case studies and some analyses, but the concept of the neo-patrimonial state has not been integrated in theoretical models that examine pastoralists' relations with the state in Africa.

The literature on pastoralists in Africa has emphasized that state centripetal forces of domination and encapsulation of nomadic pastoralists lead to their increasing marginalization (see Klute 1996). I have argued that the focus on official laws and policies of an ideal bureaucratic state misrepresents the relationships between nomadic FulBe Mare'en pastoralists and the state in the Chad Basin. My analysis of relationships of nomadic pastoralists with the neo-patrimonial state in the Chad Basin shows that the state is not all-powerful and not primarily concerned with the dominance and encapsulation of nomadic pastoralists. More important, the focus shifts from the objectives of an abstract bureaucratic "state" to the interests of the agents who make up the neo-patrimonial state and how they instrumentalize the apparent disorder for personal gain. In this analysis, it becomes clear why state agents prefer partial incorporation of nomadic pastoralists in the state. From the agents' perspective, it is better if nomadic pastoralists are not registered, do not have identity papers, do not pay taxes, and are not schooled or settled because this leaves them more easily exploited. The actions of nomadic pastoralists are also not necessarily characterized by centrifugal forces, as they have more to gain from incorporation in the state. In fact, FulBe Mare'en actively seek to be integrated in the neo-patrimonial state through the clientelistic networks of laamiiBe, wealthy absentee owners, and government officials in order to secure access to grazing lands and ensure their personal safety.

My analysis examines the state as viewed and experienced by nomadic pastoralists. In the eyes of nomadic pastoralists, the state does not manifest itself as an ideal bureaucratic state but rather as the informal politics and networks of its bureaucrats and elites with whom they come in contact.[18] To understand how the "state" affects the lives and livelihoods of nomadic pastoralists, we must focus on these informal relationships.

ACKNOWLEDGMENTS

I would like to thank Elliot Fratkin, Avinoam Meir, and Leslie C. Moore for their comments on an earlier version of this chapter.

NOTES

1. Meir (1988) applies the concepts of centripetal and centrifugal forces to examine spatial conflicts between Bedouin nomads and the Israeli state, but the concepts also aptly summarize socio-political conflicts between nomads and the state in the "pastoralist literature."

2. How nomadic pastoralists experience and view the state in Africa is shaped by these everyday encounters with agents of the state. In fact, pastoralists' view of the state probably more accurately describes the reality of the African state than do models and conceptions in the pastoralist literature.

3. Often, a distinction is made between the conventional basin (967,000 km^2), which comprises the states that border Lake Chad (Niger, Nigeria, Cameroon, and Chad), and the hydrological basin (2,335,000 km^2), often referred to as the greater Chad Basin, which also includes Libya and the Central African Republic.

4. In the Far North of Cameroon they call themselves FulBe, but they are also known under the name Fulani in the Anglophone literature or Peul in the Francophone literature.

5. Lamidat is the commonly used French Cameroonian word for the territory governed by the laamiiDo; the FulBe refer to these provinces as lesdi (singular for "land" or "territory"; plural: lesDe) (Seignobos and Tourneux 2002).

6. The social organization of nomadic FulBe has been described as a fragmentary lineage system (Dupire 1970), suggesting greater flexibility than in the ideal model of segmentary lineage systems (Evans-Pritchard 1940), as FulBe have continually changed lineage and clan affiliations in response to changes in transhumance patterns (Stenning 1960). Nomadic FulBe in the Far North of Cameroon are "organized" by sub-ethnic groups, which consist of multiple clans and lineages that are endogamous. Members of sub-ethnic groups generally follow the same transhumance route and have a number of cultural traits in common, such as dialect, ceremonies, cattle breed, and tents (Burnham 1996). The sub-ethnic groups of the Mare'en FulBe, for example, consist of multiple clans, some of which are descendants of Arab and FulBe groups (e.g., FulBe Kessu'en), while others were sedentary agropastoralists (e.g., FulBe Ngara'en), but they all keep mahogany zebu cattle and live in the same oval-shaped tents.

7. Mare'en leaders also act as brokers in economic ties with the outside world. They maintain contacts with "absentee herd owners" who split their herds and keep part in the village and entrust the remainder to nomadic pastoralists. These animals are entrusted to a kaliifa, generally an arDo who in turn entrusts the animals to herders, most often his sons or resident kin (Moritz 2003, 2012). The kaliifa has the final responsibility over the herd and answers to the absentee owners. Herders have usufruct rights over the animals and receive a monthly salary. Although, there are no direct material benefits for the kaliifa, their position as intermediary allows them to support their kin and maintain their client network.

8. Barfield (1993:17) has noted that pastoral political organization generally mirrors the complexity and "sophistication of the organization of the neighboring sedentary people with whom they interacted." This is also true of nomadic FulBe Mare'en in the Far North of Cameroon whose socio-political organization mirrors that of the patron-client networks of the neo-patrimonial state.

9. The colonialization of the Chad Basin consolidated the power of some FulBe lesDe and laamiiBe, e.g., Maroua and Mindif, but diminished the power of others, e.g., Binder, which was subjugated and then subdivided by German and French colonial administrations (Mohammadou 1988).

10. These measures would allow pastoralists to stay in the Mindif-Moulvoudaye region throughout the year instead of going on transhumance to the Logone floodplain or south into Chad. Many pastoralists from the Mindif-Moulvoudaye region did indeed cease to go on transhumance but for different reasons: civil war and insecurity in Chad and cattle theft and insecurity in the Logone floodplain.

11. Ndiyam Shinwa literally means "Chinese water" in Fulfulde. Shinwa comes from the French Chinois.

12. It's not that Abdu had not traveled. He has been on transhumance in Cameroon and Chad with his father's herd. Abdu knew the bush well and had crossed the Logone River, which forms the border with Chad in the Far North, multiple times.

13. The bandits (pasoowo, fasooBe in Fulfulde) are not a new phenomenon (Issa 2004).

14. Roitman (2004) argues that in the border regions, non-paid custom agents, military, gendarme, and other armed personnel have become *douanierscombattans* and that they are accepted as a regulatory force in the area (in the absence of the state). This has resulted in combinations of official and nonofficial taxes, such as *le taxe d'entree* and *le taxe du coupeur* (ibid.:20). Roitman (ibid.) argues that these douanierscombattans have a certain legitimacy because they do provide security and redistribute wealth, but I am not sure whether I agree since it remains unclear how much is redistributed and how much security is provided.

15. This discussion of rangeland access draws from our paper in *Human Ecology* (Scholte et al. 2006) and from a paper in the *Canadian Journal of African Studies* (Moritz 2006).

16. This means that most "negotiations" and coordination occur primarily among pastoralists. Niamir (1990) calls this "passive coordination" or "choreography" of movements

in which no formal agreements are made between pastoralists but where coordinated movements result from individual decision-making. Galaty (1994:187) adds that this coordination is a progressive and continuous process "whereby the movement of herds is effectively rationalized through progressive adjustments made by herding groups in response to the presence and trajectory of one another."

17. Through their association with the FulBe laamiiBe, FulBe Mare'en were also integrated in the bureaucratic structure of the colonial and post-colonial state. They are, for example, inscribed on the roles of the municipality in the lesdi, where they spend the rainy season and pay their poll taxes (rather than in the dry-season transhumance area of the Logone floodplain).

18. In the eyes of nomadic pastoralists, the state also does not manifest itself in the form of development projects, which are correctly associated with expatriates.

REFERENCES

Abubakar, Saad. 1977. *The Lamibe of Fombina: A Political History of Adamawa 1809–1901.* Zaria, Nigeria: Ahmadu Bello University Press.

Amnesty International. 1998. Cameroon: Extrajudicial Executions in North and Far-North Provinces. http://web.amnesty.org/library/Index/ENGAFR170161998?open&of =ENG-CMR, accessed February 9, 2004.

Azarya, Victor. 1978. *Aristocrats Facing Change: The Fulbe in Guinea, Nigeria, and Cameroon.* Chicago: University of Chicago Press.

Azarya, Victor. 2001. "The Nomadic Factor in Africa: Dominance of Marginality." In *Nomads in the Sedentary World*, ed. Anatoly M. Khazanov and Andre Wink, 250–84. Richmond, UK: Curzon.

Barfield, Thomas J. 1993. *The Nomadic Alternative.* Englewood Cliffs, NJ: Prentice-Hall.

Bayart, Jean-François. 1993. *The State in Africa: Politics of the Belly.* Harlow, UK: Longman.

Bayart, Jean-François, Stephen Ellis, and Beatrice Hibou, eds. 1999. *The Criminalization of the State in Africa.* London: International African Institute [in association with James Currey, Bloomington: Indiana University Press].

Beauvilain, Alain. 1989. *Nord-Cameroun: Crises et Peuplement*, vol. 1 and 2. Manches, France: Imprimerie Claude Bellee.

Berry, Sara S. 1993. *No Condition Is Permanent: The Social Dynamics of Agrarian Change in Sub-Saharan Africa.* Madison: University of Wisconsin Press.

Burnham, Philip. 1979. "Spatial Mobility and Political Centralization in Pastoral Societies." In *Pastoral Production and Society*, ed. L'equipe Ecologie et Antropologie des Societies Pastorales, 349–60. Cambridge: Cambridge University Press.

Burnham, Philip. 1996. *The Politics of Cultural Difference in Northern Cameroon.* Washington, DC: Smithsonian Institution Press.

Callaghy, Thomas M. 1984. *The State-Society Struggle: Zaire in Comparative Perspective*. New York: Columbia University Press.

Casimir, Michael J. 1992. "The Determinants of Rights to Pasture: Territorial Organization and Ecological Constraints." In *Mobility and Territoriality: Social and Spatial Boundaries among Foragers, Fishers, Pastoralists, and Peripatetics*, ed. Michael J. Casimir and Aparnu Rao, 153–204. New York: Berg.

Chabal, Patrick, and Jean-Pascal Daloz. 1999. *Africa Works: Disorder as Political Instrument*. Oxford: International African Institute [in association with James Currey, Bloomington: Indiana University Press].

Cleboski, Linda D. 1985. *Development and Installation of Range Management Plan: Grazing Block 1 (Maoudine-Gagadje-GayGay)*. Minneapolis: Experience Incorporated.

Diallo, Youssouf. 1999. "Dimensions Sociales et Politiques de l'Expansion Pastorale en Zone Semi-humide Ivorienne." In *Pastoralists under Pressure? Fulbe Societies Confronting Change in West Africa*, ed. Victor Azarya, Anneke Breedveld, Mirjam de Bruin, and Han van Dijk, 211–36. Leiden: Brill.

Dominik, Hans. 1908. *Vom Atlantik zum Tschadsee: Kriegs-und Forschungsfahrten in Kamerun*. Berlin: Ernst Siegfried Mittler und Sohn.

Dupire, Marguerite. 1970. *Organisation Sociale des Peuls*. Paris: Librairie Plon.

Evans-Pritchard, Edward E. 1940. *The Nuer: A Description of the Modes of Livelihood and Political Institutions of a Nilotic People*. New York: Oxford University Press.

Fisiy, Cyprian F. 1992. *Power and Privilege in the Administration of Law: Land Law Reforms and Social Differentiation in Cameroon*. Leiden: African Studies Centre.

Fratkin, Elliot. 1997. "Pastoralism: Governance and Development Issues." *Annual Review of Anthropology* 26 (1): 235–61. http://dx.doi.org/10.1146/annurev.anthro.26.1.235.

Galaty, John G. 1994. "Rangeland Tenure and Pastoralism in Africa." In *African Pastoralists Systems: An Integrated Approach*, ed. Elliot Fratkin, Kathleen A. Galvin, and Eric Abella Roth, 185–204. Boulder: Lynne Rienner.

Issa, Saibou. 1998. "Laamiido et Securite dans le Nord-Cameroun." *Annales de la FALSH de l'Université de Ngaoundéré* 3: 63–76.

Issa, Saibou. 2004. "L'Embuscade sur les Routes des Abords Sud du Lac Tchad." *Politique Africain* 94: 82–104.

Issa, Saibou, and Hamadou Adama. 2000. *Vol et Relations entre Peuls et Guiziga dans la Plaine du Diamare (Nord- Cameroun)*. Ngaoundere, Cameroon: Department of History, University of Ngaoundere.

Iyebi-Mandjek, Olivier, and Christian Seignobos. 2000. "Evolution de l'Organisation Politicoadministrative." In *Atlas de la Province Extrême-Nord Cameroun*, ed. Christian Seignobos and Olivier Iyebi-Mandjek. Paris: Institut de Recherche pour le Developpement (IRD) and Ministere de la Recherche Scientifique et Technique (MINREST).

Khazanov, Anatoly M. 1994. *Nomads and the Outside World*. Madison: University of Wisconsin Press.

Kirk-Greene, Anthony Hamilton Millard. 1958. *Adamawa Past and Present: An Historical Approach to the Development of a Northern Cameroons Province*. London: Oxford University Press.

Klute, Georg. 1996. "Introduction." *Nomadic Peoples* 38: 3–10.

Krings, Matthias, and Editha Platte. 2004. "Living with the Lake: An Introduction." In *Living with the Lake: Perspectives on History, Culture and Economy of Lake Chad*, ed. Matthias Krings and Editha Platte, 11–40. Koln: Rudiger Koppe Verlag.

Lacroix, Pierre-Francis. 1953. "Materiaux pour Servir a l'Histoire des Peuls de l'Adamawa (Suite et Fin)." *Études Camerounaises* 6 (39-40): 4–40.

Lenhart, Lioba, and Michael J. Casimir. 2001. "Environment, Property Resources and the State: An Introduction." *Nomadic Peoples* 5 (2): 6–20. http://dx.doi.org/10.3167/082279401782310817.

Loth, Paul, ed. 2004. *The Return of the Water*. Gland, Switzerland: International Union for the Conservation of Nature.

Meir, Avinoam. 1988. "Nomads and the State: The Spatial Dynamics of Centrifugal and Centripetal Forces among the Israeli Negev Bedouin." *Political Geography Quarterly* 7 (3): 251–70. http://dx.doi.org/10.1016/0260-9827(88)90015-8.

Mohammadou, Eldridge. 1976. *L'Histoire des Peuls FerooBe du Diamaré: Maroua et Pétté*, vols. 1–3. Tokyo: Institute for the Study of Languages and Cultures of Asia and Africa (ILCAA).

Mohammadou, Eldridge. 1988. *Les Lamidats du Diamare et du Mayo-Louti au XIX siecle (Nord Cameroun)*, vol. 22. Tokyo: Institute for the Study of Languages and Cultures of Asia and Africa (ILCAA).

Moritz, Mark. 1995. "Minin jogi geeraaDe: A Study of the Marginalization of Mbororo'en in Northern Cameroon." MA thesis, Department of Anthropology, Leiden University, UK.

Moritz, Mark. 2003. "Commoditization and the Pursuit of Piety: The Transformation of an African Pastoral System." PhD diss., Department of Anthropology, University of California, Los Angeles.

Moritz, Mark. 2006. "The Politics of Permanent Conflict: Herder-Farmer Conflicts in the Far North of Cameroon." *Canadian Journal of African Studies* 40 (1): 101–26.

Moritz, Mark. 2012. "Individualization of Livestock Ownership in Fulbe Family Herds: The Effects of Pastoral Intensification and Islamic Renewal." In *Who Owns the Stock? Collective and Multiple Forms of Property in Animals*, ed. Anatoly Khazanov and Günther Schlee, 193–214. Oxford: Berghahn.

Moritz, Mark, Paul Scholte, and Saidou Kari. 2002. "The Demise of the Nomadic Contract: Arrangements and Rangelands under Pressure in the Far North of Cameroon." *Nomadic Peoples* 6 (1): 124–46. http://dx.doi.org/10.3167/082279402782311013.

Niamir, Maryam. 1990. *Community Forestry: Herders' Decision-Making in Natural Resource Management in Arid and Semi-arid Africa*. Rome: Food and Agriculture Organization of the United Nations.

Niamir-Fuller, Maryam. 1999a. "Towards a Synthesis of Guidelines for Legitimizing Transhumance." In *Managing Mobility in African Rangelands: The Legitimization of Transhumance*, ed. Maryam Niamir-Fuller, 266–90. London: Intermediate Technology. http://dx.doi.org/10.3362/9781780442761.011.

Niamir-Fuller, Maryam, ed. 1999b. *Managing Mobility in African Rangelands: The Legitimization of Transhumance*. London: Intermediate Technology. http://dx.doi.org/10.3362/9781780442761.

Njeuma, Martin. 1989. "The Lamidats of Northern Cameroon, 1800–1894." In *Introduction to the History of Cameroon: Nineteenth and Twentieth Centuries*, ed. Martin Njeuma, 1–31. London: Macmillan.

Reiss, Denis. 2000. "La Gestion du Foncier Pastoral: Une Experience au Nord-Cameroun." *Intercoopérants-Agridoc* 12: 38–40.

Roitman, Janet. 2004. "Les Recompositions du Bassin du Lac Tchad." *Politique Africaine* 94: 7–22.

Rothchild, Donald S. 1985. "State Ethnic Relations in Middle Africa." In *African Independence: The First 25 Years*, ed. Gwendolen M. Carter and Patrick O'Meara, 71–96. Bloomington: Indiana University Press.

Salih, M.A. Mohamed. 1990. "Introduction: Perspectives on Pastoralists and the African States." *Nomadic Peoples* 25–27: 3–6.

Scholte, Paul, Saidou Kari, and Mark Moritz. 1996. "The Involvement of Nomadic and Transhumance Pastoralists in the Rehabilitation and Management of the Logone Flood Plain, North Cameroon." *IIED Drylands Programme Issues Paper* 66: 1–21.

Scholte, Paul, Saidou Kari, Mark Moritz, and Herbert Prins. 2006. "Pastoralist Responses to Floodplain Rehabilitation in North Cameroon." *Human Ecology* 34 (1): 27–51. http://dx.doi.org/10.1007/s10745-005-9001-1.

Seignobos, Christian, and Olivier Iyebi-Mandjek, eds. 2000. *Atlas de la Province Extrême-Nord Cameroun*. Paris: Institut de Recherche pour le Developpement (IRD) and Ministere de la Recherche Scientifique et Technique (MINREST).

Seignobos, Christian, and Henry Tourneux. 2002. *Le Nord Cameroun a Travers ses Mots-Dictionnaire de Termes Anciens et Modernes*. Paris: Karthala.

Stenning, Derrick J. 1960. "Transhumance, Migratory Drift, Migration: Patterns of Pastoral Fulani Nomadism." In *Cultures and Societies of Africa*, ed. Simon Ottenberg and Phoebe Ottenberg, 139–59. New York: Random House.

van de Walle, Nicolas. 2001. *African Economies and the Politics of Permanent Crisis, 1979–1999*. Cambridge: Cambridge University Press.

van den Berg, Adri. 1997. *Land Right, Marriage Left: Women's Management of Insecurity in North Cameroon*, vol. 54. Leiden: Research School CNWS, Leiden University.

van Raay, Hans G.T. 1971. *Land, Power, and Pastoral Development in Northern Nigeria*. The Hague, Netherlands: Regional Development Planning, Institute of Social Studies.

Weber, Max. 1964 [1947]. *The Theory of Social and Economic Organization*. New York: Free Press.

Young, Crawford. 1994. "Zaire: The Shattered Illusion of the Integral State." *Journal of Modern African Studies* 32 (2): 247–63. http://dx.doi.org/10.1017/S0022278X0001274X.

Young, Crawford. 2004. "The End of the Post-Colonial State in Africa? Reflections on Changing African Political Dynamics." *African Affairs* 103 (410): 23–49. http://dx.doi.org/10.1093/afraf/adh003.

8

Flexibility in Navajo Pastoral Land Use

A Historical Perspective

Lawrence A. Kuznar

The Navajo people, or Diné as they call themselves, are numerous and occupy a region within US territory the size of West Virginia. As with most Native American reservations, the average Navajo lives in extreme poverty compared with mainstream Americans, and Aberle (1983) argues that economic development on the reservation has been impeded to the benefit of large mining and oil interests. However, despite poverty and unemployment, the Navajo have enjoyed more success, in terms of territorial control and overall population growth, than most Native American groups in the face of colonialism and the expansion of a worldwide economic system into their lands. In the 1990s, the Navajo numbered 219,000 on a reservation of 25,351 square miles (Rogers 1993a:6, 1993b:2); the latest US Census (2010) lists their population as 332,000. Traditional dwellings, such as log and earth hogans, and traditional livestock, such as sheep, are found throughout Navajo Nation lands; and more mainstream border towns are increasing in size while modern communities are sprouting in the Nation's interior. The Navajo Nation also earns millions from the lease of mining and drilling rights on its mineral- and oil-rich lands. My thesis is that the Navajos' historical reliance on pastoralism resulted in their being economically more secure and reproductively more numerous and is a major reason for their successful survival as a people and a culture relative to other Native Americans.

My aim in this chapter is to summarize the work of historians, archaeologists, and paleo-environmental researchers and to integrate this emergent body of data into two theoretical contexts: core-periphery theory following the work of

DOI: 10.5876/9781607323433.c008

Wallerstein (1974) and Chase-Dunn and Hall (1991) and human behavioral ecology (Harpending, Draper, and Rogers 1987; Cronk 1991; Durham 1991; Borgerhoff-Mulder 1994; Smith and Winterhalder 1994; Winterhalder and Smith 2000).

THEORIES AND METHODS

I propose two measures to monitor the impact of historical and environmental factors on Navajo adaptability: reproductive success, as measured by population size and intrinsic growth rates, and economic success, as measured by livestock numbers. Both measures have salience for anthropological theories as well as for Navajo values.

Human behavioral ecologists have used reproductive success to understand human behavior in a wide variety of cultural settings, including forager, horticultural, pastoral, and peasant economies, and the full range of socio-political situations (Irons 1979; Hames 1990; Cronk 1991, 1993; Chagnon 1992; Hawkes 1993; Borgerhoff-Mulder 1994; Buss 1994; Winterhalder and Smith 2000; Kuznar 2007). Focusing on reproductive success allows one to use evolutionary theories that explain human behavior, as well as to develop testable theories of how people will respond to changing social and environmental conditions. Reproductive success also has emic salience for Navajo studies, given the Navajo religious emphasis on family and fertility; Navajo traditionally consider the creation of conditions that favor the reproduction of human beings a sacred duty (Reichard 1950:29, 80–81; Adams 1963:68; Downs 1972:22, 90, 97; Bailey 1980:93, 118; McNeley 1981:33; Chisholm 1983:57; Farella 1984:73–76; Frisbie 1993). Chisholm (1983) found that reproduction was not only highly valued in the Navajo cultural system but that people genuinely behaved in accordance with those values, allowing him to utilize evolutionary theory to explain Navajo child-rearing practices in a way that was both etically significant for anthropologists and emically relevant to Navajo culture.

Economic success is necessary for people's maintenance and is a measure relevant to important anthropological theoretical concerns, as well as to Navajo values. One theoretical approach that incorporates a consideration of economic factors in the face of imperial expansion, as has happened to the Navajo, is world-systems, or core-periphery, theory. Early work by Wallerstein (1974) and Wolf (1982) emphasized the expansion of the capitalist economic system in the past 500 years and its subsequent impact on people of the non-Western world. This work has expanded to include studies of non-Western empires (Blanton and Feinman 1984; Chase-Dunn and Hall 1991, 1995; Sanderson 1995; Kardulias 1999; Kuznar 1999a; Stein 1999; Wells 1999; Hall, Kardulias, and Chase-Dunn 2011), as well as specific studies of Western expansion in the Navajo region (Hall 1989). My analysis of Navajo history includes consideration of the larger, global economic processes, especially the economic development of the US nation, that impacted the local economy of the Navajo people.

In this chapter I focus on the Navajo use of herding as an adaptive economic strategy, since the Navajo people have relied in part or in whole on livestock for their material well-being for about 300 years. This dependence has resulted in a complex of Navajo practices and values surrounding pastoralism. Obviously, successful herders possess wealth that in its own right has material value, leading to the goal of obtaining large herds (Downs 1972:49). Success in herding goes beyond simple profit motives as well, with the successful herder regarded as fulfilling Navajo ethics of industry, skill, and knowledge (Kluckhohn and Leighton 1962:299), as well as fulfilling the mandates of the Holy People to care for and increase the gift of livestock (Howard Gorman, in Roessel and Johnson 1974:66). Therefore, Navajo would have historically valued their herds and would have considered forces that impacted the herds important.

Despite my justification of these measures of adaptive success, two methodological issues problematize them: scale of analysis and historiography. Ultimately, the question is how historical and environmental factors impacted individuals, but the historical data available are aggregate numbers of livestock and people for the Navajo tribe, or Nation, as a whole. Obviously, these are two very different levels of analysis, and aggregate data risk masking important individual differences in reproductive and economic success. I argue that despite the limitations of aggregate data, fluctuations in population size and herd size among the Navajo through history were so great as to partially alleviate this masking. For instance, fluctuations in herd size that go from tens of thousands to millions of animals, such as what we see in Navajo history, are so great that the majority of Navajo would have been impacted in the manner indicated by the aggregate data. Consequently, I use the intrinsic rate of increase, measured as per annum population growth, as an indirect measure of individual reproductive success, or human fertility, at any one time. Herd numbers are used directly to measure economic success.

Another issue is the accuracy of historical data. It is obvious that, given the difficult conditions of early population estimates, the lack of training of those making observations, and the personal agendas of those counting, population estimates and herd sizes recorded will not be accurate and precise. However, the dramatic shifts of Navajo history once again act to ameliorate these historiographic problems. Population sizes and livestock counts fluctuate widely in Navajo history. When independent counts are available, these relative fluctuations are preserved, indicating that even if counts are off by orders of magnitude, they probably still reflect trends of increase and decrease through time. Some of these historiographic problems have been dealt with by historians of the Navajo, and I rely on the figures they present (see table 8.1; Bailey 1980; Bailey and Bailey 1986; Kelley and Whitely 1989). In addition to raw counts, I also provide log transformations of the data to deemphasize the actual numbers (which are certainly imprecise) and emphasize relative changes through time. As a practical matter, in cases where researchers do not agree on livestock numbers or human population figures, I use median figures. Also, I caution the reader to regard all figures presented here as first approximations to actual population and herd sizes, as well as intrinsic rates of increase.

While it would be far more desirable to have precise data on individuals' reproductive success and economic wealth in livestock throughout history for the Navajo, no such data are available today or are likely to ever be available. Faced with using either flawed data or no data at all, I argue that a conservative use of the available data, considering its limitations, will yield further insights into Navajo history and the behavior of Navajo people and a deeper historical perspective on contemporary Navajo conditions.

EARLY NAVAJO

The original Navajo appear to have been Athapaskans who migrated south to the New Mexican Rockies and the San Juan Basin between AD 1300 and 1500 (Kelley and Whitely 1989:5; Perry 1991:128). This region is known as the Navajo homeland, or Dinétah (see figure 8.1). These Athapaskans diverged into the various Apachean peoples of the Southwest, including the Navajo (Perry 1991). After looking at archaeological, ethnohistorical, and ethnographic data, Kelley and Whitely (1989:12) argue that Apachean Navajo economy at this time was focused on foraging supplemented by maize horticulture (ibid.). Wetter climate conditions during the 1500s led to migration out of the Rockies, down to the San Juan River valley, and onto Mount Taylor, New Mexico, by AD 1583 (ibid.:8). At this time Spanish chroniclers, such as Antonio de Espejo, noted that the Navajo economy was based on a combination of hunting and gathering and maize horticulture, as well as the trade of animal hides with Puebloan peoples (Brugge 1983:491). In time, the Spanish began to raid Navajo and other Apachean groups for slaves, leading to retaliatory raids by the Navajo and their cousins (Bailey 1980:58; Kelley and Whitely 1989:11). Sheep and cattle captured on raids introduced these animals into Navajo society. Kelley and Whitely (ibid.:13) suggest that Navajo raiders primarily ate their booty at this time, although Navajo herds probably originated from these captured animals.

Spanish depredations upon Native Americans, including religious oppression, widespread slavery, forced labor, and concubinage, in the Southwest resulted in the

TABLE 8.1 Historic population figures, livestock numbers, and intrinsic population growth rates for the Navajo people.

Time Period	*Sheep/Goat Herd Size, 1000s*		*Resident Population, 1000s*		*Population Intrinsic*
AD	*Count*	*Log*	*Count*	*Log*	*Growth Rate**
1743	3	3.5	2–4	3.3–3.6	1.8
1805	8	3.9	8–10	3.9–4	0.6
1863	520	5.7	10–15	4–4.2	−0.4
1868	15	4.2	8–18	3.9–4.3	1.6
1900	401.9	5.6	20	4.3	2.4
1930	1304	6.1	39–43	4.6	1.1
1937	435	5.6	44.3	4.6	2.4
1960	390	5.6	77.3	4.9	2.6
1985	317.4	5.5	146.2	5.2	

Sources: Figures drawn from Bailey and Bailey (1986:appendix A); Kelley and Whitely (1989:tables 3 and 4).

* Intrinsic growth rate estimated by calculating all possible growth rates given the various population estimates provided for the time period and then calculating the median growth rate among estimates.

famed Pueblo Revolt of 1680. Most Puebloan peoples and some Apachean bands took part in expelling their Spanish overlords from the Southwest (Bailey 1980:60). The vacuum created by the expulsion of the Spanish enabled the Navajo to extend their territory westward to Canyon de Chelly, Arizona, as well as their regional economic influence through trading (Kelley and Whitely 1989:15). More important, the Spanish backlash in 1693–98 forced many Puebloans, perhaps as many as 1,500, to flee their fixed settlements. Many of these refugees sought refuge with the Navajo, whose own numbers may have only been around 1,500 at that time (Bailey 1980:65). This influx of immigrants swelled Navajo numbers, and some of the present-day Navajo clans appear to have been derived from these Puebloan immigrants (Brugge 1983:493). Culturally and economically, the Spanish re-conquest had a cataclysmic effect on the Navajo, who adopted many elements of Puebloan religion and economic life at this time, including Puebloan ceremonial lore, intensive agriculture, and settled village life (ibid.).

Navajo settlement shifted by 1710. Typical settlements, known today as *pueblitos*, consisted of stone structures arranged in small towns, often in defensive locations with watchtowers. The economy also became focused more intensively on maize agriculture at this time (Bailey 1980:66; Brugge 1983:494; Kelley and Whitely 1989:20). Also, with this settling down and increase in population levels, sheep could be acquired and their numbers increased. The acquisition and enlargement

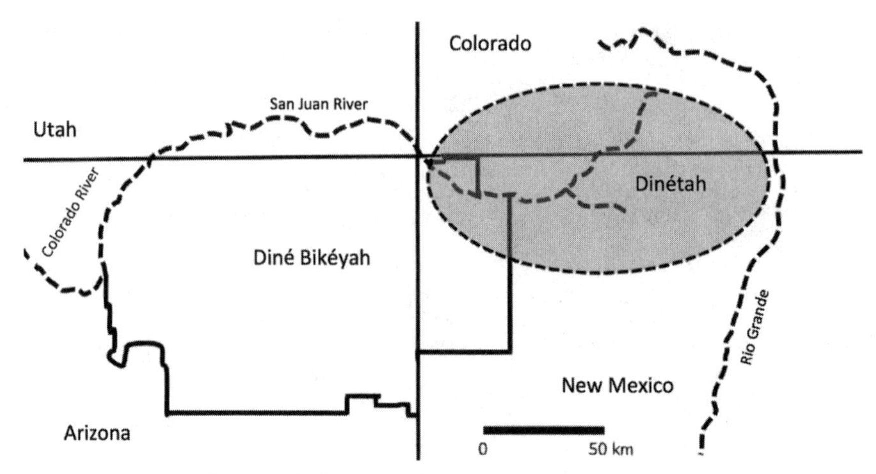

FIGURE 8.1. Map showing the location of Navajo land in the Four Corners region of the US Southwest.

of herds likely transformed these animals, once simply eaten, into a form of wealth (Bailey 1980:67; Kelley and Whitely 1989:29; see Ingold 1980 for a similar argument regarding the evolution of reindeer herding in Eurasia).

Historian Lynn Bailey (1980:77) has noted that given the high intrinsic growth rate of sheep and the fixed pasture forage capacity of Dinétah, the Navajo would have been forced to leave Dinétah by the early 1700s, partly explaining their westward movement from Dinétah to the lands they inhabit today. In addition, the Navajo population continued to grow throughout the 1700s, a time of relative peace for the Navajo people (ibid.:90). However, climate would intercede and further perturb the newly emergent Navajo agricultural system. Droughts in 1755–65 and again in 1772–82 led to increased reliance on mobile pastoralism (Kelley and Whitely 1989:27).

The sedentary agricultural system collapsed, and the Navajo picked up their approximately 3,000 animals and began moving around (ibid.:208). Regional political events also spurred this movement, as the Navajo's neighbors to the north, the Ute, began raiding in 1773, probably at Spanish urging (ibid.:24). These perturbations ironically turned out to benefit the Navajo as a whole. The forced reliance on livestock and migration to the west, where better pastures were available, provided the Navajo people with a more productive economic base, and they thrived. By the late eighteenth century, Navajo livestock numbers were estimated at almost 8,000, and the intrinsic rate of increase of the Navajo population was about 1.8 percent per annum (see table 8.1).

PASTORAL NAVAJO

During the late 1700s and early 1800s, our cultural image of mobile Navajo families following their herds southwest across the desert became a reality (Bailey 1980:115). The Navajo largely abandoned their intensive agricultural focus while continuing to practice maize horticulture and spread themselves even wider throughout the Southwest, often moving further west to the grasslands and steppes of northern Arizona to take advantage of better forage for their animals. The displaced Navajo eventually came to claim this land, which comprises much of the area of the present-day reservation, and it is known as Diné Bikéeyah, or the "Land of the Navajo" (see figure 8.1). Despite almost constant warfare and raiding by Ute and Spanish settlers, Navajo population numbers and especially livestock numbers increased, and the Navajo became prosperous, expansionary pastoralists (ibid.; Brugge 1983:495). In addition to animals captured during raids, economic events of global significance, such as the Mormon settlement of Utah in 1847 and the California Gold Rush of 1849, provided markets for Navajo mutton that further strengthened the Navajo pastoral economy (Bailey 1980:181). Navajo livestock numbers rose to 670,000 by 1846 (Kelley and Whitely 1989:208).

Despite the economic successes of this period, the near constant warfare of the time came at a demographic cost, with the Navajos' intrinsic rate of population growth falling to 0.6 percent per annum (see table 8.1). The precise reasons for the fall in Navajo numbers are not known. However, high mobility and its possible resultant decrease in women's fertility (Lee 1979; Konner and Worthman 1980; Harpending and Wandsnider 1982; Borgerhoff-Mulder 1994), as well as losses of women and children as a result of slave raids, may have accounted for the demographic downturn. Nonetheless, Navajo economic success continued because of their ability to move across a rugged terrain and because of the generally martial material culture possessed by pastoralists: horses, guns, dogs, movement in small units, and similar factors (Goldschmidt 1979).

Navajo economic and demographic success began to wane in the 1860s, indicating that while mobile pastoralism can be a successful adaptation, it is not without limits. The US takeover in 1846 introduced political and demographic factors that would ultimately spell doom for continued Navajo prosperity. Not only did Americans enter the region with the intent to bring cattle and settle their own growing population, but they also introduced a more effective political system that could be used to martial forces against indigenous peoples. As a result of pressure from Hispanic inhabitants of the new US acquisition, the US Army set up militia units that, in coordination with the Ute, conducted a series of devastating slave and livestock raids against the Navajo (Bailey 1980:212). The relentless pressure exerted by the militias resulted in a reduction of Navajo livestock holdings from 800,000 in

1846 to 500,000 by the 1860s, earning this period the Navajo name *Nahondzod*, or fearing time (ibid.:211; Kelley and Whitely 1989:208). Ironically, one effect of the raiding was to spread the Navajo people even further west to the Grand Canyon and the San Francisco Peaks, extending the territory they would someday claim (Bailey 1980:215). Nahondzod culminated with Kit Carson's military roundup of Navajo in the winter of 1864 and their brutal relocation to Bosque Redondo in southeast New Mexico. Not only did the Navajo lose their livestock, but they actually experienced negative population growth at this time (−0.4 per annum; see table 8.1). This decline resulted from malnutrition, abuse, deaths along the winter march from western to eastern New Mexico in 1864, and Comanche raids on the defenseless Navajo settlements near Bosque Redondo (ibid.:266; Kelley and Whitely 1989:43).

In late 1868, after five years of devastating pestilence, poisoned crops, raids by the Comanche, and the waste of millions of dollars on what was apparently a failed project, the US government returned the Navajo to Diné Bikéeyah, on the border of Arizona and New Mexico (Bailey 1980:262; Bailey and Bailey 1986:26; Kelley and Whitely 1989:44). Beginning with an allotment of 15,000 sheep and goats and adding scattered stock from former years, the Navajo embraced their pastoral adaptation more strongly than ever before and by 1880 once again possessed at least 1 million head of livestock (Bailey and Bailey 1986:73, appendix A; Kelley and Whitely 1989:48, 208). Navajo population numbers rebounded as well, and the intrinsic rate of growth reached 1.6 percent per annum during this period of reconstruction (see table 8.1).

The extremely rapid rebound in the numbers of Navajo livestock and population is a testament to the adaptive flexibility of mobile pastoralism. The Navajo people not only restored their economy but continued to spread throughout their former territory (Bailey and Bailey 1986:80), thereby regaining some of their former political importance.

In 1881, the construction of the Atlantic and Pacific Railroad introduced another influence on the Navajo social environment. The railroads caused towns such as Farmington, Gallup, and Flagstaff to develop and brought traders from the East and West Coasts (Kelley and Whitely 1989:65–66). The towns' increased populations created a demand for mutton, and traders created a demand for Navajo weaving and silversmithing (ibid.:78). These economic demands further strengthened the Navajo economy and provided for its increased expansion while simultaneously more fully bringing the Navajo into the world economic system (Kelley 1986:79; Kelley and Whitely 1989:68).

During this period of demographic, economic, and territorial expansion, the Navajo ironically benefited from their encapsulation within the US government and the world economic system. However, the Navajo were also politically weak, compared to the US government and its constituency of oil barons and large stock owners. Navajo prosperity would continue only to the extent that it was convenient for these larger forces, and Navajo pastoral expansion eventually came to an end. The formation of a Tribal Council in 1923 for the purpose of granting drilling rights to oil companies, while initially providing the Navajo people with revenues and representation in the larger society, also paved the way for the influence of US federal programs on the Navajo people (Kelley and Whitely 1989:71–73).

The successful Navajo pastoral adaptation came to an abrupt end during the 1930s with federally imposed stock reductions. Federal authorities worried that overgrazing would lead to the silting of the Colorado River and therefore threaten the new Boulder (Hoover) Dam, which, in turn, would threaten the electricity supply of growing California cities such as Los Angeles (Aberle 1982:55; Kelley and Whitely 1989:74). As a remedy to a problem that some have argued may not have existed (Aberle 1982; Fanale 1982), federal authorities forcibly slaughtered half of the Navajo livestock holdings and imposed restrictions on the movement of the remaining livestock (Aberle 1982:72; Fanale 1982:218, 226; Bailey and Bailey 1986:190; Kelley and Whitely 1989:109; Kuznar 2008).

The reduction of livestock alone may not have been fatal to the Navajo pastoral system. Sheep and especially goats, if properly managed, can rebound quickly, as they had done for the Navajo in the past. However, the restriction of mobility was fatal to the Navajo system of mobile pastoralism (Fanale 1982; Kuznar 2008, 1999b). The traditional herding system required occasional long-distance movements to alleviate grazing pressure on land. Ironically, the new system designed by federal officials to save the Navajo environment undermined the mobility on which the system depended by forcing livestock to remain in restricted areas of degraded forage. The end result was the destruction of Navajo pastoralism as a viable economic system. Navajo consider the period of the stock reduction one of the darkest times of their history (Roessel and Johnson 1974). Not only did livestock numbers decrease, but Navajo fertility likewise decreased precipitously, from a high of 2.4 percent per annum during the 1920s to a low of 1.1 percent per annum during the stock reduction (see table 8.1).[1]

CONTEMPORARY PASTORALISM

The era of industrialization (Kelley and Whitely 1989) ushered in after the stock reduction spelled the end of a sustainable and economically significant Navajo herding economy. In addition, the population boomed after the stock reduction, having an intrinsic growth rate of 2.4 percent per annum (see table 8.1). Population growth increased despite the erosion of the Navajo economic base because of improvements in health care introduced by the US government during this period (Chisholm 1983:57). Even if herding regulations could be changed and stock prices increased, herding could never sustain the large population of Navajo on their present lands. Nevertheless, herding is still practiced on the reservation today, although very few families derive a significant income from their flocks (Bailey and Bailey 1986:262; Kelley and Whitely 1989:149). Pastoralism remains important to the Navajo for several reasons. In a society where cash is short, unemployment hovers near 50 percent,[2] and people desire reliable transportation and health care, animals serve as cash on the hoof that can be liquidated when financial needs arise (Ruffing 1976; Kuznar 2008, 1999b). Cattle are increasingly important as people travel off-reservation for whatever jobs can be found and cannot be present for the required daily tending of sheep (Downs 1972:88; Kelley and Whitely 1989:150). Cattle do not require daily care and they also represent a larger amount of cash on the hoof compared with sheep, so many Navajo maintain their traditional use rights to rangeland and become cattle ranchers in increasing numbers.

Navajo people maintain livestock for social and cultural reasons as well. Many Navajo continue to utilize the services of *hataatłii*, or medicine men, for healing. Some ceremonies involve large groups of relatives and supporters who must be fed, and herders report to me that livestock are important in the reciprocal exchange of animals for religious purposes. Even among Christianized Evangelical and Pentecostal Navajo with whom I have worked, the exchange of animals to support Christian revivals, known as camp meetings, is important. So, animals have retained their centrality in the social exchanges among Navajo people, despite the animals' loss of economic centrality and the religious conversion of some Navajo.

Also, the Navajo intrinsically value the raising of livestock. Common phrases are "*dibé iiná nilínii át'é*," meaning "Sheep is [*sic*] life," and "*dibé wolgheíi nimá át'é*," meaning "Those called sheep are your mother" (Witherspoon 1973:1442). Navajo feel that raising children as herders instills in them a proper respect for animals, the natural world, and the supernatural and in general serves as a microcosm of life (ibid.; Schoepfle, Burton, and Begishe 1984). Sheep also reflect Navajo personal identity and social standing (Witherspoon 1983:528).

CONCLUSION

The global quest for empire brought the Spanish to the American Southwest 400 years ago, initiating contact between the Apachean people we now know as the Navajo and a world economic system. An agropastoral economic system developed out of contacts among the Navajo, Puebloans, and the Spanish as an unintended consequence of raiding, repression, and conquest. Still, the Navajo were primarily agricultural until a series of severe droughts began in the mid-1700s, making a more fully pastoral lifestyle more feasible than one focused on agriculture. This pastoral life-way withstood all but the most severe environmental and social perturbations for 150 years. The Mexican Revolution of the 1820s and even the devastating influenza epidemic in 1918 did little to alter the success of the Navajo. Also, once the defeated and impoverished Navajo were returned to their lands in 1869, they immediately flourished by virtue of their pastoral system, which was actually aided initially by increased capital intrusion into their land. The perturbations that irrevocably changed the Navajo pastoral system included major social transformations of a global scale, such as the invasion of the US Army in 1846 and the accompanying ethnic cleansing of Diné Bikéeyah, as well as the imposition of the full power of the federal government in reducing livestock and forever limiting stock movements and numbers. Despite these setbacks, Navajo tenaciously hold on to their livestock because animals continue to provide some economic support in an impoverished local economy, as well as to serve cultural ends for the Navajo people (Witherspoon 1973; Schoepfle, Burton, and Begishe 1984).

The implications of the Navajo example for the development of core-periphery, or world-systems, theory are twofold. First, the Navajo case is a clear example of how, with an emergent capitalist system and its attendant demands for raw materials and the conquest of land, the Navajo came into contact with the Spanish and, later, Americans. The Navajo were predictably exploited as slaves and later displaced from lands eventually occupied by ranchers whose production could more efficiently be assimilated into a world economic system (Hall 1989). Finally, despite successful efforts to stave off the onslaught of this world-system, the Navajo were finally fully incorporated into it, and today, multinational corporations draw raw materials (oil, coal, lumber) from the reservation. On the other hand, the Navajo case demonstrates the subtleties of core-periphery interaction. The Navajo were originally a small band of foragers in a region politically dominated by Puebloan peoples (Bailey 1980; Bailey and Bailey 1986; Kelley and Whitely 1989; Perry 1991). As a result of the Spanish introduction of sheep and cattle and the American introduction of profitable markets for pastoral products, the Navajo were able to take advantage of a pastoral lifestyle and persevere in the face of seemingly overwhelming odds. They did this not only by selling off raw materials but also by manufacturing their own value-added products (e.g., blankets and silver work) for sale to the core elite. Such opportunism in using refined products stands in contrast to Wallerstein's (1974) original monolithic thesis of core exploitation of peripheral populations for raw materials.

Finally, Navajo economic history illustrates that environmental and social factors can impact the reproduction of people in an ethnic group and that these people's struggles are often not only to acquire the material and wealth to sustain themselves but to increase their numbers. The Navajo system of matrilineal lineages and clans is a potent social symbol and reality in Navajo society, and reproducing themselves is a sacred blessing and goal traditionally valued by Navajo individuals (Reichard 1950:80; Chisholm 1983:58; Frisbie 1993:377). The tenacity with which the Navajo people apparently capitalized on historical conditions to increase their numbers and the current result of the populous Navajo Nation is a testimonial to the mutual reinforcement of human biology and Navajo cultural values.

NOTES

1. Even in a modern economy, economic downturns are associated with decreased fertility, as happened during the 2008–11 global recession (Sobotka, Skirbekk, and Philipov 2011).

2. Navajo Nation Department of Agriculture statistics note a 48.5 percent unemployment rate in the Navajo Nation in 2012. http://www.agriculture.navajo-nsn.gov/, accessed March 3, 2012.

REFERENCES

Aberle, David. 1982. *The Peyote Religion among the Navajo*, 2nd ed. Norman: University of Oklahoma Press.

Aberle, David. 1983. "Navajo Economic Development." In *Handbook of North American Indians,* vol. 10: *Southwest*, ed. Alfonso Ortiz and William C. Sturtevant, 641–58. Washington, DC: Smithsonian Institution Press.

Adams, William Y. 1963. "Shonto: A Study of the Role of the Trader in a Modern Navaho Community." In *Smithsonian Institution Bureau of American Ethnology Bulletin* 188. Washington, DC: Smithsonian Institution.

Bailey, Garrick, and Roberta Glenn Bailey. 1986. *A History of the Navajos*. Santa Fe, NM: School of American Research Press.

Bailey, Lynn. 1980. *If You Take My Sheep: The Evolution and Conflicts of Navajo Pastoralism, 1630–1868*. Pasadena: Westernlore.

Blanton, Richard, and Gary Feinman. 1984. "The Mesoamerican World System." *American Anthropologist* 86 (3): 673–82. http://dx.doi.org/10.1525/aa.1984.86.3.02a00100.

Borgerhoff-Mulder, Monique. 1994. "Reproductive Decisions." In *Evolutionary Ecology and Human Behavior*, ed. Eric Alden Smith and Bruce Winterhalder, 339–73. New York: Aldine de Gruyter.

Brugge, David. 1983. "Navajo Prehistory and History to 1850." In *Handbook of North American Indians,* vol. 10: *Southwest*, ed. Alfonso Ortiz and William C. Sturtevant, 489–501. Washington, DC: Smithsonian Institution Press.

Buss, David. 1994. *Evolution of Desire: Strategies of Human Mating*. New York: Basic Books.

Chagnon, Napoleon. 1992. *Yanomamö*, 4th ed. New York: Holt, Rinehart and Winston.

Chase-Dunn, Christopher, and Thomas D. Hall. 1991. "Conceptualizing Core/Periphery Hierarchies for Comparative Study." In *Core/Periphery Relations in Precapitalist Worlds*, ed. Christopher Chase-Dunn and Thomas D. Hall, 5–44. Boulder: Westview.

Chase-Dunn, Christopher, and Thomas D. Hall. 1995. "Cross–World-System Comparisons: Similarities and Differences." In *Civilizations and World Systems: Studying World-Historical Change*, ed. Stephen K. Sanderson, 109–35. Walnut Creek, CA: AltaMira.

Chisholm, James S. 1983. *Navajo Infancy: An Ethological Study of Child Development*. New York: Alsine.

Cronk, Lee. 1991. "Human Behavioral Ecology." *Annual Review of Anthropology* 20 (1): 25–53. http://dx.doi.org/10.1146/annurev.an.20.100191.000325.

Cronk, Lee. 1993. "Parental Favoritism toward Daughters." *American Scientist* 81: 272–79.

Downs, James F. 1972. *The Navajo*. New York: Holt, Rinehart and Winston.

Durham, William H. 1991. *Coevolution: Genes, Culture and Human Diversity*. Stanford, CA: Stanford University Press.

Fanale, Rosalie. 1982. "Navajo Land and Land Management: A Century of Change." PhD diss., Department of Anthropology, Catholic University, Washington, DC.

Farella, John R. 1984. *The Main Stalk: A Synthesis of Navajo Philosophy*. Tucson: University of Arizona Press.

Frisbie, Charlotte J. 1993. *Kinaaldá: A Study of the Navaho Girl's Puberty Ceremony*. Salt Lake City: University of Utah Press.

Goldschmidt, Walter. 1979. "A General Model for Pastoral Social Systems." In *Pastoral Production and Society: Proceedings of the International Meeting on Nomadic Pastoralism*, 15–27. Paris, December 1–3, 1976. Cambridge: Cambridge University Press.

Hall, Thomas D. 1989. *Social Change in the Southwest, 1350–1880*. Lawrence: University Press of Kansas.

Hall, Thomas D., P. Nick Kardulias, and Christopher Chase-Dunn. 2011. "World-Systems Analysis and Archaeology: Continuing the Dialogue." *Journal of Archaeological Research* 19 (3): 233–79. http://dx.doi.org/10.1007/s10814-010-9047-5.

Hames, Raymond. 1990. "Sharing among the Yanomamö: Part I, The Effects of Risk." In *Risk and Uncertainty in Tribal and Peasant Economies*, ed. Elizabeth Cashdan, 89–106. Boulder: Westview.

Harpending, Henry, Patricia Draper, and Alan Rogers. 1987. "Human Sociobiology." *Yearbook of Physical Anthropology* 30 (S8): 127–50. http://dx.doi.org/10.1002/ajpa.1330300509.

Harpending, Henry, and Lu Ann Wandsnider. 1982. "Population Structures of Ghanzi and Ngamiland! Kung." In *Current Developments in Anthropological Genetics*, ed. Michael H. Crawford and James H. Mielke, 29–50. New York: Plenum. http://dx.doi.org/10.1007/978-1-4615-6769-1_2.

Hawkes, Kristen. 1993. "Why Hunter-Gatherers Work: An Ancient Version of the Problem of Public Goods." *Current Anthropology* 34 (4): 341–61. http://dx.doi.org/10.1086/204182.

Ingold, Tim. 1980. *Hunters, Pastoralists and Ranchers*. Cambridge: Cambridge University Press. http://dx.doi.org/10.1017/CBO9780511558047.

Irons, William. 1979. "Cultural and Biological Success." In *Evolutionary Biology and Human Social Behavior*, ed. William Irons and Napoleon A. Chagnon, 257–72. North Scituate, MA: Duxbury.

Kardulias, P. Nick. 1999. "Multiple Levels in the Aegean Bronze Age World-System." In *World-Systems Theory in Practice: Leadership, Production, and Exchange*, ed. P. Nick Kardulias, 179–201. Lanham, MD: Rowman and Littlefield.

Kelley, Klara B. 1986. *Navajo Land Use: An Ethnoarchaeological Study*. New York: Academic.

Kelley, Klara B., and Peter M. Whitely. 1989. *Navajoland: Family Settlement and Land Use*. Tsaile, AZ: Navajo Community College.

Kluckhohn, Clyde, and Dorothea Leighton. 1962. *The Navajo*. Garden City, NY: Anchor Books and Doubleday.

Konner, Melvin, and Carol Worthman. 1980. "Nursing Frequency, Gonadal Function, and Birth Spacing among Kung Hunter-Gatherers." *Science* 207 (4432): 788–91. http://dx.doi.org/10.1126/science.7352291.

Kuznar, Lawrence A. 1999a. "The Inca Empire: Detailing the Complexities of Core/Periphery Interaction." In *World-Systems Theory in Practice: Leadership, Production, and Exchange*, ed. P. Nick Kardulias, 223–40. Lanham, MD: Rowman and Littlefield.

Kuznar, Lawrence A. 1999b. "Traditional Pastoralism and Development: A Comparison of Aymara and Navajo Grazing Ecology." In *Ethnoecology: Knowledge, Resources and Rights*, ed. Ted L. Gragson and Ben Blount, 74–89. Athens: University of Georgia Press.

Kuznar, Lawrence A. 2007. "Rationality Wars and the War on Terror: Explaining Terrorism and Social Unrest." *American Anthropologist* 109 (2): 318–29. http://dx.doi.org/10.1525/aa.2007.109.2.318.

Kuznar, Lawrence A. 2008. *Reclaiming a Scientific Anthropology*, 2nd ed. Walnut Grove, CA: AltaMira.

Lee, Richard B. 1979. *The !Kung San*. Chicago: University of Chicago Press.

McNeley, James K. 1981. *Holy Wind in Navajo Philosophy*. Tucson: University of Arizona Press.

Perry, Richard J. 1991. *Western Apache Heritage: People of the Mountain Corridor*. Austin: University of Texas Press.

Reichard, Gladys. 1950. *Navajo Religion*. Princeton: Princeton University Press.

Roessel, Ruth, and Broderick Johnson, comps. 1974. *Navajo Livestock Reduction: A National Disgrace*. Chinle, AZ: Navajo Community College Press.

Rogers, Larry, ed. 1993a. *1990 Census: Population and Housing Characteristics of the Navajo Nation*. Window Rock, AZ: Division of Community Development, Navajo Nation.

Rogers, Larry, ed. 1993b. *Chapter Images: 1992 Edition General Facts on Navajo Chapters*. Window Rock, AZ: Division of Community Development, Navajo Nation.

Ruffing, Lorraine T. 1976. "Navajo Economic Development Subject to Cultural Constraints." *Economic Development and Cultural Change* 24 (3): 611–21. http://dx.doi.org/10.1086/450900.

Sanderson, Stephen K., ed. 1995. *Civilizations and World Systems: Studying World-Historical Change*. Walnut Creek, CA: AltaMira.

Schoepfle, Mark, Michael Burton, and Kenneth Begishe. 1984. "Navajo Attitudes toward Development and Change: A Unified Ethnographic and Survey Approach to an Understanding of Their Future." *American Anthropologist* 86 (4): 885–904. http://dx.doi.org/10.1525/aa.1984.86.4.02a00040.

Smith, Eric Alden, and Bruce Winterhalder, eds. 1994. *Evolutionary Ecology and Human Behavior*. New York: Aldine de Gruyter.

Sobotka, Tomáš, Vegard Skirbekk, and Dimiter Philipov. 2011. "Economic Recession and Fertility in the Developed World." *Population and Development Review* 37 (2): 267–306. http://dx.doi.org/10.1111/j.1728-4457.2011.00411.x.

Stein, Gil J. 1999. "Rethinking World-Systems: Power, Distance, and Diasporas in the Dynamics of Interregional Interaction." In *World-Systems Theory in Practice: Leadership, Production, and Exchange*, ed. P. Nick Kardulias, 153–77. Lanham, MD: Rowman and Littlefield.

United States Census Bureau. 2010. *The American Indian and Alaska Native Population 2010*. Washington, DC: United States Department of Commerce, Economics and Statistics Administration. http://www.census.gov/prod/cen2010/briefs/c2010br-10.pdf.

Wallerstein, Immanuel. 1974. *The Modern World-System: Capitalist Agriculture and the Origins of the European World-Economy in the Sixteenth Century*. New York: Academic.

Wells, Peter. 1999. "Production within and beyond Imperial Boundaries: Goods, Exchange, and Power in Roman Europe." In *World-Systems Theory in Practice: Leadership, Production, and Exchange*, ed. P. Nick Kardulias, 85–101. Lanham, MD: Rowman and Littlefield.

Winterhalder, Bruce, and Eric Alden Smith. 2000. "Analyzing Adaptive Strategies: Human Behavioral Ecology at Twenty-Five." *Evolutionary Anthropology* 9 (2): 51–72. http://dx.doi.org/10.1002/(SICI)1520-6505(2000)9:2<51::AID-EVAN1>3.0.CO;2-7.

Witherspoon, Gary. 1973. "Sheep in Navajo Culture and Social Organization." *American Anthropologist* 75 (5): 1441–47. http://dx.doi.org/10.1525/aa.1973.75.5.02a00150.

Witherspoon, Gary. 1983. "Navajo Social Organization." In *Handbook of North American Indians,* vol. 10: *Southwest*, ed. Alfonso Ortiz, 524–35. Washington, DC: Smithsonian Institution Press.

Wolf, Eric. 1982. *Europe and the People without History*. Berkley: University of California Press.

9

Accidental Dairy Farmers

Social Transformations in a Rural Irish Parish

Mark T. Shutes

The basic theme of this book is to find some common ground by which to compare societies wherein some form of animal husbandry plays a crucial role in their production strategies. As an anthropologist who works in a rural southwestern Irish parish (see figure 9.1), where some combination of cattle and cows has played such a role for thousands of years, I have a clear interest in finding a broader comparative base from which to understand their behavior and history. Yet, in my initial review of the basic literature on what is commonly referred to as pastoralist groups, I began to realize that we anthropologists seem to have shot ourselves in the theoretical foot not once but multiple times in investigating this particular topic. So, I come limping to my more experienced and knowledgeable colleagues with multiple questions and some extremely modest suggestions about how we might make this area of study more inclusive and useful.

To begin, there seem to be two basic and interactive problems with our examination of groups dependent on animal husbandry: (1) an overemphasis on strictly ecological parameters as the delimiters of the field of comparative study, which has led to the development of confusing and overlapping categorical types ranging from pure nomadism to ranching and to a fruitless debate over which kinds of groups should or should not be included in the various categories (see Barfield 1993:3–9); and (2) the naive assumption that animal husbandry groups that experience economic transformations that entail more sedentary, market-oriented "agricultural" strategies will automatically undergo social transformations as well and eventually

DOI: 10.5876/9781607323433.c009

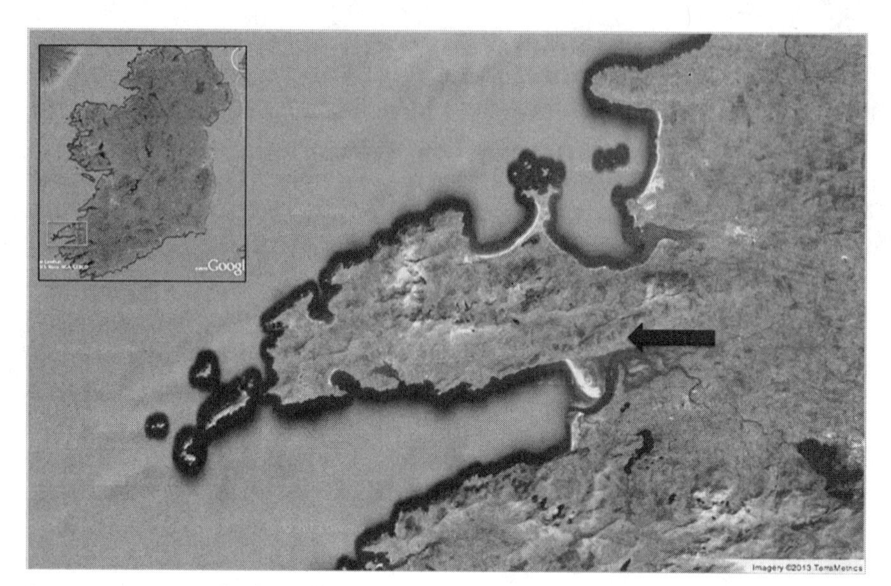

FIGURE 9.1. Map of Ireland showing the location of County Kerry.

become culturally indistinguishable from groups that have always pursued sedentary farming strategies (Barfield [ibid.:210–14] and Salzman [2004:137–55] provide more nuanced views). Generally, this latter assumption is also accompanied by some implicit or explicit value judgment on the part of the researcher that such transformations are inherently negative, since they are imposed upon the groups by more powerful outside forces and ultimately result in economic exploitation and the loss of freedom, mobility, and choice for the members of those groups.

With regard to the first problem, and keeping in mind our goal of inclusion rather than exclusion, it seems to me that precisely the wrong questions are being asked. Rather than emphasizing the ecological differences of such groups, should we not be inquiring about their social similarities? Perhaps a better question might be: What kind of system of social relationships would be required of all such groups that would give them the ability to accommodate the wide variety of conditions that can exist in the maintenance of herding animals, including multiple economic goals, ranging from pure subsistence orientation to pure market orientation; multiple mobility goals, ranging from transhumant to sedentary; and multiple production goals, ranging from total dependence on livestock maintenance to some combination of livestock with other productive elements? The system of social relationships I have in mind would have to be highly fluid, capable of emphasizing the central importance of those relationships within the minimal productive unit and yet able to telescope outward, whenever necessary, and embrace any necessary number of other such units. This "telescoping flexibility" is the key element, and it can be and has been accomplished by animal husbandry groups in a number of creative ways: agnatic segmentary lineages in Africa (Evans-Pritchard 1940), the matrilineal clan structure of the Navajo, the cognatic clans of the ancient Celts, the labor-dependent relations between neighbors through *cooring* and *meithal* groups of the Irish, and the use of immediate family, distant kin, and hired labor among Greeks (Campbell 1964; Chang 1997). Certainly, in re-framing the question in this manner, not only do we broaden the base of our empirical inquiries into the characteristics of human groups that depend on herding animals, but we also bring the nature of our investigations more in line with the historical realities that have confronted such groups, for I agree with Salzman (1972) and others that the vast majority of them have always had to contend with these kinds of highly variable, multiple goals.

The second problem points to another serious flaw in our theories concerning change among such groups. Specifically, we have tended to view them as having no active role in managing the changes they have encountered; that is, to merely be the passive recipients of externally generated change processes. This theoretical assumption has produced, therefore, an artificial dichotomy of possibilities with respect to the outcome of change: (1) either these producers totally accept the changes from the outside and thus initiate the inevitable process of transformational value changes, or (2) their traditional values make it impossible for them to manage such changes and they gradually wither away, or are driven, to extinction. Certainly, a significant part of the literature I have thus far encountered documents disruptive economic and political transformations taking place among such groups. But if we view them as active participants in the acceptance or rejection of such changes, as negotiators of their own lives, if you will, trying to retain that which is most significant to them while at the same time struggling to accommodate that which cannot be retained, how does that change our view of the number of possible outcomes? Clearly, they are multiple rather than simply dichotomous. It should be possible to imagine such groups initiating major economic and social changes in some parts of their production strategy while at the same time retaining older elements in other parts. And haven't such groups always had to confront such challenges, and haven't they exhibited precisely this kind of variability and flexibility? It is even possible to imagine that such groups could retain a significant part of their values even if they no longer kept any livestock. The system of fluid social relationships is a product of their initial herding strategies, and these crucial relationships must certainly be strongly value-bound to the animals they keep and to the lifestyle that keeping them demands, but I suggest that it is the fluidity and adaptability of their social relationships that is the key to their survival rather than the raw fact that they may no longer depend on herding animals.

To trace the development of the issues outlined here, first I review the historical literature on the earlier patterns of transhumance in Western Irish culture and identify the pattern of social flexibility as inherent in two specific areas of social relationships: *cooring*, a pattern of limited sharing of resources between two close neighbors, and the *meithal*, a broader-based labor coalition of close neighbors. Subsequently, using data from my own field site in County Kerry (see figure 9.1), I then show that these same patterns of social relationship endured well beyond the period of transhumant cattle rearing and became the basis for later twentieth-century integration into a market economy based on dairying. Finally, I conclude by suggesting that it is not ecology/economics alone that are the crucial variables in understanding animal husbandry strategies but rather the ways in which these two elements interface in those strategies to produce fluid and flexible social relationships. It is this flexible social strategy that is at the heart of an animal husbandry strategy and the key to a broader-based comparison of societies that depend on such strategies for their maintenance.

THE EARLY MODERN PRODUCTION STRATEGY, 1850–1959

A number of elements of the local production strategy during this period are relevant to our present discussion.

1. Overall, the strategy can be referred to as "mixed-farming," since it involved elements of both grassland (cows, cattle, and sheep) and tillage (oats and other cereals, potatoes, other root crops, and cabbage) enterprises. A farmer at any level had to maintain a mix of these enterprises, and the technology and knowledge base to maintain them, in order to survive.

2. The strategy involved a mixture of both cash and subsistence production. Despite numerous reciprocal arrangements with neighbors and tradesmen, cash was a vital part of the farm enterprise. It was needed for the payment of rent to landlords and later, after the land transfer acts of the early twentieth century, for farm buyout payments. Maintenance of farm equipment, local services, dowry and emigration expenses, and payments to hired labor constituted the other areas of the cash economy.

The major source of necessary cash came from the sale of livestock, typically at one or more of the timely local fairs. The vast majority of the livestock sold were yearling cattle, or "store cattle," which were bred predominantly from the milking stock in the spring and kept for one winter. The major buyers of this stock were the owners of large farms in the Midlands and eastern Ireland. They would buy these underweight animals at the cheapest possible price and then fatten them, or "finish them off," on their own rich lands for final sale as beef to the British market. They had the lucrative end of the market, for few farms in the west of Ireland had the resources to fatten cattle directly for final sale. In addition, local farmers had to compete with numerous other localities throughout the west of Ireland for this market, and it was a rare and lucky day when the big buyers couldn't find enough of the "store cattle" they needed.

The sale of sheep and wool was important to a number of parishes to the west and east, where soil quality was dramatically poorer than that found in Kilcastle. Local farmers never got involved with this market to any great degree. Only a few mountain farmers whose farms were predominantly rough grazing land ever relied on sheep to any significant extent, and their production was relatively small. Additional smaller sources of cash might come from sales of grain surpluses, if any, and the sale of eggs, poultry, and pigs.

The sale of milk to the local creamery was also an important source of cash income, but it varied according to farm size. Milking cows ate approximately one acre of good-quality grazing land per cow per year, and the production of milk was dependent entirely on hand labor. In addition, milking cows had to be grazed near the milking parlor to facilitate the twice-a-day milking process. A large number of milking cows would therefore severely limit the amount of good tillage land near the main farm and potentially threaten the crucial subsistence production areas on even the largest local farms. Finally, during this period the price of milk at the creamery remained between one and two pence a gallon for decades and only exceeded seven pence a gallon after World War II.

Much of the information on milk prices prior to 1925 comes from local sources, since official agricultural statistics were not kept until that year. But a brief examination of table 9.1 will partially support these claims. If, for example, we take the average number of cows kept in the local parish between 1925 and 1940 and divide it by the 66 farmers active during that period, it amounts to only 7 cows per farmer. Certainly, the smallest farms (under 30 acres) could not afford to keep anywhere near that average and still have sufficient good land available for tillage. Local informants report that the largest herds in the parish probably numbered somewhere between 10 and 15 milking cows during most of this period. That number is consistent with the above findings and would certainly justify sale of milk to the creamery. Dairying was therefore probably a major source of cash only for the larger farms.

As might be expected, the cash economy was intricately interwoven with the subsistence economy. With the exception of the hay and oat crops, which were used

Table 9.1 Price of milk (pence per gallon) and number of cows in Kilcastle Parish, 1925–70. Source: Kerry Co-operative Creameries Ltd.

Years	*Pence/Gallon*	*Number of Cows*
1925	2.00	390
1930	4.00	434
1935	4.00	577
1940	7.00	677
1945	11.17	693
1950	15.12	702
1955	17.98	636
1960	18.97	657
1965	21.97	768
1970	23.94	826

exclusively to feed cows, cattle, sheep, and farm horses, all of the other gardening and tillage enterprises were needed to sustain human as well as animal life. This is particularly true of potatoes and other root crops. Poultry and pigs were also fed from these garden products and were principally designed for on-farm consumption, although periodic surpluses could be transferred to the cash sphere.

This kind of interdependency gave local farmers the ability to move in and out of the cash economy sphere without threatening the survival of the farm itself. Understanding and predicting fluctuating market prices, as we have seen, posed a major problem for the majority of farmers, who relied almost exclusively on the experience of their few larger farm neighbors in these transactions. Consider this example:

> A "strong" farmer was usually a good judge of cattle, and he would usually be called upon by the small farmer at a fair to help the man judge a fair price for animals . . . The big farmer would go to every fair whether he was trading or not, and so he had more experience and was not as tied down as the small man, who could not afford to go to every fair because he was tied down to the daily work that the big farmer had hired men to do. The "strong" farmer would also know the men that would be buying the cattle, and he would be more likely to know what the man would settle for. Also, the small farmer wouldn't be the type that would muscle his way around the fair, like. He'd only go if he had something to sell, and then he'd stay out of the way, like. The small fellow would follow the line of the big man in breeding, buying stock, and the like, for the big man would be the first to try some new type of feed or the like. (Farmer #17; from field notes, Winter 1976)

Even this experience, while certainly much greater than that of the majority of farmers, was limited to the short term, since it was based on the frequency with which these farmers visited the fairs. While their larger-volume sales may have given them more influence with outside buyers, the owners of larger farms had no more long-term control over the pricing structure than the others.

3. Much of the work described here was accomplished with nothing more than horse-drawn equipment or hand tools, so the entire strategy was heavily labor-intensive. This is particularly true regarding the production of potatoes and root crops, crucial for human and animal subsistence, as well as "turf" production. No farm, regardless of size, could accomplish all of the tasks necessary for its survival without access to some form of cheap outside labor. Indeed, the larger the farm, the greater the need for such labor, and the higher the potential risks if such labor were not forthcoming.

Managing access to this labor was the single most critical aspect of the farming enterprise during this period. As with many other factors at this time, the problems of labor management differed significantly with the size of the farm. Let us examine a series of statements about labor relationships between "big," or "strong," farmers and "small" farmers drawn from interviews with older farmers in 1971:

> A small farmer would usually have no major equipment of his own. He might have one horse, and so he would coor with another small farmer, and they would do the two farms with one horse coming from each house. The rest would be just like the big farms, on a scaled-down basis, like; he'd have a garden, mangels, turnips, and all the rest . . . A lot of the bigger farms would not be wealthy at all because they would be more likely to get into big trouble with labor and other things. What they really had that made them big was that they were "strong" in the community. A big farmer was admired and talked about by the small ones because he had a lot, and he might give you the loan of his machinery. (Farmer #32; from field notes, Summer 1971)

> The choice was based on nearness, for the small farmer would need the loan of the big farmer's equipment, and the big farmer would have the advantage of the small farmer's labor. The big farmer would also own a bull, and the small farmer would need its services, as they could not afford to keep one themselves. The small farmer would not pay in cash for the services of the bull but would usually come in [give labor] for the sowing of the potatoes, or he would help with the hay or the corn [wheat, barley, oats]. The small farmer probably paid his debt three times over with all this labor, but it was the handiest thing to do because it meant that he didn't have to keep and feed a bull. Even if he did keep one, it probably couldn't be of the fine

> quality that the big farmer could [afford], so the small farmer would get better cattle and more milk. But the small farmer was in a bad position after he was granted the bull's services because in order to keep up the quality of his stock, he had to be available when the big farmer wanted him. So it worked one way in terms of labor . . . the small farmer might be called upon to help the big farmer cut turf, but the big farmer would not help the small [one] to bring in his. The small farmer didn't have an awful [lot of] choice. He had to have some arrangements with the big farmer, so he didn't count the cost to himself. (Farmer #26; from field notes, Summer 1971)

> Some big farmers would have a different system altogether. To save them from the obligations to small farmers or to any farmer, they would never lease the services of their bull or let out equipment. If they needed an extra hand, they would go out and pay cash for the labor. But few around here could afford that. The usual way [for a big farmer] was to hire one man on a cash basis for the season but not to go out and hire additional labor for certain times because he [the big farmer] had these small farmers to call upon for these times. All the time the small farmer would be watching the big farmer, so that if it looked as if the big fellow was too hard on him, he [the small farmer] would be looking for ways to get out from under [the big farmer's thumb]. At the same time the big farmer would be looking at the small one in terms of the labor he was getting for his favors, and if this wasn't satisfactory, he'd [the big farmer] be looking for ways out. They all kept each other on target. (Farmer #49; from field notes, Summer 1971)

As can be seen from these statements, farmers at all size levels practiced exactly the same mix of farming strategies, but their labor strategies varied by size. A big farmer earned sufficient income to hire at least one male farm laborer for the season (and usually a female farm servant to assist the big farmer's spouse) or to maintain a permanent laboring family in a cottage on his land. The rest of his labor force came from fellow members of his meitheal and from small farmers, who gave labor at crucial times in return for services and equipment loans. Obviously, the larger the farm at this level, the greater the number of options open to the big farmer, but there were no farms in the parish that could afford to hire all the labor they needed. They had to rely on reciprocal arrangements with other members of the parish to accommodate at least some of their greater labor needs.

A smaller farmer earned insufficient income to pay for outside labor of any sort. Nor could he afford to own much of the equipment or services on which his scaled-down farm enterprise depended. He could coor with another neighbor to get the other horse necessary for tillage work, and he could depend on the members of his meitheal for the big cooperative labor projects. But he had the additional burden of having to plan how to divide his labor time between his own enterprise and that of a big farmer. This was particularly difficult, since the outside demands on his labor occurred at precisely the same time as the heaviest labor demands on his own farm.

4. In a very real sense, the survival of the smaller farms depended on the economic success and stability of the "strong" farms. These farms, in turn, depended on the ready availability of cheap hired labor to carry out a significant part of their production strategy. The labor surplus created by the change to single inheritance after the great famines provided that availability. However, what began as a relatively large surplus continually declined as more and more young non-inheritors acquired the means to emigrate. The result was a steadily aging farm population with a decreasing birth rate, which created a decreasing pool of available surplus labor. Locally, this process reached a climax in 1955 with the closing of the labor exchange in Killorglin.

SOCIAL FACTORS AFFECTING PRODUCTION, 1850–1950

The nature and extent of social interaction between farm households depended largely on farm size. Farmers with larger holdings, the "big" or "strong" farmers, had much greater freedom of choice in their establishment of neighborly relations, and although they did depend heavily on the labor of small farmers, they clearly had the upper hand in such relationships. They moved more freely within the parish and had knowledge of, and regular contacts with, the larger world outside. In the words of one older small farmer: "The small and big farmer could be fierce friends, with the small man keeping the big informed about what people said about him and what his workmen were up to. It was all on a first-name basis, like, but the big farmer would confine his chats to just a few small farmers, and then he wouldn't have that much time for the others . . . this way he could avoid having to give out favors to too many small farmers" (Farmer #53; from field notes, Summer 1971).

In contrast, farmers with smaller holdings had little or no such freedom of choice with regard to neighborly interactions. They were just as dependent on their neighbors with small farms as they were on those with larger ones, for both provided them with services that were vital to their survival. In addition, small farmers were in direct competition with each other for the valuable favors of the larger farmers.

So, they moved more cautiously within the parish, being careful to make no public mistakes in their interactions with any of their fellow farmers so they wouldn't be "caught out," the phenomenon of publicly revealing your feelings about your neighbors or their actions. The labor on their farms was unrelenting, and so small farmers' knowledge of and contact with the world outside the parish was minimal. Their behavior during times when they were outside the parish was simply a greater magnification of their public behavior inside the parish, which made them cautious to the point of becoming almost invisible.

These obvious differences in status and public behavior were balanced, however, by the great common need for labor. The large farmer would fall faster and harder if he had no "free" labor to depend on during highly intensive work periods. The work of the small farmer would be almost impossible without the loan of labor, equipment, or a horse. As long as the two groups were mutually dependent on each other for labor, the status differences remained blurred. Nowhere was this leveling of status through mutual labor dependency more evident than in the neighbor-oriented social and labor subgroups known as meitheals. Here are a few comments about meitheals from some local farmers:

> [Speaking about the meitheal during the time the thresher was in the parish. The names have been replaced by numbers.] In ours there would be Farmer #1, Farmer #2, Farmer #4, Farmer #5, Farmer #6, Farmer #7, and ourselves. It was based on neighbors and distance, you see, so when the thresher got to Farmer #5's, Farmers #1 and 2 would start to drop out, and Farmer #5's neighbors on the other side would start to come in. It just found its own level, unless it was known that someone outside of your normal range was shorthanded, but in that case they would be setting it up in advance and making arrangements as it moved along to their place . . . Once it got started, all the men would be doing each other's fields and there would be no room for another thresher to come in. (Farmer #3; from field notes, Summer 1971)

> Football [Irish] was very important to the young lads when the days got long again. It was usually made up of the ones from the same meitheal group, who would challenge each other. They also formed a social group, as those same team members would go to the fairs together . . . they were a strong faction agent. It [the meitheal] was the most important group to those lads, closer than the relatives. They would all be at church on Sunday, outside and in. They grew up together. (Farmer #61; from field notes, Summer 1971)

> The meitheal members all knew each other's habits perfectly. They all knew the fellow that wouldn't go to Mass, and they all knew the fellow that might be after the local girls, and they all sort of kept an eye on each other. It ended up so that rarely could a fellow go wrong because even if he was given to bad habits, he would end up getting a hammering from his own gang. So everybody was kept under control. (Farmer #10; from field notes, Summer 1971)

Beginning with early childhood, therefore, the meitheal was a mechanism for enforcing egalitarian relations between neighbors, regardless of size of farm. So the delicate balance was maintained between status differences and the universal need for labor in a situation that could otherwise have led to more serious exploitation of the small by the large.

These fundamental relationships between large and small farmers also permeated every other type of interaction within the parish. Local shopkeepers could ill afford to deny credit to the larger farmers, since their small profit margin depended on higher-volume purchases. They were very careful to cater to the larger farmers, in terms of both physical goods and overt recognition of their higher status. The small farmers would never encounter such obsequiousness in their dealings with shopkeepers (quite the opposite in some cases).

For their part, the larger farmers could not afford to trample on the shopkeepers' status, for they were far more dependent on the credit than were their smaller neighbors, and an angry shopkeeper could more easily ruin a large farmer than he could a small one. By the same token, shopkeepers could not overplay their status with smaller farmers, for they usually paid in cash and the shopkeepers depended on this trade to meet their short-term monetary needs. The net result was, again, a delicate balance between status and dependency accomplished through public behavior that was fastidiously egalitarian in nature.

Similar behavioral rules also existed for the landless permanent farm laborer. Consider this:

> The origin of laborers' cottages on the farms started when a farm laborer, who always came back to the same farm to work, wanted to get married. And so he let a piece of land from the farmer and would get the farmer's help in building a house on it. This usually worked out fine, but sometimes the son of the laborer would not take up the trade of the father, and so the cottage would be wasting the land of the farmer with none in the house supplying labor to him. After the cottage was built, the farmer had to be careful in the way he treated the laborer, and so it all depended on the way he sized up the man in the first place. And the same was true with the laborer, for it was the only way he could stay and be married and make a life for himself. (Farmer #70; from field notes, Summer 1971)

Here again we have an example of the balance that needed to be struck between labor dependency and status, where a landless laborer had leverage in his interactions with those of clearly higher status, provided he managed his public behavior carefully, and where a larger farmer could exercise extreme influence on the work habits of his laborer and guarantee his labor needs for the future, provided he treated the worker in an overtly fair and equitable manner.

Such dependencies produced a lifestyle within the parish that was both overtly egalitarian yet highly regulated. The production strategies of the larger farms were meticulously copied by the smaller farms to maintain the balance of favors and labor. Perhaps most important, this social system also regulated access to the cash economy. Although not "economic barons" by any means, the "strong" farmers acted as the conduit to the outside market for the other farms and as a buffer against the vagaries of world capital. Because of their own labor dependencies, however, "strong" farmers could be expected to act in an extremely conservative fashion with regard to capital and not use their position to significantly exploit their smaller neighbors in unexpected ways. In sum, to borrow the words of one farmer quoted earlier, "they all kept each other on target."

RELATIONSHIPS BETWEEN PARISH FARMS

By the early 1970s, as a result of nearly two decades of increases in milk prices and the virtual disappearance of farm laborers through emigration, most of the larger farms in the parish had made the switch to some form of specialized dairying, and the social relationships attached to the older patterns of mixed farming were in a state of flux. With this change came the virtual elimination of the meitheal and the asymmetrical labor exchanges, since these were based primarily on the labor needs of the large farm household when they were pursuing mixed-farming. Occasions for interaction between farms of any size were greatly diminished with the elimination of their mutual labor dependencies, and, as the small farms moved more into dairying and cattle, even the practice of cooring was rapidly disappearing. The great social occasions of local village fairs were now preempted by auction marts.

During my first fieldwork in the parish in 1971, I would regularly hear private complaints from farmers with small holdings that "the big men haven't even the time of day for us now" and from farmers with larger holdings that "the small fellow is always looking for something from you." There was also a growing sense of isolation among parish farmers. Consider this example of a lament commonly expressed to me by parish farmers during this time: "I haven't been in that fellah's house for nearly three years. Oh, we see each other at Mass and in the village all the time, but we don't seem to have time for the odd visit, what with the milking and all. When we were young, I spent as much time in his father's kitchen as in my own. His grandmother practically reared the both of us" (from field notes, July 1971).

Clearly, it is not the frequency of the interaction between farm households that is changing but the nature and quality of those interactions. The labor-dependent system of the early modern period offered numerous opportunities for spontaneous, informal, and highly enjoyable social interactions, often involving music and dance and storytelling (what locals call the "crack," as in "the crack was mighty, boy, you should have been there," or "ah, it was great crack altogether"), and it was the lack of these opportunities in the emerging labor-independent system that was at the source of the forms of lament. The social rules governing relationships between farm households were being significantly altered, and the expected social rewards for appropriate behavior were no longer assured.

REFERENCES

Barfield, Thomas J. 1993. *The Nomadic Alternative*. Englewood Cliffs, NJ: Prentice-Hall.

Campbell, John K. 1964. *Honour, Family, and Patronage: A Study of Institutions and Moral Values in a Greek Mountain Village*. Oxford: Clarendon.

Chang, Claudia. 1997. "Greek Sheep, Albanian Shepherds: Hidden Economies in the European Community." In *Aegean Strategies: Studies of Culture and Environment on the European Fringe*, ed. P. Nick Kardulias and Mark T. Shutes, 123–39. Lanham, MD: Rowman and Littlefield.

Evans-Pritchard, Edward E. 1940. *The Nuer, a Description of the Modes of Livelihood and Political Institutions of a Nilotic People*. Oxford: Clarendon.

Salzman, Philip C. 1972. "Adaptation and Change among the Yarahmadzai Baluch." PhD diss., Department of Anthropology, University of Chicago, IL.

Salzman, Philip C. 2004. *Pastoralists: Equality, Hierarchy, and the State*. Boulder: Westview.

10

Real Milk from Mechanical Cows

Adaptations among Irish Dairy Cattle Farmers

Mark T. Shutes

One day in the summer of 1986, a local farmer from a small parish in southwestern Ireland found himself the subject of a front-page story in the *Irish Times* and a sixty-second news spot on Irish national television. He and a few close friends had constructed from scratch a strikingly realistic life-size model of a Frisian/Holstein cow, complete with nodding head, switching tail, sound effects, and—most important—a mechanical udder capable of delivering stirred and warmed cow's milk to real calves through a set of rubber teats. The farmer brought this model to a regional agricultural fair in County Cork in a last-ditch effort to market the calf-feeding device, which was his own invention.

It was the novelty of the event that made the news that day: a mechanical cow that gave real milk. But the circumstances that led both to the invention itself and to the unusual marketing strategy are the subject of this present inquiry, for they offer an excellent opportunity to examine the ways political-economic, ideological, and linguistic analyses in anthropology are interdependent, in the complementary fashion suggested by Friedrich (1989), so as to provide a more comprehensive understanding of the processes of rural community change.

Friedrich (ibid.) identifies two established approaches to understanding human behavior: the analytic-scientific and the emotional-ethical. Although there is some overlap in these approaches, the basic structure of each can be readily identified. He defines the analytic-scientific approach as follows: "It is rational, intellectual, and cognitive, and focuses upon constructing rigorous scientific models and empirical,

DOI: 10.5876/9781607323433.c010

operational, or at least the insight-yielding methods to go with them. It is rooted in analysis and, ultimately, the drive to know" (ibid.:295).

The emotional-ethical approach, in contrast, is "more concerned with the emotions, motivation, and issues of right and wrong—often with exploitation and oppression, the dominance of one individual class, or national polity over another, as in the case of colonialism. Social justice and individual liberation loom large, and the student may be driven by a sense of social criticism, even outrage . . . The approach is rooted in identification and affinity with one's fellow human" (ibid.).

In Friedrich's view, the two approaches

> *may* exclude each other, and neither is reducible to the other. But most of the time the two approaches are essential to each other, for both are critical and concerned with values, although the meanings of these terms differ greatly in context: a scientific criticism is always implicitly ethical to a significant degree, and an ethical criticism is almost always scientific to some extent. In other words, the scientific approach is primarily rooted in the cognitive (e.g., the logic of experiment) and is concerned with diverse levels of knowledge, and the ethical-emotive is rooted in the affective as well as being overtly focused on such phenomena, but it is also commonplace for a cognitive analysis to arise from an ethical concern, and for an analysis of affect to arise from the *libido cognosciendi* (the drive to know). Barring the extreme of certain professional economists—who seem a-ethical—or the poets innocent of economic analysis, most radical theory will exhibit both approaches. (ibid.:296; emphasis in original)

For Friedrich, the two approaches cannot be segregated, and so the question then becomes: How are we to perceive and articulate their conjunction and, indeed, the conjunction of both approaches and theory/practice (ibid.)? His answer is to examine the ways in which the political-economic, linguistic, and ideological analyses of a group interdepend to produce a more coherent understanding of the human behaviors in question (ibid.:296–97).

Friedrich's approach seems particularly relevant to the case of agricultural producers within the European Union (EU), who have experienced dramatic social, economic, and political changes in their rural communities as a result of their participation in the Common Agricultural Policy (CAP) of the EU (cf. Fennell 1979; Bowler 1985; Wilson and Curtin 1989; Wilson and Smith 1992). The planning and implementation of the CAP have mostly been governed by development theory, which pays little or no attention to the active role played by the rural producer in the acceptance or rejection of change elements (cf. Mouzelis 1978; Goodman and Redclift 1982; Clout 1984; Curtin 1986; Long et al. 1986; Shutes 1989). The results of the CAP have therefore been uneven at best and potentially disastrous at worst, particularly in the smaller, more peripheral areas (cf. Hill 1984; Kelleher and O'Mahony 1984; Shutes 1991, 1992).

By Friedrich's standards, this is clearly a case where a scientific-analytic model of political economy (the CAP) has been applied without concern for the ethical-emotional aspects of the problem, which would entail a precise understanding of the roles played by linguistic and ideological factors in the local interpretation of the policy and the changes contained therein. As social scientists doing research in European rural communities, we are perfectly positioned to ensure that our own work both meets the standards of empirical scientific analysis and does justice to the lives and values of the people with whom we work by making certain that we examine the interplay among political economy, ideology, and linguistic analysis within the context of the local community. In this fashion, we can satisfy our need to know and our concerns for fellow humans. What follows is an example of this kind of approach applied to a change event.

BACKGROUND

The community in which these events took place is located in County Kerry in the Republic of Ireland (see figure 9.1). The community has a total population of approximately 380, the vast majority of whom (55 families) are farmers who live on their holdings. The village center contains three shops, four pubs, a post office, and a collection point for the Kerry Co-op dairy cooperative, to which all local farmers belong. The average size holding within the community is approximately 50 acres, with most farms falling into the 30- to 50-acre category. Only about 30 percent of the farms are larger than 50 acres, with the largest 110 acres. They are principally livestock farms, relying chiefly on the production of milk and, to a far lesser extent, cattle to meet their cash needs. There is no cereal grain production of any sort or any significant tillage, save an occasional small garden consisting of potatoes, turnips, and cabbage. The vast majority of farm families now purchase these root crops from local stores (for a more thorough general description of this parish, see Shutes 1987).

In this community lives the farmer mentioned earlier, known hereafter by the fictitious name of Declan. Married with three teenage sons, the forty-five-year-old Declan farmed 75 acres of good-quality valley land after inheriting the farm from his father in the 1960s. Declan's family has held this same acreage for over 400 years.

Like most of the farmers with larger holdings in the parish (known locally as big fellahs, or strong farmers), Declan had always relied on the sale of both milk and year-old cattle as the two sources of farm income. For over a hundred years this two-product strategy had protected local farms from the vagaries of the marketplace. Although milk was their primary product, they always kept some number of cattle to meet emergency expenses or as a hedge against bad weather and disease, which could severely curtail their income from dairying. In this manner, they maintained control over their farming decisions, and this control was a vital and important component of their self-image.

In early 1983, nearly ten years after Ireland's entry into the EU, Declan began to feel that he and his neighbors were losing that control. For reasons discussed below, they were being forced to abandon cattle raising entirely and concentrate exclusively on the production of milk as the sole source of their income. Except for those calves intended as replacements for older dairy cows, most were now sold a month after birth (called sucks), and the money earned from the sale of these skinny little calves was hardly equal to that previously earned through the sale of yearling cattle. Indeed, it was barely enough to pay for the artificial feed they had consumed since their birth. In addition, because all these calves were marketed at about the same time, the glutted market worked in favor of continually lower prices.

Declan believed it was very dangerous for him to rely solely on milk sales, since the market price was heavily subsidized by the EU, and quotas on production seemed inevitable. He began to look for a way to regain a two-product strategy to maintain his control over his enterprise. This concern led him to invent the calf feeder. His reasoning at that time can be summarized as follows:

1. His goal was to use his primary product, raw whole milk, in such a way as to provide him with more than one source of income.
2. The market for sucks was depressed, and they could be bought at well under their real market value, quickly fattened, and then sold at a later period when fewer calves were on the market and their greater weight brought a price comparable in total to that previously gained by the sale of cattle.
3. Feeding raw whole milk would increase the calves' weight far faster and less expensively than feeding artificial formulas, but the feeding process had to be automated or the labor time spent hand-feeding them would be cost-prohibitive.
4. Automated feeding could be accomplished in a very small area and would not necessitate a reallocation of prime grass fields used to graze milking cows and produce winter silage. The size of the dairy herd, therefore, could remain constant.
5. The entire process could be initiated slowly, with only minor reductions in the amount of whole milk delivered to the creamery, and the loss of that income would be more than made up by the profits from the sale of the older calves.
6. In this manner, Declan would regain control over his short-term production decisions. If the price for whole milk declined or if the EU initiated quotas on milk production, then Declan could shift more of his whole milk into calf-

fattening and increase his calf herd accordingly. If the price for calves declined, he could sell more of his whole milk to the creamery. Only if both markets collapsed simultaneously would he be in trouble, but that was a familiar risk.

Declan enlisted an unemployed engineer and an underemployed machinist to assist him in carrying out the design he had in mind. The machine had to be able to keep the milk at a constant warm temperature and had to continually stir the raw milk to prevent separation. In addition, it had to be able to deliver the milk to calves through rubber teats so the calves would be able to feed themselves on demand. The only human labor involved would be to keep the tank of the machine filled with raw whole milk. In effect, the machine would operate much like a real cow, and whole milk delivered in such a fashion would assure that the calves would gain substantial weight in a minimum amount of time.

After about a year of tinkering with various designs, Declan and his friends came up with a prototype that seemed to accomplish all of these objectives. Declan set up three of the new machines in his own farmyard, using part of a small field formerly used to fatten a few sheep but that was not part of his dairy herd fields. He began with a modest purchase of fifteen sucks and devoted part of his milk product to their feeding. It took a few days for the calves to respond to the feeding system, but they fed vigorously afterward. Declan kept careful records of their weight gains over the next six months and sold them for a profit that was slightly more than he would have made had he sold the extra milk to the creamery.

The next year, 1985, Declan increased the number of sucks purchased to thirty, since this was the first year of the EU-imposed quota on milk and he would have had to cut production in his dairy herd if he had not had this other option. The profits from that year's calves were about the same as the first, and Declan's invention accomplished what he had intended. But Declan was encouraged by his two assistants to market the device. Declan finally agreed, partly because doing so would help him recover the initial construction costs but also because he sincerely believed the device would be of great benefit to other small dairy farmers like himself.

Declan spent the winter of 1985–86 traveling to other parts of Ireland, trying to get a few larger farmers to independently test his invention and thus verify the weight-gain results he had attained. At least two farmers agreed to do so and got results very similar to Declan's. In the spring of 1986, Declan tried to interest both government agencies and the local cooperative in his invention, using his own statistics and those of the two outside farmers as evidence of the device's potential. His attempts were a failure, with the agencies either choosing to ignore his evidence altogether or claiming that a much larger sample was necessary to demonstrate the efficacy of the device. It was then that Declan decided to build the life-size mechanical cow and take the device in its innards to the regional agricultural fair in County Cork.

The media success of the mechanical cow at the fair enabled Declan to make a few sales that very day, and he was able to market the product a few at a time, although he lacks adequate financing to produce them in great numbers. He still uses the device himself and is pleased with the results. Other local farmers have begun to experiment with it as well.

ANALYSIS

After examining the basic details of Declan's activities over seven years, certain questions crucial to this inquiry arise: (1) What caused the community-wide shift from a two-tiered production strategy based on milk and cattle sales to a one-tiered strategy based solely on milk production? (2) How did Declan's perceptions about his role in the community and about the community itself affect his reactions to these changes? (3) Why did Declan finally decide on the mechanical cow as a way of marketing his device, and why did such a strategy succeed?

Answering each of these questions is essential to fully understand the change-event described above, yet each demands a different sort of analysis. Question #1 requires the kind of analysis found under the general heading of political economy: an empirical, quantifiable account of the changes in political and market forces that led to a shift in local production strategies. Question #2 requires an analysis of the ideology held by community members with respect to strong farmers and their importance to community life. Question #3 involves what might be referred to as a linguistic or symbolic analysis, wherein we examine the manner in which Irish small farmers may manipulate, or be manipulated by, the language and symbols of power.

Each type of analysis requires different kinds of community data, and any single one of the approaches would offer only a partial explanation for the event described. They are therefore complimentary, in that each provides a component that is crucial to our understanding of the event. Let us now examine each component and its contribution to the total analysis.

Political-Economic Component

Ireland's entry into the EU in 1973 introduced dramatic changes to the farms throughout Declan's parish. Milk price subsidies, grants, and low-interest loans made available through EU membership and administered by a local dairy cooperative transformed the two-tiered income system based on milk and cattle into a one-tiered system based solely on milk production.

Now, on all but the smallest farms, dairy herds have increased dramatically. Non-milking stock is regularly sold a few weeks after birth to accommodate the increased herds. Outside contractors prepare the vast majority of the winter feed as local farmers attempt to increase the efficiency and reliability of food production for their dairy herds. Tillage of any sort is practically nonexistent because of the labor-intensity of such activities. The parish farmers have become specialized milk producers, more or less, with all of the positive and negative elements such a transformation implies. From the standpoint of a political-economic analysis, the reasons for these changes can be readily identified.

First, the continuing downturn in the world demand for beef throughout the 1970s, coupled with EU milk price subsidies that nearly doubled the amount per gallon received by farmers between 1973 and 1986, forced local farmers to reduce their cattle herds and replace them with milking stock. In 1976, the total number of cows in the community numbered 938 head. From 1977 to 1986, the size of cow herds expanded by nearly 30 percent, to 1,323 head. In 1976, the total number of cattle over one year old was 994. During the following decade, that number decreased to 318 head, a reduction of 68 percent. By the late 1980s, 90 percent of all calves were sold one month after birth, with the money used to purchase fertilizer and possibly feed concentrates for the cows if the weather prohibited them from going out to graze in the early spring.

Second, the increase in the size of dairy herds placed an excessive strain on the labor supply available to community farms, necessitating an unprecedented increase in the acquisition of labor-saving equipment and processes, which were acquired through EU-subsidized grants and low-interest loans. For example, in 1986, nearly all the farms were using the central road and paddock system, which allowed a farmer to drive the cows to and from the milking parlor to pre-selected and fertilized grazing plots while continually re-fertilizing plots used on previous days. Only one such system was operating in 1976. In 1986, nearly all the farms were hiring contractors to cut and prepare self-feeding grass silage. In 1976, only about 21 percent of the farms had such a feeding system. In 1976, only 14 percent of the parish farms had milking parlors. By 1986, nearly 80 percent of the farms had some form of automated milking equipment and parlors. In 1976, only 21 percent of the farms used bulk milk refrigeration and transport tanks, compared with 80 percent in 1986. The bulk of these expansions and acquisitions took place during what local farmers call the boom period, between 1977 and 1981.

Third, the move to single-commodity production was greatly facilitated by the fact that the community was already involved in an increase in milk production prior to Ireland's entrance into the EU and thus saw these latest changes as part of a familiar pattern. Since a large part of the expansion was fueled by grants and low-interest loans that could easily be met by the increased price received for their milk, no farmers believed they were taking an unwarranted or unfamiliar risk. They saw themselves as following the same sort of pragmatic, short-term strategy they had always followed, which was, in the words of one farmer, taking it all into account and doing what we thought was best for us in any one particular year. No one dropped their entire stock of cattle overnight. They simply reduced it gradually over the ten-year period between 1976 and 1986. As far as they could see, it was merely an extension of the process in which they had already been involved prior to 1973, and they thought they could always increase their cattle production at some later time when the market recovered.

Fourth, the boom period was quickly replaced by the bust cycle of the early 1980s, which saw the emergence of EU milk quotas that fixed the price at 1984 rates to recover from the massive milk and butter surplus the EU nations had produced. In addition, the community suffered three years of excessive rainfall between 1981 and 1984, which severely curtailed the milk output and fixed the 1984-based production quota at rates well below their normal capacity. In addition, in each of the three years of the flood, local farmers were obligated to the silage contractors through arrangements made the previous year, even though most of their fields were too wet to be cut. This resulted in massive losses in income over the three-year period and the establishment of an unprecedented debt among local farmers, who were forced to purchase foodstuffs for their milking herds. Finally, the inflationary cycle that had been developing during the boom period severely curtailed any further increases in labor-saving techniques and equipment and forced a number of farmers who were in the middle of their expansion to more than double their payments for the work-in-progress.

Fifth, given the reduced production rates imposed by the quotas, few farmers in the community escaped the debt cycle, with the luckiest having only a short-term debt repayable within five years. Others were faced with repayment that would require ten years or more. No farms were lost during this period, but the debt structure necessitated that they concentrate all of their energies on dairying, which was now their sole stable source of income. Effectively, they became producers of only one commodity: milk.

Ideological Component

Between 1850 and 1950, two apparently disparate yet internally consistent ideological notions were strongly felt and collectively held by the members of Declan's community: (1) the idea of the importance of the strong farmer and (2) the idea that all relationships within the community were based on strictly egalitarian principles, or, as one farmer put it, the notion that we're all Indians here, there are no chiefs. Let us examine each in turn.

(1) The Strong Farmer

These were the farmers who were at the center of community life. In charge of highly labor-intensive farming enterprises that included a mixture of tillage, cattle, and hand-milked cows, these big fellahs traded the use of their equipment and breeding stock, their expertise and knowledge, with the market outside the parish and their general patronage and goodwill to the small farmers in exchange for labor at crucial production junctures. They also hired agricultural laborers, who were dependent on such employment to earn the price of their emigration. The volume of their purchases at local shops kept those places open and available to all. Their celebrations of births, deaths, and marriages were memorable in terms of available food, beverages, and entertainment.

During that period, these individuals' farm enterprises differed only in scale from those of their smaller neighbors. They had more acreage and hence a higher volume of productive output. That volume was just high enough to make a difference in terms of the quality and quantity of life other community members could achieve through participation in their activities. In a very real sense, the success of these farms defined the success of the entire community.

Being strong, therefore, meant being able to maintain a viable enterprise under outside market conditions that seemed incredibly complex, unpredictable, and dangerous. The ideal strong farmer was seen as a shrewd bargainer, an excellent judge of people and livestock, a worldly-wise individual not fooled by the world's guile, and—most important—a cautious, conservative, and knowledgeable evaluator of the marketplace. Such a farmer would never be the first to experiment with new techniques unless the benefits were virtually certain or no other options were available, and he would avoid any long-term debt that might threaten the loss of the land. In addition, strong farmers would be, as one local expressed it, moral and serious men, devoted to work and with no time for foolishness.

The fact that any real individual was unlikely to possesses all of these characteristics or that farmers classed in that category often acted in ways that were inconsistent with the ideal pattern in no way diminished the notion itself, for, as Geertz (1972) has suggested, it acted as both a model of and for the world of this rural community. As a model of the world, it identified the need for extreme caution in the marketplace and a careful year-by-year appraisal of farming strategies. As a model for the world, it identified the behavioral guidelines community members must attempt to follow to remain in control of their lives, given the uncertainty and power of the outside world. By focusing on the few farmers who both bridged the gap between local community and marketplace and were essential to the quality of life within the community, the notion of the strong farmer tapped the perfect source for both aspects.

Guided by this ideological model, the larger farmers felt constrained to exercise extreme caution in their economic decisions and could be subject to extreme local criticism if they deviated too far from this conservative strategy. They also truly believed they were ultimately responsible for the community's welfare. Their rewards for proper behavior included prestige and the access to cheap labor that guaranteed them higher profits than their smaller neighbors. This same model constrained the smaller farmers to both yield respect and sacrifice labor to the larger farmers, but in the process they gained access to the equipment and expertise that allowed them to base their own economic decisions on those of the larger farmers without substantial risk or the need for increased knowledge on their own part.

(2) Egalitarian Relationships

The linchpin of the entire process just described was labor. The larger farmers were ultimately dependent on access to the donated labor of the smaller farmers if they were to maintain the scale of their labor-intensive enterprises. The threat behind small farmers' criticism of the larger farmers' community behavior was the withdrawal of their labor support. For their part, smaller farmers could not afford to lose access to the larger farmers' equipment and knowledge for any length of time. The threat behind the larger farmers' criticism of any aspect of their community behavior was the withdrawal of that access and knowledge. In addition, small farmers relied on each other's labor for certain activities (*cooring*, discussed in chapter 9), and geographically proximate farms, regardless of size, depended on each other's labor for collective activities such as threshing (*meithal*, discussed in chapter 9). Clearly, these delicate dependencies precluded the development of any sort of ideological system that would have based social interaction within the community strictly on differences in wealth and prestige, in what is typically referred to as a system of social stratification. Something more subtle was obviously required.

It was precisely at this point that the idea of egalitarianism came into play. The strongly held notion that all community members are equals and ought to be treated as such provided the perfect guidelines for social interaction under these circumstances. No farmers, large or small, risked a potentially dangerous misinterpretation of their motives or economic goals when their behavior with each other in public was carefully egalitarian. Indeed, such public statements as "we're all Indians here," "we all only work to pass the day," or "the only difference between a big fellah and a small fellah is the size of their hats" and the use of the common greeting "lads" between males of all ages and statuses are all indications of the constraints of this egalitarianism.

The necessity and efficacy of an ideology of egalitarianism also led to an elaborate separation of public and private behaviors. Farmers rarely directly discussed their own farming enterprises and plans in public or those of any other community member. Agricultural workers could be fired for discussing their employers' actions in public, and those who became permanently attached to one farm and were allowed to build a house on that land (called cotters) were referred to as being like one of the house, who never carried tales.

The use of indirect questions and negative inquiries also reflects this concern to mask private feelings and motives while trying to discover those of others. There was also the commonly expressed local fear of being "caught out," which in general referred to the highly undesirable state of having revealed one's private feelings or motivations in public. This is not to say that individuals could not be direct and forthright regarding their private behavior, only that there were strong ideological commitments to the opposite. It was safer and sounder behavior to keep the public and private domains separate. Not all individuals were equal masters of the game of separation, but the language itself guaranteed that everyone had access to the basic rules.

Finally, it is certainly not correct to interpret these locally expressed ideological ideas of the strong farmer and egalitarianism as somehow indicative of a peasant community, wherein commitment to traditional values prohibits or impedes necessary economic change. The whole purpose of these ideas was to allow change to take place within the framework of local control, specifically through the gradual changes in strategy made first by the strong farmers and eventually borrowed in scale by all the rest. They were culturally transmitted behavioral guidelines that provided all members of the community with the ability to resist lack of control over change, not change itself.

Too frequently, the ethnographic literature concerning rural Ireland has emphasized these ideological aspects without examining the economic history of the groups in question (cf. Arensberg 1959:35–107; Arensberg and Kimball 1968:31–58; Schepper-Hughes 1979; Hannan 1982), which has sustained the peasant model. Less frequently, the opposite emphasis has been offered, which leads to an equally inappropriate class-analysis model based strictly on differences in wealth and prestige (e.g., Gibbon 1973).

Between 1950 and 2000, both an earlier locally induced change that mechanized the dairying part of the strategy and the later EU-induced change to the single-commodity production of milk virtually obliterated the labor dependencies that had characterized the previous 100 years. Insufficient time has elapsed to accommodate a major change in ideology, and, as a result, the older notions of the strong farmer and egalitarianism still prevail, although their contextual relevance is rapidly changing.

Faced with these growing inconsistencies between ideological perspective and economic reality, the most common reaction, particularly among the strong farmers, has been to resist total commitment to the single-income strategy—even though they stand to gain from it in the long run—while desperately seeking alternatives that will give them back control over what their farms will produce. Not surprisingly, the most commonly stated reason for this resistance is fear that the community itself will rapidly decline and eventually disappear.

Linguistic-Symbolic Component

Given the ideological components discussed above, it is possible to characterize the linguistic-symbolic style of the community by reference to the term *irony*, wherein what is said is not actually what is meant. Consider the relationship between the two major notions of the strong farmer and the egalitarian community. On the one hand, we are presented with the idea that some individual farmers are unique from all the rest in terms of their skills and abilities and are therefore deserving of much higher levels of respect and prestige than would be obtained by an ordinary farmer. On the other hand, we are given the notion that no such individuals exist within the community and that all members are inherently equal and to be treated with the same level of respect and prestige.

And so, when a farmer is identified as strong, it does not necessarily mean he is worthy of more respect than any other community member, and when community members insist that there are no chiefs among them, it does not necessarily mean there are no individuals who have more influence than others because of their superior skills and abilities. This is irony in the classic sense.

Irony is also clearly evident in the everyday speech of community members. Under the ideological constraints of strict separation of public and private behaviors, what is said in public typically contains such ironic elements. Even the structure of questions permits the individual to speak from the other's point of view, so that, for example, asking someone for a ride into town on the following day would be framed as "you wouldn't be going to town tomorrow, now, would you" or "I don't suppose you'll be in town tomorrow?" In this fashion, the language provides for the avoidance of being caught out.

The solution to these ironic elements in speech lies in the fact that all those who use such conventions are intimately familiar with each other's history, habits, behaviors, and emotional tendencies. There is a real exclusionary power to such uses of irony, for, although outsiders may share the same conventions, only those with such intimate knowledge really understand the complete message, with all its subtle nuances. In this sense, then, the use of irony serves as a mechanism for bonding community members to each other in a conspiracy of uniquely shared meanings.

From the perspective of the urban, bureaucratic world outside, such ironic uses are often interpreted as genuine dissimulation, wherein the speaker is seen as feigning ignorance and self-effacement to gain some advantage. Rural people who use such patterns of speech are often referred to as crafty or shrewd. Regardless of whether the dissimulation is always intended, the use of irony does give locals an edge in their dealings with bureaucracies since it tends to evoke a sort of begrudging respect in an individual who would otherwise have the total upper hand.

CONCLUSIONS

In returning to the matter of Declan and his invention, we should be able to view that event from the unified perspective in which it occurred, for we can now answer the major questions it provoked.

1. What caused the shift in strategy? As we have seen, the shift from a two-tiered to a one-tiered economy was the direct result of changes in the international markets for cattle and milk. At the local level, however, the gradual change from more milk to fewer cattle was never seen as a shift away from cattle altogether but rather as the same kind of careful yearly decision-making that had always typified the farmers' strategies. Since the market was strong for milk, they shifted more of their energies to the production of milk but were prepared to shift back if the milk market weakened.

The debt incurred by this shift could easily be handled by the increase in milk income, and so, from the farmers' perspective, it was business as usual, wherein they would pit their local expertise against the market. Because their strategies were necessarily made for the short term, it was impossible for anyone to foresee that two years of the worst weather in the county's history, followed by the imposition of a milk quota, could eliminate their two-tiered strategy entirely and force them into production for the maintenance of debt. Almost everyone was trapped by the situation. The market strategy that had sustained them through much bleaker times than the present had failed, and they were no longer in control of their production decisions.

2. How did Declan's perceptions about his role in the community and about the community itself affect his reactions to these changes? No group was more embarrassed and shocked by these changes than strong farmers like Declan. They had done

everything correctly. They had proceeded cautiously in expanding their dairy herds since the initial years of EU membership. They were not, as far as they could see, incurring any meaningful debt in the process, taking advantage of grants and relying on loans only when the debt could be repaid out of their present income. Indeed, most of them were proud of the fact that they had been able to lead their community into the prosperous times of the EU without severe disruptions to community life. The smaller farmers who, as always, followed them into this series of gradual changes were also prospering in scale and privately praised the initiatives taken by the strong farmers throughout the 1970s. In spite of all this, the bottom fell out in the early 1980s.

Declan in particular felt in some way responsible for the plight of his community. He had always been a solid, if cautious, supporter of the EU system, and he had been a leader in the early 1970s in mechanizing his dairying operation, prior to Ireland's entry into the EU. Although severely criticized for these actions then, most of the other farmers did likewise after EU membership was accomplished, and his private prestige was very high throughout the 1970s. Declan, like most of the strong farmers, was probably in an excellent position to weather these debt and quota problems and come out on the other end in a sounder economic position than before. But Declan and the others did not just view their enterprises from an economic standpoint. He was a member of a community and he was a strong farmer, and he felt the responsibility of that position. He and the other strong farmers also felt the harsh criticism of smaller farmers, with comments like "well, they're [larger farmers] free of us now, and the big fellahs won't have the time of day for us now, you'll see."

Declan's entire plan for recapturing some form of two-tiered income through the invention of a calf-feeder was motivated by this sense of responsibility to all the farmers in the community and by his conviction that his strength and value as a farmer in that community rested in his own ability to control his production decisions. Had he and his fellow strong farmers been motivated solely by concerns for their own economic gains, such an invention would not have been forthcoming, for time was on their side in a single-income situation.

3. Why did Declan decide on a mechanical cow as a way of marketing his invention, and why did this succeed? When Declan was encouraged by his urban-based co-workers to market his device, the initial strategy involved getting the facts and figures to support the claims of weight gain. Without such evidence, reasoned his allies, the attempt to sell the device would fail. In fact, this evidentiary approach failed miserably, and Declan could interest none of the powers in his device. Declan had very little advantage in presenting bureaucratic data to bureaucrats, and the invention he was offering had the potential to take the control over the milk supply

out of their hands and put it back into the hands of farmers. It was a question of power, and the data Declan offered could easily be managed to their advantage by saying there was simply not enough of it or ignoring it altogether.

It was only when Declan called upon his substantial sources of symbolic irony that he met with any success in marketing the device. He recreated the image of the crafty farmer on a grand scale, and his message had a poignancy that could not be ignored. Who could not respect, even begrudgingly, the image of an uneducated farmer having the skills to produce not only such a perfect mechanical model but the device that was in its innards? And who could miss the irony of a farmer having to produce mechanical symbols of his livelihood to be able to make better use of the real objects? He attained the small advantage irony has always given the members of his community and was able to market his device and recover his investment in his strategy. What is of much greater significance to Declan, however, is that the device is now used by other farmers in and around his community, for he believes it will help them regain control of their enterprises and hence over their lives and community.

Control over their lives is the unifying factor in this entire presentation, for these are the political stakes involved in the changes taking place in rural communities like Declan's. In emphasizing one individual's attempts to regain that control, this chapter has shown the absolute necessity of an approach to change that incorporates within a single analysis the political-economic, ideological, and linguistic-symbolic components, for individual behavior is always a unity of these components. From a theoretical standpoint, it is impossible to understand the individual acts involved in what we call social change without such a combined approach. From an ethical standpoint, we cannot possibly hope to work for fair and equitable changes within such communities if we do not understand how individual members view their lives and their work.

REFERENCES

Arensberg, Conrad M. 1959. *The Irish Countryman*. Gloucester, MA: Peter Smith.

Arensberg, Conrad M., and Solon T. Kimball. 1968. *Family and Community in Ireland*. Cambridge: Cambridge University Press. http://dx.doi.org/10.4159/harvard.9780674729469.

Bowler, Ian R. 1985. *Agriculture under the Common Agricultural Policy: A Geography*. Manchester: Manchester University Press.

Clout, Hugh. 1984. *A Rural Policy for the EEC?* London: Methuen.

Curtin, Chris. 1986. "The Peasant Family Farm and Commoditization in the West of Ireland." In *The Commoditization Debate: Strategy and Social Network*, ed. Norman

Long, Jan Douwe van der Ploeg, Chris Curtin, and Louk Box, 58–76. Papers of the Department of Sociology17. Wageningen, Netherlands: Agricultural University.

Fennell, Rosemary. 1979. *The Common Agricultural Policy of the European Community: Its Institutions and Administrative Organisation*. London: Granada.

Friedrich, Paul. 1989. "Language, Ideology, and Political Economy." *American Anthropologist* 91 (2): 295–312. http://dx.doi.org/10.1525/aa.1989.91.2.02a00010.

Geertz, Clifford. 1972. "Religion as a Cultural System." In *Reader in Comparative Religion*, 3rd ed., ed. William A. Lessa and Evon Z. Vogt, 167–78. New York: Harper and Row.

Gibbon, Peter. 1973. "Arensberg and Kimball Revisited." *Economy and Society* 2 (4): 479–98. http://dx.doi.org/10.1080/03085147300000023.

Goodman, David, and Michael Redclift. 1982. *From Peasant to Proletarian: Capitalist Development and Agrarian Transitions*. New York: St. Martin's.

Hannan, Damian. 1982. "Peasant Models and the Understanding of Social and Cultural Change in Rural Ireland." In *Ireland: Land, Politics and People*, ed. P. J. Drudy, 141–66. Irish Studies 2. Cambridge: Cambridge University Press.

Hill, Brian E. 1984. *The Common Agricultural Policy: Past, Present and Future*. London: Methuen.

Kelleher, Carmel, and Ann O'Mahony. 1984. *Marginalisation in Irish Agriculture*. Socio-Economic Research Series 4. Dublin: Economics and Rural Welfare Research Centre of the Agricultural Institute.

Long, Norman, Jan D. van der Ploeg, Chris Curtin, and Louk Box. 1986. *The Commoditization Debate: Strategy and Social Network*. Papers of the Department of Sociology 17. Wageningen, Netherlands: Agricultural University.

Mouzelis, Nicos P. 1978. *Modern Greece: Facets of Underdevelopment*. New York: Holmes and Meier.

Schepper-Hughes, Nancy. 1979. *Saints, Scholars and Schizophrenics: Mental Illness in Rural Ireland*. Berkeley: University of California Press.

Shutes, Mark T. 1987. "The Role of Agricultural Production in Social Change in a Rural Irish Parish." *Social Studies* 9 (2): 17–28.

Shutes, Mark T. 1989. "Changing Agricultural Strategies in a Kerry Parish." In *Ireland from Below: Social Change and Local Communities*, ed. Chris Curtin and Thomas Wilson, 186–206. Galway: Galway University Press.

Shutes, Mark T. 1991. "Kerry Farmers and the European Community: Capital Transitions in a Rural Irish Parish." *Irish Journal of Sociology* 1: 1–17.

Shutes, Mark T. 1992. "Rural Communities without Family Farms? Family Dairy Farming in the Post-1993 EC." In *Culture Change and the New Europe: Perspectives on the European Community*, ed. Thomas M. Wilson and M. Estellie Smith, 123–42. Boulder: Westview.

Wilson, Thomas M., and Chris Curtin, eds. 1989. *Ireland from Below: Social Change and Local Community*. Galway: Galway University Press.

Wilson, Thomas M., and M. Estellie Smith, eds. 1992. *Culture Change and the New Europe: Perspectives on the European Community*. Boulder: Westview.

11

Island Pastoralism, Isolation, and Connection

An Ethnoarchaeological Study of Herding on Dokos, Greece

P. Nick Kardulias

As interest in the nature of the ancient Greek economy increased during the second half of the past century, anthropologists and archaeologists focused a significant amount of attention on pastoralism past and present. Campbell (1964) provided one of the first detailed ethnographic accounts of Balkan pastoralists in his study of the Sarakatsani, linking cultural values and kinship structure to the herding of sheep and goats in central Greece. Koster (1976, 1977, 2000) examined in detail the ecology of herding in several locations and, among other important insights, demonstrated that contrary to Hardin's Tragedy of the Commons thesis, pastoral groups do not overexploit the communal grazing lands but rather utilize these areas carefully (Koster 1997). Herzfeld (1985) explored the ways in which herders on Crete expressed deep cultural values and personal identity through the animals they tended. Of more direct relevance to the interests of archaeologists has been the seminal ethnoarchaeological research of Claudia Chang (1981, 1993, 1994, 1997, 2000; Chang and Koster 1986, 1994). With her detailed observations about the material culture of modern pastoralists in the Peloponnesos and Macedonia, Chang has contributed significantly to our understanding of how to recognize pastoral sites in the archaeological record and the periods when various forms of herd management (e.g., transhumant movement) could have been developed. The present study builds on this earlier work, especially that of Chang and Koster, to comprehend the role pastoralism has played on small islands in the Aegean. At a more general level, the concern with pastoralism's role in the Greek economy reflects its position in

DOI: 10.5876/9781607323433.c011

Figure 11.1. Map showing the location of Dokos in relation to other islands and the southern Argolid in Greece.

the broader Mediterranean economy (Braudel 1972:85–102) and demonstrates the type of core-periphery interactions that form the basis of world-systems analysis (Wallerstein 1974; Chase-Dunn and Hall 1991, 1997).

In 1996 the Ohio State University initiated a study of human occupation on the island of Dokos, located in the Aegean Sea off the northeastern coast of the Peloponnesos (figure 11.1). This research was part of a long-term project whose goal is to examine the nature of human interaction with a dynamic landscape. In doing so, the project brings together scholars from a variety of disciplines in a coordinated effort to comprehend the mutual effects of human activity and natural forces in the Korinthia and surrounding regions of Greece. Among the experts who participate in this collaborative endeavor are archaeologists, cultural anthropologists, physical anthropologists, geomorphologists, ethnoarchaeologists, and historians (see Tartaron et al. 2006). The purpose of the present study is to document the use of marginal land by contemporary herders as a response to variable economic, political, and social conditions in modern Greece. In addition, the ethnoarchaeological information provides insights that can inform our reconstruction of ancient habitation on Dokos and other small islands lacking natural sources of drinking water. I suggest that a mixed strategy of herding, farming, and commercial employment makes possible the use of such islands to expand economic opportunities in an area where good land is at a premium. The present report focuses on only one family and

FIGURE 11.2. View of the narrow channel between Dokos (*left*) and the mainland (southern Argolid; *right*).

is thus somewhat limited in scope, but it does provide important information on adaptations to a specific location.

The rocky island of Dokos lies on the eastern edge of a sea lane through which ships passed between Cape Malea (and then on to the west) and Athens and from there on to the northern Aegean and eventually Constantinople (figure 11.2). Gregory (1997) argues that this passage was one of the main routes of transport and communication between Byzantium and the west. Kyrou (1995) suggests that Dokos may be a place mentioned in the tenth-century narrative of Paul of Monemvasia. In the extant Arabic translation of that document, Paul tells the story of the remains of three patron saints of Barcelona—Valerius, Vincent, and Eulalia—which materialized miraculously at a castle called Ashab al Bakar; on linguistic and geographic grounds, Kyrou (1995) identifies the fortified settlement, or *kastro*, on Dokos as the castle mentioned by Paul.

In 1996, a team of six did a surface collection and an architectural survey in the area of the Early Byzantine fortified settlement at the north end of Dokos. The kastro circuit walls are 575 m long, enclose an area of 1.65 ha, and were built in three phases. The initial construction occurred in the late sixth century, followed by an extension in the mid-seventh century. Gregory (1997) associates the last phase with the recorded refurbishing of the kastro in the 1680s by the Venetian Francisco Morosini, best known for his artillery attack against the Turks entrenched on the Akropolis of Athens in 1687 during which the Parthenon sustained considerable damage. The OSU team also recorded a number of buildings in the interior of the fortified area and outside the walls along the slopes of the steep hill. Many cisterns stored water for the Byzantine inhabitants; what seems to have been a spring, which some locals said supplied abundant fresh water at the beginning of the twentieth century, lies near the north gate of the fortifications.

The goal of the 1997 expedition was to examine the area around the small chapel of Agios Ioannis O Theologos (St. John the Theologian) on a saddle below the fortified area to determine if an early Christian basilica may have existed there (figure 11.3). Some aligned stones suggest the possibility of older walls, while ceramics on the surface date to the seventh century AD. In addition, the current chapel contains a number of features that are clearly ancient or medieval in date. As an adjunct to this excavation, our team had the opportunity to undertake an ethnoarchaeological investigation that involved interviewing resident herders and recording a number of agricultural and domestic features. The herders represent the last permanent residents who extract their livelihood directly from the land. These people offer a fascinating glimpse into a complex lifestyle that requires careful balancing of various activities in order to survive on an island with no permanent water source.

ISLANDS: BIO-GEOGRAPHY, ISOLATION, AND CONNECTION

Over forty years ago, Evans (1973) argued that islands can act as laboratories for the study of culture change. By their very nature, he suggested, islands have conditions of particular interest to archaeologists, that is, they often provide significant shelter from external contacts while simultaneously isolating the various biotic communities. Water transport is often a viable mechanism for maintaining contacts, depending on the vagaries of weather and available technology. Evans pointed out that in the Mediterranean, for example, sea travel in antiquity was limited by such factors. Ancient mariners stayed close to shore, hugging coastlines to maintain bearings; they avoided open passages whenever possible (Frost 1997:29). In addition, much sea travel was linked to shifting seasonal wind patterns. One is reminded of Hesiod's (1983:lines 618–26, 663–81) prescription to stay put during the winter months when strong winds in the Aegean make sailing a hazardous venture. An important addendum to Evans's point is that these rules were true until recently. The advent of steam, gasoline, and diesel engines has made passages possible in all seasons, although the rough winter waters of the Aegean during that time of the year make for diminished travel schedules by passenger ships.

A second important trait Evans finds of great potential use to archaeological analysis is that islands have only a limited range of resources. Scholars can learn a great deal by examining the ways humans adapt to the circumscribed conditions;

FIGURE 11.3. Chapel of Saint John the Theologian near the Douskos family complex.

insular innovations are often peculiar to the specific settings. The lack of diversity also helps clarify instances of outside contact, that is, it is relatively easy to see intrusive materials in the archaeological record. Furthermore, this dearth of resources encourages islanders to seek connections with other regions and helps establish the exchange networks that define world-systems.

Third, Evans argues that island societies often exhibit pronounced evolution of certain cultural traits, frequently of a ceremonial or symbolic nature. The development of such trends on the mainland is often truncated as a result of the intercession of outside influences—a cultural trait is not allowed to reach its fullest expression because some new element appears and leads matters in a different direction. An example of this process is the situation on prehistoric Easter Island, with the production of the large stone statues and the attendant ritual elements.

Fourth, Evans (1973:519) notes that clusters of islands can provide a setting to study "the development of discrete communities in close proximity, [where one can] see the mutual effects of contact." Evans cites the Cyclades as an example of such an island group, and one can easily extend that characterization to the islands of the Saronic Gulf (Aegina, Poros) and along the shores of the Argolid peninsula (Idhra, Dokos, Spetses).

Evans makes the case that islands offer appropriate conditions to undertake the processual dictum of "archaeology as anthropology." The conditions specified earlier offer a level of control not attainable on the mainland, so that islands can serve as laboratories for archaeological investigations. He also notes that small islands tend to meet the conditions better than larger ones. Among the factors to consider are (ibid.:520):

1. The geographic and geological structure. For the present study, the relationship between these factors and local hydrology, the presence of sheltered inlets and bays, and defensible positions are all important in determining the nature of past and present occupation and land use.
2. The variety of eco-niches and resources available. Evans implies that the range of specific eco-zones provides the environmental parameters, that is, what humans have to work with.
3. The size and structure of the population.
4. The cultural diversity of the group(s) in question.

From this list, one can draw certain basic conclusions. The most significant perhaps is the link between environment and culture. Specifically, the more uniform the ecological conditions between islands, the more homogeneous the forms of cultural adaptation/expression. While this is not a great revelation, Evans did suggest that islands are the best places to study the phenomenon.

Several scholars working in the eastern Mediterranean have continued the theoretical development of island archaeology with the delineation of concepts particular to insular studies (Cherry 1981, 1985, 2004; Broodbank 2000, 2006; Knapp 2008:13–65). Held (1990, 1993) has provided some key concepts, such as configuration, defined as the relation of an island to nearby islands and the mainland. He places such work in an evolutionary framework (Held 1990). First, adaptation to island conditions (often quite different from those on the mainland) requires unique adjustments. Second, one must consider the ramifications of genetic drift on island populations after they are established; the physical traits of the initial colonists dominate the gene pool and often lead to significant differences with the parent group from which the island population derives. The primary goal of such analysis is to comprehend the boundary effect, which refers to the impact of resource scarcity on a given population and which varies according to the degree of openness of the system in question. This approach has direct applicability to Dokos, which, because of its size, experiences constraints on its resources. This, in turn, influenced the trajectory of cultural development.

As a counter to the processual approaches of Evans and Held, Broodbank (2000) and Knapp (2008) emphasize the concepts of agency and insularity, suggesting that we must see island occupants as active participants in formulating distinctive identities that govern their actions in the world and not simply as subject to the material conditions that surround them. Knapp (2008:18) provides a useful definition of insularity: "The quality of being isolated as a result of living on islands, or of being somewhat detached in outlook and experience. Insularity can result from personal, historical or social contingency." Since islanders clearly also interact with others, Knapp argues that this insularity "is contingent in both space and time, and thus may be adopted or adapted as individual or wider social concerns dictate" (ibid.). In this way, he accommodates the factors that can at times isolate and in other instances connect islands with the places beyond their shores.

An important approach in studying island ecology is bio-geography, which helps define the relationships between various plant and animal populations. MacArthur and Wilson (1967:185), in reference specifically to island conditions, defined it as an examination of the distribution of organic species over the face of the earth. Biogeography is concerned with the limits and geometric structure of individual species populations and with the differences in biotas at various points on the earth's surface. The local, ecological distribution of species, together with such synecological features as the structure of the food web, are treated under bio-geography only insofar as they relate to the broader aspects of distribution.

In contrast, human bio-geography is "the study of the size, distribution, and population structure of, and the interactions among, human populations found in similar or divergent habitats, and of the conditions and events leading to the development and maintenance of similarities and differences among human populations living at various points on the earth's surface" (Terrell 1977:7; see also 39–40). The use of population size, distribution, structure, and interaction as key concepts demonstrates the affinity of this approach to general systems theory (ibid.:8). The perspective is important because it assesses the biological, cultural, and physiographic elements that determine human interaction with the environment and forces us to think in terms of specific features of that interaction.

Farming and pastoralism are two forms of subsistence that people have adapted to islands. While there is some evidence on Mediterranean islands for human visitation and occupation in Late Pleistocene and Early Holocene times (and perhaps as early as the Middle Pleistocene [see Strasser et al. 2010]), most permanent habitation took place during the Neolithic (Cherry 1990). There is clear evidence that people visited Melos to acquire obsidian at the end of the Upper Palaeolithic (Perlès 1987), and recent evidence demonstrates the presence of Mesolithic foragers on Cyprus at around 10,000 BP (Simmons 1991, 1999), but long-term successful habitation did not occur until the migration of farmers, accompanied by the full suite of domesticated species, provided ample human numbers and a sufficient subsistence base to support those people (Broodbank and Strasser 1991). The impact of pastoralism on island ecologies must have been considerable. Critical to understanding the nature of this impact is knowing how ancient herders managed their flocks. What factors determined whether to increase or decrease herd numbers? How and when did they dispose of surplus secondary products, such as wool and milk? What was the nature of the connection between island pastoralists and their neighbors on other islands and the mainland? The present study has as a key goal to investigate modern herding practices on Dokos as a way to understand how past inhabitants of this and other, similar islands may have acted and under what kinds of economic circumstances (i.e., in periods of general prosperity and expansion or of decline and retrenchment).

To gain such comprehension, it is necessary to begin with a general model of pastoralism. Koster and Chang (1994:8) define pastoralists as "those who keep herd animals and who define themselves and are defined by others as pastoralists." They argue that a pastoral lifestyle requires people to structure their activities around the needs of the animals on which they depend. This symbiotic relationship "must be measured analytically in terms of time, labor, production levels, number of household members involved, assets invested, and income derived" (ibid.:9–10). To make such a system work, pastoralists often develop multifaceted economic strategies in which they utilize "livestock, agricultural, and artisanal production; wage labor; foraging; trade and smuggling; and, often, predation" (ibid.:4). This risk-reduction approach permits pastoralists to maintain some degree of control even in capitalist economies that might otherwise dominate their economic fortunes. On Dokos, we had a unique opportunity to examine a herding family in operation and thus to provide some empirical support for the model Koster and Chang have generated.

A general approach that can aid in pulling together the various elements discussed earlier into a coherent framework is world-systems analysis (WSA). Initiated through the work of Frank (1967) and Wallerstein (1974) and modified by many others, WSA emphasizes the ways different social units are interconnected. Wallerstein argued that there is an exploitative relationship between dominant cores and dependent peripheries, with semiperipheries often playing the role of middlemen. Critical to the approach is the process of incorporation, which can vary from weak to strong; at the weak end, peripheries maintain substantial autonomy and individuals have more options that diminish as the bonds of incorporation tighten, leading from core-periphery differentiation to core-periphery hierarchy (Chase-Dunn and Hall 1997:59–65).

Over time, world-systems can expand and contract, becoming more and less integrated. What the world has witnessed over the past two centuries in particular is a massive expansion of this system, a period of intense integration many people now identify as globalization. In this process, many groups have lost local control of essential economic structures and, consequently, of their political systems. We see evidence of this hegemony in the intensive exploitation of Africa and Asia in the eighteenth and nineteenth centuries by European states, followed by the control of local economies by multinational corporations in the twentieth century. However, even as the pendulum of incorporation swung toward the strong end, there were numerous instances of both resistance and efforts to take advantage of certain situations by those people on the margins (Carlson 2012; Hall 2012). Historically, people on the peripheries of world-systems have been able to gain some benefits through a process of negotiation in which they selectively accept or reject certain aspects of intrusive cultures (Kardulias 2007; Hall, Kardulias, and Chase-Dunn 2011:255–61). Furthermore, in the modern world-system, with its complex economic network, people can move back and forth between occupations in an effort to garner the most benefits under various circumstances. These shifting economic personae can take several forms. In some cases, people openly and legally hold down two or more jobs as a way to supplement income. Others may participate in a shadow economy by taking odd jobs for which they receive payment "off the books" and thus pay no taxes, in addition to having a position in the formal economy. In these ways, individuals attempt to increase the liquidity of their labor, and globalization offers certain opportunities for such activity in addition to the restrictions it imposes in other ways.

The discussion that follows presents some examples of how individuals attempt to mix elements of traditional subsistence with those of the modern formal economy. Furthermore, this flexible strategy is the way people strike a balance between the insularity and connectivity that define island life; these elements form a sliding scale, with more or less expression of one trait depending on the prevailing conditions that can include local, regional, and international economic conditions; political circumstances; physical distances between landforms; and—of particular importance in the present study—immediate subsistence needs.

ETHNOARCHAEOLOGY

For the purposes of this study, ethnoarchaeology is defined as "the subfield of anthropology in which an archaeologist (or one sufficiently attuned to archaeologists' problems) does ethnographic fieldwork with the ultimate goal of providing ethnographic information of particular use to the archaeologist . . . Ethnoarchaeology may be done at various levels of complexity depending on the ultimate use to which the data [are] to be put" (White 1974:107). This definition is a composite of notions expressed by Gould (1968, 1971), Binford (1967, 1968), Watson (1979), and others. The "archaeologists' problems" referred to are the difficulties associated with translating inanimate objects into a model of behavior. Although the problems are substantial, they are not insuperable.

There are three levels of complexity, as noted by Gould (1971:175) and redefined by White (1974). The initial stage (Gould's practical level, White's Level One) involves the use of informants to determine such things as where to dig, the age of a site, known occupants, and similar factors. Such supplemental information "can increase the efficiency and scope of archaeological survey" (Gould 1971:175). The second stage (Gould's specific level, White's Level Two) attempts "to gain insights into the manufacture, function, classification, etc. of material culture items" (White 1974:102). At its most complex (Gould's level of general interpretation, White's Level Three), ethnoarchaeology is concerned with "broad interpretations of culture history" (Gould 1971:175), which can include aspects such as social organization, exchange networks, and ritual activity. The present study operates at all three of these levels. While the focus of the fieldwork was on the collection of specific information about the residents and their pastoral activities on the island, the larger goal is to comprehend the interlocking economic system of which such activity is a part at present and what that may tell us about the exploitation of marginal environments in the past. In addition, the present study demonstrates some of the ways in which ethnoarchaeology specifically aids an understanding of ancient pastoral societies (see Kuznar 1995).

ETHNOARCHAEOLOGICAL RESEARCH

During the four days on the island, we became acquainted with Nikos Douskos and his wife, who keep a herd of goats and sheep in a farm complex in the saddle where the chapel is located (see figure 11.4). Their willingness to talk to us about their lifestyle became the impetus for an ethnoarchaeological and ethnographic study. The focus of this investigation is the nature of adaptation to the circumscribed resources of Dokos (and by implication other, similar islands). The residents, past and present, have had to and continue to deal with the lack of a natural water source on the island, limited amount of land suitable for agriculture, and the delicate balance between human and animal populations in terms of the availability of food. The advantages of the island are that it is close to Idhra and the coast of the southern Argolid and sits astride a major transportation and communication route for the eastern seaboard of Greece. Key research questions include: (1) How do the residents manage their resources to make life viable on the island? What are the key factors in making what are obviously strategic decisions about the disposition of manpower and other resources? (2) Under what conditions did the residents come to the island and why did they stay, considering the numerous difficulties associated with life on Dokos? (3) To what degree are they self-reliant? Can they get by without frequent (how often?) contact with larger populations on other islands and the mainland? (4) What archaeological and historical implications are there for past residents of similar

FIGURE 11.4. Nikos Douskos demonstrating the use of a glass sherd as a wood scraper.

places (e.g., Makronisos, Evraionisos)? (5) Do the residents preserve traditional practices and technologies to a greater extent than their neighbors? How has mechanization affected their lives? (6) What is the nature of the land tenure system on the island? What role do social relations play in this system? (7) What is the future of such economic activity on the island? How will or do other aspects of the economy (such as tourism, fishing) affect their activities and plans for the future?

During our stay on the island, we had a number of impromptu conversations with the Douskos family; we also conducted a long formal interview in which we systematically recorded data. In addition, we were able to observe them conducting a number of activities directly related to herding. Finally, we drew plans of the agricultural and pastoral complex in the area of the saddle. We plotted all the buildings, locations of special features (i.e., animal folds, cheese-making shed, trash pits, threshing floor), and the distribution and kinds of artifacts around some of the locations. It became clear that, although modern technology and transportation have ameliorated conditions on Dokos, life on the island is still difficult and requires a finely balanced accommodation to local and regional conditions. Successful adaptation to the often harsh conditions requires careful planning of activities. It is also clear that success requires integration into a large regional interaction sphere for the acquisition of important commodities, sale of various products (cheese, animals for meat, olive oil), and the periodic use of other land.

FIGURE 11.5. The saddle at the north end of Dokos, where the chapel and Douskos house are located.

Nikos Douskos came to Dokos with his family in 1945 at age fifteen. At the time, there were twenty-two families on the island. The people cultivated wheat and olives in the rocky earth and raised goats and sheep, which could graze freely over most of the island (see figure 11.5). The men in two of the families were sponge divers and were gone for part of each year collecting sponges throughout the Aegean Sea and elsewhere. Sponge fishing is an excellent example of a local activity that blossomed into a major industry through globalization; sponges from the eastern Mediterranean became important commodities in the world-system in the nineteenth century and brought significant wealth to various islands such as Kalymnos in the eastern Aegean (Bernard 1976).

Most of these people went to Dokos from Idhra to have some means of survival in the lean years of World War II and its aftermath, when the Greek economy recovered slowly. Douskos married the daughter of a local landowner and practiced a mixture of herding and agriculture with his in-laws for many years. Douskos and his wife raised three children on the island; all three now reside on Idhra and have offered their parents a place to stay on the big island. However, Douskos and his wife choose to stay on Dokos and look after their herds, despite the complaints about the hard nature of the work.

From the interviews and our observations, we noted that the Douskos family has about 140 goats, 35 sheep (I counted 90 goats in one pen and 27 sheep in another

FIGURE 11.6. Mules and donkeys grazing on an abandoned agricultural terrace.

pen), a small flock of turkeys and chickens (about 30 total), 2 dogs, 2 mules, and 2 donkeys (figure 11.6). In the past, they used to have several oxen for plowing. They have pens for the sheep and goats, where they keep the animals from about sunset to dawn. Early each morning they release the herd, which scampers down the hillside toward the sheltered bay (probably the Early Byzantine harbor) on the west side of the island. At the base of the hill is a large aboveground stone cistern that gathers runoff from the surrounding hills and provides drinking water for both the herders and their animals. Douskos or his wife opens the faucet and fills a variety of basins, which include an old metal bathtub, to water the animals. After that, the herd is allowed to roam freely to graze during the rest of the day. He supplements their diet with cornmeal, which he pours into feeding troughs outside the pens.

Douskos milks the goats and sheep on a regular basis. This is typically an early-morning chore. On the day when I observed him, he herded the animals into an oval-shaped roofed pen. After selecting an animal with swollen udders, he chased it around the enclosure until he caught it by the neck with the crook of his shepherd's staff. He pulled in the animal until he could take hold of a hind leg and then the udder, then he positioned the metal milk pail beneath the udders; milking each animal took about one to two minutes. He continued this procedure until he had filled the pail, milking a total of about eight animals in the process. The family members consume some of the milk themselves and sell some when they go to Idhra, but the bulk of it is used for the production of cheese. The main milking season begins in February, after the lambs and kids are weaned. Birthing takes place in May. Until recently, they produced about 100 kilos of *kefalotiri* and a small amount of feta each year, most of which they sold to merchants on Idhra or the mainland. Douskos stated that sheep milk is fatter and thus better for making cheese. They produced the cheese in a small shed near the animal pens. The shed is made of flat stones stacked with no mortar. A fireplace in one corner serves to heat the milk to the necessary 30–32°C. The equipment in this facility did not appear to have been used recently.

The other main product the Douskos family sells is meat. Greek law does not permit them to butcher the animals themselves to sell to the public. Periodically, they will take a group of 20–30 animals to Idhra, where they sell them to a butcher named Gounaris; they tend to get a better price (100+ drachmas more per kilo) if they sell the animals on Idhra as opposed to Ermioni on the mainland. It was clear that he understood the value of the animals. In the past, when his father-in-law was still alive, the extended family would take their herd to the mainland near Ermioni, where they would rent pasture for a period of 5–6 months between early spring and mid-fall (ca. March-April–October-November). They often rented hillsides where the animals could graze. They tended to pay the rent in various animal products—milk, cheese, and some wool. During this time, they lived in huts they built on the rented land. This practice is similar to that of the Valtetsiotes, who in the past alternated between the highlands of Arkadia and the coastal area around Ermioni in the southern Argolid (Koster 1976). For the sixteen years or so during which the Douskos family took their herd to the mainland, they were members of a milk cooperative. One benefit of the cooperative was an immediate outlet for their milk that could provide cash and removed the burden of marketing their product. Another benefit was the prevention of overgrazing of the vegetation on Dokos, so that when they returned to the island for the late fall and winter, the animals would have adequate forage.

The Douskos family lives in a complex that dates to the 1940s (figure 11.7). They reside in a two-story stone house built in 1946. Another house about 30 m to the southwest, with a stone courtyard, was built by another family but has not been inhabited for a number of years. Adjoining this house to the west and south is a series of animal pens, some roofed and others open to the sky. The cheese-making shed is 20 m south of the main residence. The entire complex is set on several terraces at the base of the large hill crowned by the Early Byzantine *kastro* on the northern edge of the saddle (i.e., it has a southern exposure). The modern chapel of Agios Ioannis, surrounded by a rubble wall surmounted by a wire fence to keep out the goats, lies 100 m south-southeast of the main residence on the flattest part of the

FIGURE 11.7. Douskos complex on Dokos.

saddle. The Douskos family maintains the chapel and brings a priest several times a year to celebrate the liturgy on special occasions. Scattered around the houses and pens are several garbage dumps that consist of a wide range of debris. On the terrace in front (south) of the main house, the trash included metal camping gas canisters, part of a metal anchor, many glass bottles, metal cans, a metal chair frame, and some animal bones. A deep pit on the north side of the unoccupied house was originally built as a cistern, but the project was never completed. This rectangular hole is over 2 m deep and serves as a major refuse pit. At the far northwest corner of the complex, one of the Douskos sons, a building contractor on Idhra, built a new house for use as a country home on visits to the island.

The next topic of concern is the diversified economic strategy of the Douskos family. While the family has considered Dokos its primary residence for over fifty years and has lived on the island year-round for some time, the members have exhibited a rather flexible settlement strategy linked to economic pursuits. As mentioned, for sixteen years the clan spent a significant part of each year on the mainland near Ermioni. In addition, Nikos Douskos spent sixteen years in the merchant marine and traveled to Brazil on a number of occasions. While it was not possible to get precise dates, it seems that he did this work from the early to mid-1950s to the late 1960s or early 1970s, during which time he spent about half the year onboard ship and the rest of the time resident on Dokos. His retirement pay from the merchant marine amounts to 100,000 drachmas ($1US = 270 drachmas in 1997) per month. It is in part because his pension is not very large that Douskos and his wife continue to herd animals on Dokos. It is clear from this brief history that the Douskos family and its members have been involved in both the traditional and modern economies for essentially all of their time on Dokos. One might view the shifting balance among herding, farming, work as a merchant seaman, and selling animal products as the way these people have negotiated their economic status for more than half a century. The issues of insularity and connectivity come into clear view in this strategy with multiple branches.

In the mid-1970s, Douskos says he used his boat to take tourists to and from various locations in the area, for example, from Idhra to Ermioni and from Kosta to Spetses. By the 1990s regular water taxis handled the tourist and local traffic. At the time of the interviews, he owned two small boats primarily for personal travel between Dokos and Idhra.

Another aspect of the Douskos family's economic strategy was the effort to be self-sufficient. In the past, it was important to provide as many products as possible for family consumption because of the poverty (Nikos emphasized on several occasions how poor he was in the early days on Dokos) everyone experienced, especially in the difficult years of the 1930s and 1940s. The family practiced a mixture of subsistence and commercial farming and herding. While they sold animals and various by-products, the family also produced wheat for bread, olives for oil, and peas and several other vegetables for family consumption. In addition, they used to manufacture their own soap and some of their own clothing.

Still, these were clearly not subsistence farmers. As difficult as conditions might have been, they were involved in the regional economy and to some extent in the international system. This embedded form of incorporation required the shifting of labor efforts but did spread economic risk. That strategy has undergone alteration with the next generation. While the Douskos couple has stayed on the island, their children are enjoying a middle-class existence on Idhra. In addition, Mrs. Douskos's brother, who also has claim to much of the land on the north part of the island, owns and runs a successful tavern on the sheltered bay, which many Greek and foreign tourists visit. Furthermore, the Douskos family is clearly aware of subsidies the European Union provides to Greek pastoralists. They also receive visits from the agricultural agent based on the island of Poros and a veterinarian; they get prescription drugs for the animals from Ermioni. In brief, these hardy people following a rigorous life on an island with no natural source of fresh water have lived and even thrived as a result of their connections to neighboring islands and the mainland. The relationship is dynamic. The Douskos could live on Dokos only because the other communities offered markets for their produce. Conversely, the people of Idhra and the region around Ermioni have benefited from their relationship with the residents of Dokos. The latter have clearly not been just passive recipients in this exchange. Indeed, the Douskos, and others on the island in the past, made pastoralism a viable economic alternative because they carefully integrated it with a number of other activities. This ingenious use of land and other resources has made seemingly inhospitable settings places where people have earned a livelihood for at least three millennia. It remained a successful strategy because it was one that did not take undue risks. What Kuznar (1995:18) says about Andean herders is appropriate for Aegean islanders: "Awatimarka herders focus on animals that not only produce income, but that are particularly resilient in the face of hazards such as drought, disease, and predation. This indicates that pastoral behavior tends to be risk-averse, and one would expect herders to make economically conservative decisions in the management of their herds."

ARCHAEOLOGICAL IMPLICATIONS

This ethnoarchaeological project has a number of implications for ancient life on Dokos and other so-called desert islands around the Peloponnesos. Hood (1970) suggested that people moved to these islands, which lack springs, only under the duress of the Slavic invasions in the sixth century AD. More recent work on islands in the Korinthian Gulf (Gregory 1986) and the Saronic Gulf (Kardulias, Gregory, and Sawmiller 1995) has proved that occupation of such marginal zones actually began in the Bronze Age, with a series of peaks in Classical, Late Roman, and Late Byzantine times. My colleagues and I have suggested that people moved to these dry islets not to seek refuge but to exploit additional land in periods of economic expansion (see figure 11.8).

The work on Dokos helps us envision the nature of such an economic system. Dokos was intensively inhabited from early in the Bronze Age, with substantial communities at several times, including the Late Roman and Early Byzantine eras (fourth–seventh centuries AD), as the large fortified settlement at the north end of the island demonstrates. The current study provides some clues about how such a rather large population could survive in an austere arid environment. Many of the agricultural terraces on Dokos may be quite old and suggest intensive cultivation. The Douskos family grew wheat, peas, and olives on the slopes and in the saddle around their domestic complex. They threshed and winnowed the grain on a floor on the north slope. Douskos indicated that in the 1940s and early 1950s they could grow and process two to four tons of wheat annually, enough to meet most of their needs. It is clear, however, that herding was critical to survival. Not only did

FIGURE 11.8. A Late Roman to Early Byzantine period cistern built around a natural cave on the island of Evraionisos in the Saronic Gulf.

the sheep and goats provide meat and milk for immediate consumption, but they also were the source of cheese the residents sold to people on Idhra and the mainland. The infusion of cash allowed for the purchase of certain manufactured goods, additional food, and construction materials (such as roof tiles, cement). While the Douskos family was tied into the regional, national, and even international economies by virtue of their various activities, the basic level of interdependence was most likely a feature that reflects a very old pattern of adaptation. In the past, the meager resources on islands such as Dokos could support communities only if the residents maintained close links to other islands and the neighboring mainland. The practice of moving herds of sheep and goats by boat between various islands and the mainland to avoid overgrazing must have required contractual arrangements with the occupants of those other regions.

While there is no direct evidence of herding by early residents of Dokos, ancient literary sources attest to such practices on a variety of islands. For example, Longus's novel *Daphnis and Chloe,* from the second to third centuries AD, about herding on Lesbos discussed this process. Bremmer (1986) argues that several ancient authors, including Homer, mention "goat islands" in the Aegean that most likely provided pasture for caprids; the practice seems to have been common in the central and western Mediterranean as well. Bean and Cook (1957:133) identified the small islet of Gaidouronesi near Kalymnos in the eastern Aegean as one such place where shepherds periodically brought their herds in the mid-twentieth century. Greaves (2010:56) notes that small islands near the coast of Ionia in western Asia Minor have been used in this manner in modern times, and he suggests that exploitation of these areas for grazing was important in antiquity as well. It is clear that this practice of moving herds to small, often uninhabited islands was a common practice for a number of centuries and demonstrates some of the linkages that existed in world-systems.

One goal of any future archaeological research would be to provide details of such subsistence on Dokos or other, similar islands. I envision a fluid economic system, with people exploiting a variety of eco-niches. Herding was in some ways the key linchpin in this system, since it offered mobile wealth and required the inhabitants of Dokos to make numerous connections with people and regions beyond the confines of their island. I believe this model of an integrated regional economy applies equally well to ancient and medieval residents of Dokos and similar islands.

CONCLUSION

The viability of archaeological reconstruction of past life-ways depends largely on the development of appropriate analogies. Since we interpret ancient artifacts, sites, and patterns of activity by means of comparison with similar contemporary phenomena, archaeologists must strive to build convincing models of behavior. Careful attention to ethnographic detail aids immensely in this important endeavor. The purpose of the study on Dokos was to enhance our ability to explain how herding of domesticated animals played, and to some extent still plays, an important role in adaptation to areas in which good land for grazing and farming is at a premium. The present inhabitants of Dokos, and by implication its past residents, have adopted a flexible subsistence pattern in which herding provides important food items and commodities for exchange. In addition, important social and economic relationships often derive from the requirements of a herding lifestyle as played out in this insular setting. An intricate web of such connections ties the herders to people on other islands and the mainland, in a system of mutual benefit. Such studies make archaeologists aware of the range of behavior they must consider in their attempts to explain past societies. Finally, one gains a greater appreciation for the ingenuity, persistence, and ability to cope with difficult conditions, which many "traditional" peoples exhibit. The peasants of the Greek countryside have proved to be resilient and resourceful in the face of the encroaching world-system.

ACKNOWLEDGMENTS

Timothy E. Gregory of Ohio State University, the supervisor of the investigations on Dokos, graciously provided the opportunity to undertake the research. The Greek Ministry of Culture and Science granted the permit to conduct the fieldwork. Lita Diacopoulos, my collaborator on this project, helped frame the questions and conduct the interviews and prepared drawings of the several locations on the site; unfortunately, the plans did not survive the return trip to the mainland. Stelios Nomikos assisted us very ably in the field.

REFERENCES

Bean, George E., and John M. Cook. 1957. "The Carian Coast III." *Annual of the British School at Athens* 52: 58–146. http://dx.doi.org/10.1017/S0068245400012909.

Bernard, H. Russell. 1976. "Kalymnos: The Island of the Sponge Fishermen." In *Regional Variation in Modern Greece and Cyprus: Toward a Perspective on the Ethnography of Greece*, ed. Muriel Dimen and Ernestine Friedl, 289–307. Annals of the New York Academy of Sciences 268. New York: New York Academy of Sciences.

Binford, Lewis R. 1967. "Smudge Pits and Hide Smoking: The Use of Analogy in Archaeological Reasoning." *American Antiquity* 32 (1): 1–12. http://dx.doi.org/10.2307/278774.

Binford, Lewis R. 1968. "Methodological Considerations of the Archaeological Use of Ethnographic Data." In *Man the Hunter*, ed. Richard B. Lee and Irven Devore, 268–73. Chicago: Aldine.

Braudel, Fernand. 1972. *The Mediterranean and the Mediterranean World in the Age of Philip II*, vol. 1. Trans. Siân Reynolds. Berkeley: University of California Press.

Bremmer, Jan N. 1986. "A Homeric Goat Island (Od. 9.116–41)." *Classical Quarterly* 36 (1): 256–57. http://dx.doi.org/10.1017/S0009838800010703.

Broodbank, Cyprian. 2000. *An Island Archaeology of the Early Cyclades*. Cambridge: Cambridge University Press.

Broodbank, Cyprian. 2006. "The Origins and Early Development of Mediterranean Maritime Activity." *Journal of Mediterranean Archaeology* 19 (2): 199–230. http://dx.doi.org/10.1558//jmea.2006.v19i2.199.

Broodbank, Cyprian, and Thomas F. Strasser. 1991. "Migrant Farmers and the Neolithic Colonization of Crete." *Antiquity* 65: 233–45.

Campbell, John K. 1964. *Honour, Family and Patronage: A Study of Institutions and Moral Values in a Greek Mountain Village*. Oxford: Clarendon.

Carlson, Jon D. 2012. "Externality, Contact Periphery and Incorporation." In *Routledge Handbook of World-Systems Analysis: Theory and Research*, ed. Salvatore J. Babones and Christopher Chase-Dunn, 87–96. New York: Routledge.

Chang, Claudia. 1981. "The Archeology of Contemporary Herding Sites in Didyma, Greece." PhD diss., Department of Anthropology, State University of New York, Binghamton.

Chang, Claudia. 1993. "Pastoral Transhumance in the Southern Balkans as a Social Ideology: Ethnoarchaeological Research in Northern Greece." *American Anthropologist* 95 (3): 687–703. http://dx.doi.org/10.1525/aa.1993.95.3.02a00080.

Chang, Claudia. 1994. "Sheep for the Ancestors: Ethnoarchaeology and the Study of Ancient Pastoralism." In *Beyond the Site: Regional Studies in the Aegean Area*, ed. P. Nick Kardulias, 353–71. Lanham, MD: University Press of America.

Chang, Claudia. 1997. "Greek Sheep, Albanian Shepherds: Hidden Economies in the European Community." In *Aegean Strategies: Studies of Culture and Environment on the European Fringe*, ed. P. Nick Kardulias and Mark T. Shutes, 123–39. Lanham, MD: Rowman and Littlefield.

Chang, Claudia. 2000. "The Material Culture and Settlement History of Agro-Pastoralism in the Koinotis of Dhidhima: An Ethnoarchaeological Perspective." In *Contingent Countryside: Settlement, Economy, and Land Use in the Southern Argolid since 1700*, ed. Susan B. Sutton, 125–40. Stanford: Stanford University Press.

Chang, Claudia, and Harold A. Koster. 1986. "Beyond Bones: Toward an Archaeology of Pastoralism." *Advances in Archaeological Method and Theory* 9: 97–148.

Chang, Claudia, and Harold A. Koster, eds. 1994. *Pastoralists at the Periphery: Herders in a Capitalist World*. Tucson: University of Arizona Press.

Chase-Dunn, Christopher, and Thomas D. Hall. 1997. *Rise and Demise: Comparing World-Systems*. Boulder: Westview.

Chase-Dunn, Christopher, and Thomas D. Hall, eds. 1991. *Core/Periphery Relations in Precapitalist Worlds*. Boulder: Westview.

Cherry, John F. 1981. "Pattern and Process in the Earliest Colonization of the Mediterranean Islands." *Proceedings of the Prehistoric Society* 47: 41–68.

Cherry, John F. 1985. "Islands out of the Stream: Isolation and Interaction in Early East Mediterranean Insular Prehistory." In *Prehistoric Production and Exchange: The Aegean and Eastern Mediterranean*, ed. A. Bernard Knapp and Tamara Stech, 12–29. UCLA Institute of Archaeology Monograph 25. Los Angeles: UCLA Institute of Archaeology.

Cherry, John F. 1990. "The First Colonization of the Mediterranean Islands: A Review of Recent Research." *Journal of Mediterranean Archaeology* 3: 145–221.

Cherry, John F. 2004. "Mediterranean Island Prehistory: What's Different and What's New?" In *Voyages of Discovery: The Archaeology of Islands*, ed. Scott M. Fitzpatrick, 233–48. Westport, CT: Praeger.

Evans, John D. 1973. "Islands as Laboratories of Culture Change." In *The Explanation of Culture Change*, ed. Colin Renfrew, 517–20. Pittsburgh: University of Pittsburgh Press.

Frank, Andre Gunder. 1967. *Capitalism and Underdevelopment in Latin America: Historical Studies of Chile and Brazil.* New York: Monthly Review.

Frost, Frank. 1997. *Greek Society*, 5th ed. New York: Houghton Mifflin.

Gould, Richard A. 1968. "Living Archaeology: The Ngatatjara of Western Australia." *Southwestern Journal of Anthropology* 24: 101–22.

Gould, Richard A. 1971. "The Archaeologist as Ethnographer: A Case from the Western Desert of Australia." *World Archaeology* 3 (2): 143–77. http://dx.doi.org/10.1080/00438243.1969.9979499.

Greaves, Alan M. 2010. *The Land of Ionia: Society and Economy in the Archaic Period.* Malden, MA: Wiley-Blackwell.

Gregory, Timothy E. 1986. "A Desert Island Survey in the Gulf of Corinth." *Archaeology* 39 (3): 16–21.

Gregory, Timothy E. 1997. "Dokos." *Archaeological Reports for 1996–1997.* 43: 26–27.

Hall, Thomas D. 2012. "Incorporation into and Merger of World-Systems." In *Routledge Handbook of World-Systems Analysis: Theory and Research*, ed. Salvatore J. Babones and Christopher Chase-Dunn, 37–55. New York: Routledge.

Hall, Thomas D., P. Nick Kardulias, and Christopher Chase-Dunn. 2011. "World-Systems Analysis and Archaeology: Continuing the Dialogue." *Journal of Archaeological Research* 19 (3): 233–79. http://dx.doi.org/10.1007/s10814-010-9047-5.

Held, Steve O. 1990. "Sardinia to Samoa: Studying Comparative Insularity and Patterns of Prehistoric Island Colonization." Paper presented at the 89th Annual Meeting of the American Anthropological Association, November 28–December 2, New Orleans, LA.

Held, Steve O. 1993. "Insularity as a Modifier of Cultural Change: The Case of Prehistoric Cyprus." *Bulletin of the American Schools of Oriental Research* 292: 25–33. http://dx.doi.org/10.2307/1357246.

Herzfeld, Michael. 1985. *The Poetics of Manhood: Contest and Identity in a Cretan Mountain Village.* Princeton: Princeton University Press.

Hesiod. 1983. *Works and Days.* Trans. Apostolos N. Athanassakis. Baltimore: Johns Hopkins University Press.

Hood, Sinclair. 1970. "Isles of Refuge in the Early Byzantine Period." *Annual of the British School at Athens* 65: 37–45. http://dx.doi.org/10.1017/S0068245400014684.

Kardulias, P. Nick. 2007. "Negotiation and Incorporation on the Margins of World-Systems: Examples from Cyprus and North America." *Journal of World-Systems Research* 13: 55–82.

Kardulias, P. Nick, Timothy E. Gregory, and Jed Sawmiller. 1995. "Bronze Age and Late Antique Exploitation of an Islet in the Saronic Gulf, Greece." *Journal of Field Archaeology* 22: 3–21.

Knapp, A. Bernard. 2008. *Prehistoric and Protohistoric Cyprus: Identity, Insularity, and Connectivity*. Oxford: Oxford University Press.

Koster, Harold A. 1976. "The Thousand Year Road." *Expedition* 19 (1): 19–28.

Koster, Harold A. 1977. "The Ecology of Pastoralism in Relation to Changing Patterns of Land Use in the Northeast Peloponnese." PhD diss., Department of Anthropology, University of Pennsylvania, Philadelphia.

Koster, Harold A. 1997. "Yours, Mine and Ours: Private and Public Pasture in Greece." In *Aegean Strategies: Studies of Culture and Environment on the European Fringe*, ed. P. Nick Kardulias and Mark T. Shutes, 141–85. Lanham, MD: Rowman and Littlefield.

Koster, Harold A. 2000. "Neighbors and Pastures: Reciprocity and Access to Pasture." In *Contingent Countryside: Settlement, Economy, and Land Use in the Southern Argolid since 1700*, ed. Susan B. Sutton, 241–61. Stanford: Stanford University Press.

Koster, Harold A., and Claudia Chang. 1994. "Introduction." In *Pastoralists at the Periphery*, ed. Claudia Chang and Harold A. Koster, 1–15. Tucson: University of Arizona Press.

Kuznar, Lawrence A. 1995. *Awatimarka: The Ethnoarchaeology of an Andean Herding Community*. Fort Worth: Harcourt Brace.

Kyrou, Adonis. 1995. "Periplaniseis Agion Leipsanon kai Mia Agnosti Kastropoliteia ston Argoliko" [The Movement of Holy Relics and an Unknown Fortified Town in the Argolic Gulf]. *Peloponnisiaka* 21: 97–118.

MacArthur, Robert H., and Edward O. Wilson. 1967. *The Theory of Island Biogeography*. Monographs in Population Biology. Princeton: Princeton University Press.

Perlès, Catherine. 1987. *Présentation Générale et Industries Paléolithiques*, vol. I. Les Industries Lithiques Taillées de Franchthi (Argolide, Grèce). Bloomington: Indiana University Press.

Simmons, Alan H. 1991. "Humans, Island Colonization and Pleistocene Extinctions in the Mediterranean: The View from Akrotiri-*Aetokremnos*, Cyprus." *Antiquity* 65: 857–69.

Simmons, Alan H. 1999. *Faunal Extinction in an Island Society: Pygmy Hippopotamus Hunters of Cyprus*. New York: Kluwer Academic/Plenum.

Strasser, Thomas F., Eleni Panagopoulou, Curtis N. Runnels, Priscilla M. Murray, Nicholas Thompson, Panayiotis Karkanas, Floyd W. McCoy, and Karl W. Wegmann. 2010. "Stone Age Seafaring in the Mediterranean: Evidence from the Plakias Region for Lower Palaeolithic and Mesolithic Habitation of Crete." *Hesperia* 79 (2): 145–90. http://dx.doi.org/10.2972/hesp.79.2.145.

Tartaron, Thomas F., Timothy E. Gregory, Daniel J. Pullen, Jay S. Noller, Richard M. Rothaus, Joseph L. Rife, Lita Tzortzopoulou-Gregory, Robert Schon, William R.

Caraher, David K. Pettegrew, and Dimitri Nakassis. 2006. "The Eastern Korinthia Archaeological Survey: Integrated Methods for a Dynamic Landscape." *Hesperia* 75 (4): 453–523. http://dx.doi.org/10.2972/hesp.75.4.453.

Terrell, John. 1977. *Human Biogeography in the Solomon Islands*, vol. 68, no. 1: *Fieldiana*. Chicago: Field Museum of Natural History.

Wallerstein, Immanuel. 1974. *The Modern World-System: Capitalist Agriculture and the Origins of the European World-Economy in the Sixteenth Century*. New York: Academic.

Watson, Patty Jo. 1979. *Archaeological Ethnography in Western Iran*. Tucson: University of Arizona Press.

White, John R. 1974. "Ethnoarchaeology, Ethnohistory, Ethnographic Analogy, and the Direct-Historical Approach: Four Methodological Entities Commonly Misconstrued." *Conference on Historic Sites Archaeology Papers* 8: 98–111.

12

The Ecology of Herding

Conclusions, Questions, Speculations

Thomas D. Hall

Since the preface and introduction have summarized the aims of this collection and the gist of the various chapters, I will not repeat that information here. Rather, I make some general comments, draw a few conclusions, ask some questions, and speculate on topics for future studies of pastoralism. Throughout this discussion I try to link these studies, several highly localized, to larger issues in long-term macro-social change. The extreme variability and considerable volatility of herding societies, while at times frustrating, are superb for exploring complex changes in considerable detail and for making links between local, small-scale change and regional, large-scale change. I begin by noting, as have most of the authors in this collection, that what is meant by pastoralism, herding, nomadism, transhumance, and similar words is often not clear because so many writers have used the terms in so many ways. Furthermore, many of those definitions or descriptions have been heavily colored by the almost universal tendency of sedentary people to "look down their noses" at pastoral nomads, and some pastoralists respond in-kind. Sedentary people often fear raids yet often participate in trade, surprisingly often, both at the same time. When trading and raiding coincide, it is most often on a clan-by-clan or family-by-family basis rather than involving entire groups (examples in Hall 1989; Brooks 2002). Most problematic are the various ways analysts try to fit pastoralists into various larger schemes, generally evolutionary and often in support or denial of Marxist principles (chapters by Chang and Kradin, this volume).

DOI: 10.5876/9781607323433.c012

A few general conclusions—which means they fit many cases but not necessarily all—are that pastoralism is a continuum that runs from entirely dependent on animal resources, with no direct farming, to agriculture that includes some herding of animals. An analogous continuum of nomadism runs from highly localized transhumance to nearly permanent mobility. Furthermore, rooted in the typically derogatory view of nomads by sedentary peoples—which probably includes more than a modicum of jealousy—is the belief that nomads are aimless wanderers. The latter view distorts conceptions of indigenous peoples, especially Native Americans, who expectantly resent the characterization. Key here is that even for fully or nearly fully nomadic peoples, movement typically follows circuits that are mostly shaped by ecological needs but are almost never random. So, in short, "pastoral nomadism" actually refers to broad ranges of variation along several axes, as the preceding chapters collectively demonstrate.

Another common mistake is the reification of terms like nomad, pastoralist, herder, transhumant, and the like, and the use of general characteristics as applicable to all pastoralists. In addition to the axes of variation just noted, an important methodological or even epistemological error views typologies as capturing some sort of universal, if not cosmic, truth. Rather, as I and many others have argued (e.g., Hall 2009), typologies are tools. Like many tools used by archaeologists, tools that are vital for one kind of research (say, using a backhoe to remove overburden) are entirely inappropriate for others (say, deftly removing small artifacts from their surrounding matrix). What a good typology should do is divide a variety of instances into categories wherein most of the objects in one category are more similar to each other than they are to instances in other categories. Nearly always, some instances straddle the divides, hence my preference for continua. Furthermore, the criteria for a typology are more often shaped by the questions it is intended to help answer than by the objects themselves. This factor clearly demonstrates how overly simple (or even simplistic) typologies, for example, the classic steppe vs. sown, can blind researchers to much more complex relations and make it all but impossible to understand how they change through time.

That said, I turn to an account that caricatures such nuanced approaches by examining an evolutionary approach.

SOCIAL EVOLUTION AND PASTORAL NOMADS

Gerhard Lenski and Patrick Nolan (Lenski 1966; Nolan and Lenski 2011; see also Chase-Dunn and Lerro 2014) argue that there is a discernible evolutionary sequence of societal types: from foraging to horticultural to agrarian to industrial societies. They do note, however, some interesting side forms: fishing societies, herding societies, and maritime societies. While one might quibble about details, they make a useful typology for setting up comparisons. Lenski and Nolan argue that the typical foraging group is not sedentary and does not produce any significant storable surplus. Horticultural societies, however, do both. Now comes the clever comparison: fishing societies are typically sedentary but produce little surplus, whereas herding societies can produce significant surplus but are rarely sedentary, typically being nomadic or at least transhumant. They further note that *all* states (agrarian societies) emerge from horticultural societies but not from fishing or herding societies. From this they conclude that the key features of horticultural societies that allow them to sometimes transform into agrarian societies or states or civilizations is that they both produce a surplus and are sedentary. Furthermore, given that herding societies can and do become quite large and complex, rivaling Lenski's advanced horticultural societies, they argue that production of surplus is more important than being sedentary *but* that it remains that the coincidence of surplus production *and* sedentism is necessary for states to emerge. Maritime societies are actually variants of agrarian societies, which are far from self-sustaining. Rather, they are nodes in various trade networks. That is, though sedentary, they engage in minimal production but utilize extensive trade networks.

This entire analysis could be, and has been, hotly debated. Yet even if one rejects most of the Lenskian analysis, it does point to one reason why the study of herding or pastoral societies remains important. These societies underscore the coincidence of surplus production and sedentarism that allowed repeated inventions of states. They also suggest a refinement of socio-cultural evolutionary sequences. Pastoralism seems to emerge from horticultural or agrarian societies but not directly from foraging societies, although the Saami may be a salient exception to this general pattern. Once formed, whether deriving from horticultural or agrarian societies, they need to maintain some sort of relations with those societies, either through trading or raiding and most often a combination of both. In short, pastoral societies almost never exist in splendid isolation—even when they live on islands, as Kardulias shows.

There are at least five further reasons why the study of pastoralists is important:

1. Pastoralists often live on the edges, or frontiers, of sedentary societies and have complex relations with them, sometimes symbiotic, sometimes predatory, sometimes as a reserve army of labor or victims, and often all three.
2. Pastoralists are often major actors, or at least catalysts, in social change. Andre Gunder Frank (1992) argues that Central Asia is the pivot, mediated by pastoral nomads, to understand Afroeurasian development. Christopher Beckwith (2009; see also Hall 2010) also argues for a pivotal role for Central Asia.

Christopher Chase-Dunn and Thomas Hall (1997:ch. 8) extend that argument. Thomas Barfield (1989; see also Anatoly Khazanov [1983]) pointed out how steppe confederacies rose and fell in tandem with Chinese dynasties. That is, nomads did *not* cause the periodic collapses of China but rather collapsed with them (see also Barfield 2001). To be sure, their relationships with West Asians were more often of the kind that induced collapse. This leads us to the more familiar argument advanced by Ibn Khaldun, but, as Barfield points out, the valleys in West Asia are much smaller, so the sedentary-steppe interaction sphere is smaller, with considerably different consequences (Barfield 1990, 1991). Certainly, both the Diné (Navajo) and the people of Dokos fit somewhere in this range, albeit toward the smaller end of the continuum (Kuznar and Kardulias chapters, this volume).

Moritz's discussion of the FulBe Mare'en and the patrimonial state shows how groups can use the vagaries of frontier settings to their own advantage. This is a case of overly simple categories, for example, state, colonialism, and dependency, which are in fact large collections of types of relationships that can and often do change through time. Indeed, his example could be very useful in unpacking how and why pastoral nomads often play important, albeit commonly almost invisible, roles in macro-social change.

3. Precisely because pastoralists so often play pivotal roles in macro-social change, it is vitally important to study them "from the ground up." There is a need to examine how local actions, decisions, and processes condition the ways nomads influence larger social, political, economic, and cultural changes and in turn are influenced by them. A major contribution of all the chapters in this volume is the progress they make in this direction, however inchoate at this point. One can almost envision a mural being painted from both ends but the middle of which is yet to be completed. Both approaches, macro-social change and local analyses, have left loose ends that might be tied together but have not—yet. Any analysis that proceeds from one direction only will be flawed. Rather, it is the simultaneous pursuit of both that is most useful (Hall 1999, 2009, 2013a). There remains, however, a fourth set of issues and relations.
4. These derive from the different volatilities of pastoral and sedentary life noted in many chapters (also see Cribb 1991). They include both the non-coincidence of annual lean or dangerous times for pastoralists and farmers and conflicting needs by pastoralists for foraging land and by farmers for protection of ripening crops from hungry animals. The critical times for pastoralists are the times of birthing and survival of newborn animals, which are much more volatile by season and from year to year than is typically the case for agriculture. As Sidky

notes, investment risk can be controlled somewhat by diversifying herds. But this strategy is difficult for pastoralists who do not operate from a fixed base, as do the Hunzakutz. In contrast, for farmers the critical times are planting and harvest. Small changes in the first and last frost can potentially put an entire year's crop at risk. Farmers can manage investment risk by dispersing among soil types, plant varieties, and elevations. Once in a while lambing and planting coincide, putting both pastoralists and farmers at risk simultaneously.

This condition in part drives the oscillation between symbiotic trade and parasitic raiding. It also drives the phenomenon Owen Lattimore (1951 [1940], 1962), Frederik Barth (1969), and Gunnar Haaland (1969) have often noted: that individuals oscillate, migrate, and change from herder to farmer and back again with shifts in climate, ecology, and larger macro-social cycles. Just where the West Irish fit in this range is not entirely clear, not because of any failing in Shutes's presentation but because the time span he has been able to observe has been too short. Sidky's account of the Hunzakutz offers the tantalizing possibility that their mixed farming and herding strategy may have been yet another solution to these competing pressures. The fact that the region has several different languages and a seeming mix of religious rites is suggestive of an amalgamation of what were once different groups with different cultures. However, these shifts *do* sometimes occur more than once in an individual's lifetime. So, *why* do they occur at one scale or another? This is itself an interesting problem for further study and analysis. Shutes shows how the superior flexibility of a pastoral adaptation may drive such a change.

5. Herd animals that can be used for transportation, especially horses, often transform ways of making a living. Most obviously, travel over wider areas becomes more possible. When they are used in herding other animals or hunting other herd animals, changes can be drastic. The celebrated shift of Plains groups in what is now the Great Plains of the United States is probably the most familiar and dramatic example (Roe 1955; Hämäläinen 2003). But as David Anthony (2007) shows, there were tremendous consequences from the use of horses throughout Central Asia and the steppe territories generally (Honeychurch and Amartuvshin 2006), not the least of which involved movements of peoples, grains, other products, ideas, and languages. Several have argued that mounted pastoralists in Eurasia were the glue that held together larger systems (Beckwith 1991; Frank 1992; Hall 2005a, among others). They were vital to world-system formation and enlargement (Chase-Dunn and Hall 1997; Kradin et al. 2000; Kradin 2002; Kradin, Bondarenko, and Barfield 2003; Hall 2005b). Indeed, steppe pastoralists often played the role(s) of semipheripheral components in the larger world-system, the middle

> leg that stabilized the system, as well as being a transmission mechanism for goods and ideas. Peter Turchin and Thomas Hall (2003) show how even weak connections between sedentary civilizations may lead to synchronization of supposedly internal cycles. It may well be that for Afroeurasia, pastoral nomads supplied those weak connections. This, arguably, is how and why the rises and falls of West Asia/Rome and East Asia became somewhat synchronized, something Frederick Teggart (1939) pointed out long ago.

In North America, the spread of horses radically changed indigenous life-ways. Hunting buffalo (bison) became a viable way of life, leading some groups to all but abandon agriculture for horse herding and extensive buffalo hunting, especially among Plains groups (see examples in Hall 1989). The impact of horses spread far and wide, well beyond direct contact with Spanish settlers. This is why when Lewis and Clark explored the northwest territories of the expanding United States, they encountered peoples who had long since adopted horse herding and riding as a way of life, even though these explorers were probably the first Europeans the natives encountered directly (Fenelon and Defender-Wilson 2004). Also, as inter-group raiding became more intense, virtually all groups were pressured to move or take up mounted hunting to avoid being easy targets for rivals. Comanche bands indeed became "lords of the south plains" (Wallace and Hoebel 1952) precisely because they sat squarely in the middle of the exchange of horses from the Southwest to the Northeast and of guns in the reverse direction. Pekka Hämäläinen (1998, 2003, 2008) argues persuasively that Comanches were able to capitalize on this position to build a large trading "empire" that lasted until expansion of the United States preempted their trade networks. This ability to capitalize on the mobility to trade or raid with multiple partners gave nomadic groups, especially those with herds, considerable stability, since they had both multiple suppliers and multiple buyers. Thus, they had alternatives and could play one group against another. Here a bit of counterfactual thinking suggests that had a rise in industrial technology not led to definitive US conquest and domination, Comanches might have gone on to play roles similar to those played by pastoral nomads in Central Asia. While highly debatable, this is a speculation well worth pursuing to tease out just which factors, in what combination, can lead to the rise of large pastoral federations.

In regard to point 3, how local changes affect larger changes, Sidky's analysis is suggestive of how people make strategic choices to minimize risks by combining strategies. One might expect such adjustments in highly volatile or marginal environments. The Diné, as Kuznar discusses, seem to have made an analogous set of adaptations when they combined sheep herding with farming. Indeed, they often move in and out of canyons with their herds through the seasons. In both cases, trade and occasionally raiding were supplements to these strategies. One thinks of Fernand Braudel's (1979) term *bricolage*, that is, using whatever is at hand to get by. It seems that what is at work in these cases is the *interaction* of multiple local processes and pressures with wider processes and pressures. Chang shows not only that mixed-strategy economic systems are probably quite common but also that their specific configurations can have significant effects on gender roles and stratification.

This is where connections with similar processes elsewhere in time and space could prove useful. They would help determine how unusual—or typical—these specific changes are. They might also help to discern just what it is they show about such general processes. For instance, Lance Blyth (2012; see also Hall 2013b) gives a very detailed and elaborately nuanced account of Chiricahua-Spaniard interactions over several centuries in northern New Spain and what is now the southwestern United States. He paints a vivid picture of the complexity of continuing interaction. Unfortunately, he does not address larger processes of change, save to discuss communities of violence. His account is an invaluable examination of the trading-raiding oscillations.

William Robbins (1994) argues that the entire American West, in the American era at least, was shaped most forcefully by the expansion of capitalism. Similarly, Eliott West (1998) offers insight into the demise of the Plains Indians cultures under the onslaught of European invasion. I would also argue that what is described here, among other things, is a process of ethnogenesis in which the Diné (Navajo) and other groups were being invented, or at least drastically reshaped, by these processes. My argument is *not* that Diné (or other groups) were "invented" out of whole cloth. Rather, it is that their sense of identity shifted radically over the time Kuznar discusses. While virtually all these people called themselves Diné (the people or the humans), they did not share any overarching political or social unity. However, as the use of horses and sheep spread and pressure from Spaniards and then European Americans increased, their collective identity strengthened and then solidified when the United States gathered them on a reservation and forced a "tribal" (i.e., national) government on them. Anyone who has lived for any time in the Navajo Nation in the last forty years or so has seen how that identity and tribal government have become thoroughly indigenized by the Navajo people. This is an example of the formation and solidification of symbolic ethnicity Joane Nagel (1996) discusses so well. As part of these changes, Navajo sheep have come to symbolize traditional identity and practice, a point William Adams (1963) notes. Louise Lamphere (1989, 2007) also documents the important symbolic role of sheep in contemporary Navajo life without vitiating their continual roles in ecology and economy.

Thus, while of decreasing marginal economic significance, sheep and cattle remain of symbolic significance. This, too, is something that appears in many accounts of pastoral peoples.

Also quite common, and noted in Nagel's discussions, is that ethnicity is often shaped, modified, and transformed in the interaction of local social processes with assorted external pressures. This is most assuredly not a simple process (see Chang, this volume). Sorting out the mixes of internal and external pressures is a major task. It does seem, however, that pastoral nomads and mounted nomadic hunters are especially adept at manipulating these pressures to their own advantage.

Johannesson's discussions show how the complications of ethnicity can become even more complex in archaeological settings. He does make a persuasive case for exploring these processes in such a setting. Indeed, he presents good advice for ways such recondite issues as ethnic identity, especially when it is changing, can be studied in the archaeological record, as he does for the Xiongnu. In what might seem like an entirely unrelated area, Anne Ross and colleagues (2011) discuss the complex ways local indigenous groups can work with various nongovernment organizations to enhance environmental conditions. In the case studies they discuss, it is clear that the indigenous people often have a better understanding of complex ecological relations. Examination of the techniques Ross and colleagues discuss might provide further insights into how such complex relations might be explored in an archaeological context.

Here it is worthwhile to recap a thorny empirical and theoretical issue in world-systems analysis, one that probably plays a significant role in the origin of states. This is the question of when, how, and why a world-system of interacting societies with different levels of complexity, called core-periphery differentiation, can transform into a system in which the members are hierarchically ranked, called core-periphery hierarchy. Research shows that while this transformation is not rare, neither does it always occur. This transformation may be one of the keys to state formation. Notable here is that it is part of system change, not change in an individual society (for elaboration, see Chase-Dunn and Lerro 2014). Kradin's chapter discusses a number of factors that promote hierarchy. The reasons behind migrations of nomadic pastoralists are equally complex.

Sidky shows how the mixed strategy of farming and herding can be yet another solution to the macro- and micro-pressures that induce the oscillations between raiding and trading on the one hand and herding and farming on the other. Was this an amalgamation of formerly competing or occasionally symbiotic groups? Is it the result of some older core-periphery hierarchy? Indeed, if this is an instance of core-periphery differentiation *without* hierarchy, it becomes all the more interesting because long-term maintenance of core-periphery differentiation that does not turn into hierarchy is relatively rare, although not as rare, it seems, as a transition from hierarchy to differentiation (see Chase-Dunn and Hall 1997, 2012). It is for this reason that the question of how this adaptation originated—likely more than one time and in more than one way—remains so fascinating. This question is clearly beyond the scope of intention of Sidky's chapter, but I hope he or others will address this issue in future work.

Equally fascinating but again not fully developed in this chapter is how ritual gender restrictions shape adaptive strategies, ones that prove ecologically sound. What else is going on that reinforces a ritual taboo of women in the high country and also calls forth an allowance for ritual exception? This, too, is an interesting topic, deserving further work.

Kardulias returns full circle to examine how a pastoral adaptation, on Dokos at least, can only work because of the juxtaposition to sedentary societies (as was the case for the Hunzakutz). He is arguing—and I *am* putting words in his mouth—that many types of adaptation and social organization *can only* be explained when embedded in a larger, inter-societal context. Of course, world-systems analysis is not the only way to do this, but it is one of the most thought-out and tested of such generic explanations (for detailed explication of this and other points, see Hall, Kardulias, and Chase-Dunn 2011; Babones and Chase-Dunn 2012; Chase-Dunn and Lerro 2014). That said, this embedding in the analysis of Dokos remains largely implicit. I would push for more explicit analysis in this direction. Indeed, I argue that the inter-societal context of all pastoral societies is as vital as internal social dynamics to understand how they operate and change. Just what is it about the ways a pastoral society is enmeshed in a larger network that shapes, limits, or influences the details of how that society operates? Is it primarily reactive to the outside, or is it using the outside? Or both, as Kardulias (2007) has suggested elsewhere? Who is predator and who is prey? Or are things symbiotic? Or a mix? Whatever the answers, how is it that any current situation, which may appear "permanent," may only be the most recent manifestation of longer cycles of change? Dokos appears reasonably stable over a long time period. But is stability stasis? Or is it a particularly well-balanced dynamic equilibrium, a state of homeostasis, as Sidky suggests?

All these chapters have the advantage of working "from the ground up," of building on what specific actors do in specific situations. What remains unclear or at least undone is how these people sense and react to larger social processes and pressures as they adopt and adapt to changing circumstances. It would be helpful, for instance, to know just how and when they use trade outside their local region to help them adjust to unpredictable or unpredictably large changes in micro-climate. How long have current adaptations been in place? Have they changed historically? If so, under what conditions?

Finally, I want to toss out (some readers may decide that "toss out" is a singularly apt choice of words) an idea about neutral zones between groups (see Hall 2000, 2009, 2012 for more detailed discussions). I initially borrowed this concept from Melissa Meyer (1994) and David Anderson (1994). It has been amplified recently by Paul Martin and Christine Szuter (1999), who call them War Zones. The argument in a nutshell is that neutral zones between competing groups serve as areas where game that has been overharvested can be replenished. They argue that this is why Lewis and Clark found more buffalo east of the Rockies than west of them. Anderson observed similar zones in the pre-contact southeastern United States, and Melissa Meyer saw such a zone between Anishinaabe and Lakota peoples in the northeastern Great Plains. I suggest that in or at least around such zones, pastoralists, mounted hunters on the US Great Plains, are commonly found, with their volatile relations with any nearby sedentary folk. If so, why this is the case is not fully clear.

Building on the preceding comments, I suggest that pastoralists, precisely because of their mobility, have often been major vectors of change. Not only do they carry information, cultural practices, crops, animals, and diseases between peoples, they also are often the stimulus for increasing complexity in neighboring societies and for binding together heretofore isolated groups into larger—even if inchoate—systems. Increasing complexity and expansion, and at times decreasing complexity and contractions, are major processes in world-system evolution (Hall and Chase-Dunn 2006; Chase-Dunn and Hall 2012; Hall 2012; Chase-Dunn and Lerro 2014).

REFERENCES

Adams, William Y. 1963. *Shonto: A Study of the Role of the Trader in a Modern Navaho Community*. Washington, DC: Bureau of American Ethnology.

Anderson, David G. 1994. *The Savannah River Chiefdoms: Political Change in the Late Prehistoric Southeast*. Tuscaloosa: University of Alabama Press.

Anthony, David W. 2007. *The Horse, the Wheel, and Language: How Bronze-Age Riders from the Eurasian Steppes Shaped the Modern World*. Princeton: Princeton University Press.

Babones, Salvatore, and Christopher Chase-Dunn, eds. 2012. *Handbook of World-Systems Analysis: Theory and Research*. London: Routledge.

Barfield, Thomas J. 1989. *The Perilous Frontier*. London: Blackwell.

Barfield, Thomas J. 1990. "Tribe and State Relations: The Inner Asian Perspective." In *Tribe and State Formation in the Middle East*, ed. Philip S. Khoury and Joseph Kostiner, 153–82. Berkeley: University of California Press.

Barfield, Thomas J. 1991. "Inner Asia and Cycles of Power in China's Imperial Dynastic History." In *Rulers from the St'eppe: State Formation on the Eurasian Periphery*, ed. Gary Seaman and Daniel Marks, 21–62. Los Angeles: Ethnographics Press, Center for Visual Anthropology, University of Southern California.

Barfield, Thomas J. 2001. "The Shadow Empires: Imperial State Formation along the Chinese-Nomad Frontier." In *Empires: Perspectives from Archaeology and History*, ed. Susan E. Alcock, Terence N. D'Altroy, Kathleen D. Morrison, and Carla Sinopoli, 11–41. Cambridge: Cambridge University Press.

Barth, Frederik, ed. 1969. *Ethnic Groups and Boundaries*. Boston: Little, Brown.

Beckwith, Christopher I. 1991. "The Impact of the Horse and Silk Trade on the Economies of T'ang China and the Uighur Empire." *Journal of Economic and Social History of the Orient* 34 (2): 183–98. http://dx.doi.org/10.1163/156852091X00111.

Beckwith, Christopher I. 2009. *Empires of the Silk Road: A History of Central Eurasia from the Bronze Age to the Present*. Princeton: Princeton University Press.

Blyth, Lance R. 2012. *Chiricahua and Janos: Communities of Violence in the Southwestern Borderlands, 1680–1880*. Lincoln: University of Nebraska Press.

Braudel, Fernand. 1979. *Civilisation Matérielle, Économie et Capitalisme: XVe–XVIIIe Siècle*. Paris: A. Colin.

Brooks, James F. 2002. *Captives and Cousins: Slavery, Kinship, and Community in the Southwest Borderlands*. Chapel Hill: University of North Carolina Press.

Chase-Dunn, Christopher, and Thomas D. Hall. 1997. *Rise and Demise: Comparing World-Systems*. Boulder: Westview.

Chase-Dunn, Christopher, and Thomas D. Hall. 2012. "Global Scale Analysis in Human History." In *A Companion to World History*, ed. Douglas Northrop, 185–200. Oxford: Wiley-Blackwell. http://dx.doi.org/10.1002/9781118305492.ch12.

Chase-Dunn, Christopher, and Bruce Lerro. 2014. *Social Change: Globalization from the Stone Age to the Present*. Boulder: Paradigm.

Cribb, Roger. 1991. *Nomads in Archaeology*. Cambridge: Cambridge University Press. http://dx.doi.org/10.1017/CBO9780511552205.

Fenelon, James V., and Mary Louise Defender-Wilson. 2004. "Voyage of Domination, Purchase as Conaues, Skakawea for Savagery: Distorted Icons from Misrepresentation of the Lewis and Clark Expedition." *Wicazao sa Review* 19 (1): 85–104.

Frank, Andre Gunder. 1992. *The Centrality of Central Asia*. Comparative Asian Studies 8. Amsterdam: VU University Press, for the Center for Asian Studies Amsterdam (CASA).

Haaland, Gunnar. 1969. "Economic Determinants in Ethnic Processes." In *Ethnic Groups and Boundaries*, ed. Frederik Barth, 53–73. Boston: Little, Brown.

Hall, Thomas D. 1989. *Social Change in the Southwest, 1350–1880*. Lawrence: University Press of Kansas.

Hall, Thomas D. 1999. "World-Systems and Evolution: An Appraisal." In *World-Systems Theory in Practice: Leadership, Production, and Exchange*, ed. P. Nick Kardulias, 1–25. Lanham, MD: Rowman and Littlefield.

Hall, Thomas D. 2000. "Frontiers, and Ethnogenesis, and World-Systems: Rethinking the Theories." In *A World-Systems Reader: New Perspectives on Gender, Urbanism, Cultures, Indigenous Peoples, and Ecology*, ed. Thomas D. Hall, 230–70. Lanham, MD: Rowman and Littlefield.

Hall, Thomas D. 2005a. "Borders, Borderland, and Frontiers, Global." In *New Dictionary of the History of Ideas*, vol. 1, ed. Maryanne Cline Horowitz, 238–42. Detroit: Charles Scribner's Sons.

Hall, Thomas D. 2005b. "Mongols in World-System History." *Social Evolution and History* 4:2 (September): 89–118.

Hall, Thomas D. 2009. "Puzzles in the Comparative Study of Frontiers: Problems, Some Solutions, and Methodological Implications." *Journal of World-Systems Research* 15 (1): 25–47. http://www.jwsr.org, accessed December 15, 2014.

Hall, Thomas D. 2010. "The Silk Road: A Review Essay on Empires of the Silk Road: A History of Central Eurasia from the Bronze Age to the Present, by Christopher I. Beckwith (Princeton University Press, 2009)." *Cliodynamics: The Journal of Theoretical and Mathematical History* 1 (1):103–15. http://escholarship.org/uc/item/67z5m9d3, accessed December 15, 2014.

Hall, Thomas D. 2012. "Incorporation into and Merger of World-Systems." In *Routledge Handbook of World-Systems Analysis*, ed. Salvatore J. Babones and Christopher Chase-Dunn, 37–55. New York: Routledge.

Hall, Thomas D. 2013a. "Lessons from Comparing the Two Southwests: Southwest China and Northwest New Spain/Southwest USA." *Journal of World-Systems Research* 19 (1): 24–56. http://www.jwsr.org/, accessed December 15, 2014.

Hall, Thomas D. 2013b. "Reflections on Violence in the Spanish Borderlands: A Review Essay on *Chiricahua and Janos*, by Lance R. Blyth." *Cliodynamics* 4: 171–84. http://escholarship.org/uc/item/67z5m9d3, accessed December 15, 2014.

Hall, Thomas D., and Christopher Chase-Dunn. 2006. "Global Social Change in the Long Run." In *Global Social Change: Comparative and Historical Perspectives*, ed. Christopher Chase-Dunn and Salvatore Babones, 33–58. Baltimore: Johns Hopkins University Press.

Hall, Thomas D., P. Nick Kardulias, and Christopher Chase-Dunn. 2011. "World-Systems Analysis and Archaeology: Continuing the Dialogue." *Journal of Archaeological Research* 19 (3): 233–79. http://www.springerlink.com/openurl.asp?genre=article&id=doi:10.1007/s10814-010-9047-5, accessed December 15, 2014.

Hämäläinen, Pekka. 1998. "The Western Comanche Trade Center: Rethinking the Plains Indian Trade System." *Western Historical Quarterly* 29 (4): 485–513. http://dx.doi.org/10.2307/970405.

Hämäläinen, Pekka. 2003. "The Rise and Fall of Plains Indians Horse Cultures." *Journal of American History* 90 (3): 833–62. http://dx.doi.org/10.2307/3660878.

Hämäläinen, Pekka. 2008. *The Comanche Empire*. New Haven: Yale University Press.

Honeychurch, William, and Chunag Amartuvshin. 2006. "States on Horseback: The Rise of Inner Asian Confederations and Empires." In *Archaeology of Asia*, ed. Miriam T. Stark, 255–78. Malden, MA: Blackwell. http://dx.doi.org/10.1002/9780470774670.ch12.

Kardulias, P. Nick. 2007. "Negotiation and Incorporation on the Margins of World-Systems: Examples from Cyprus and North America." *Journal of World-Systems Research* 13 (1): 55–82. http://www.jwsr.org, accessed December 15, 2014.

Khazanov, Anatoly M. 1983. *Nomads and the Outside World*. Cambridge: Cambridge University Press.

Kradin, Nikolay N. 2002. "Nomadism, Evolution and World-Systems: Pastoral Societies in Theories of Historical Development." *Journal of World-Systems Research* 8 (3): 368–88. http://www.jwsr.org, accessed December 15, 2014.

Kradin, Nikolay N., Dmitri M. Bondarenko, and Thomas J. Barfield, eds. 2003. *Nomadic Pathways in Social Evolution: The Civilization Dimension Series*, vol. 5. Moscow: Russian Academy of Sciences, Center for Civilizational and Regional Studies.

Kradin, Nikolay N., Andrey V. Korotayev, Dmitri M. Bondarenko, Victor de Munck, and Paul K. Watson, eds. 2000. *Alternatives of Social Evolution*. Vladivostok: Institute of History, Archaeology and Ethnology, Far Eastern Branch of the Russian Academy of Sciences.

Lamphere, Louise. 1989. *To Run after Them: Cultural and Social Bases of Cooperation in a Navajo Community*. Tucson: University of Arizona Press.

Lamphere, Louise. 2007. *Weaving Women's Lives: Three Generations in a Navajo Family*. Albuquerque: University of New Mexico Press.

Lattimore, Owen. 1951 [1940]. *Inner Asian Frontiers*, 2nd ed. Boston: Beacon.

Lattimore, Owen. 1962. *Studies in Frontier History: Collected Papers, 1928–58*. London: Oxford University Press.

Lenski, Gerhard. 1966. *Power and Privilege: A Theory of Social Stratification*. New York: McGraw-Hill.

Martin, Paul S., and Christine R. Szuter. 1999. "War Zones and Game Sinks in Lewis and Clark's West." *Conservation Biology* 13 (1): 36–45. http://dx.doi.org/10.1046/j.1523-1739.1999.97417.x.

Meyer, Melissa L. 1994. *The White Earth Tragedy: Ethnicity and Dispossession at a Minnesota Anishinaabe Reservation, 1889–1920*. Lincoln: University of Nebraska Press.

Nagel, Joane. 1996. *American Indian Ethnic Renewal: Red Power and the Resurgence of Identity and Culture*. New York: Oxford University Press.

Nolan, Patrick, and Gerhard Lenski. 2011. *Human Societies: An Introduction to Macrosociology*, 11th ed. Boulder: Paradigm.

Robbins, William G. 1994. *Colony and Empire: The Capitalist Transformation of the American West*. Lawrence: University Press of Kansas.

Roe, Frank G. 1955. *The Indian and the Horse*. Norman: University of Oklahoma Press.

Ross, Anne, Kathleen Pickering Sherman, Jeffrey G. Snodgrass, Henry D. Delcore, and Richard Sherman. 2011. *Indigenous Peoples and the Collaborative Stewardship of Nature: Knowledge Binds and Institutional Conflicts*. Walnut Creek, CA: Left Coast.

Teggart, Frederick J. 1939. *Rome and China: A Study of Correlations in Historical Events*. Berkeley: University of California Press.

Turchin, Peter, and Thomas D. Hall. 2003. "Spatial Synchrony among and within World-Systems: Insights from Theoretical Ecology." *Journal of World-Systems Research* 9 (1): 37–64. http://www.jwsr.org, accessed December 15, 2014.

Wallace, Ernest, and E. Adamson Hoebel. 1952. *Lords of the South Plains*. Norman: University of Oklahoma Press.

West, Eliott. 1998. *The Contested Plains: Indians, Goldseekers, and the Rush to Colorado*. Lawrence: University of Kansas Press.

About the Contributors

Claudia Chang (PhD, SUNY-Binghamton) is professor of anthropology at Sweet Briar College in Virginia. Her research interests include the ethnography and archaeology of pastoralism, social evolution, and environmental reconstruction in the Mediterranean and Central Asia. She has conducted archaeological and ethnoarchaeological fieldwork in Greece and Kazakhstan supported by grants from the National Science Foundation. In 1994 she co-edited *Pastoralists at the Periphery: Herders in a Capitalist World* (Arizona) with Harold A. Koster.

Michelle Negus Cleary (PhD, University of Sydney) focused her dissertation on the Late Iron Age fortified sites of western Central Asia from a landscape archaeological perspective, using remote sensing, GIS, geophysical, and extensive survey techniques. She has conducted archaeological fieldwork in Uzbekistan, Turkey, Georgia, and Australia and is currently involved in several research projects in these countries. Her main research interests are landscape archaeology, architecture, archaeological survey and prospection, GIS, mapping and geophysical survey, ancient Central Asia, and the archaeology of the Caucasus.

Thomas D. Hall (PhD, University of Washington) is professor emeritus in the Department of Sociology and Anthropology, DePauw University, Greencastle, Indiana. His interests include indigenous peoples, ethnicity, comparative frontiers, and world-systems analysis. A recent book is *Indigenous Peoples and Globalization: Resistance and Revitalization*, with James V. Fenelon (Paradigm, 2009). His most recent article is "Lessons from Comparing the Two Southwests: Southwest China and Northwest New Spain/Southwest USA," *Journal of World-Systems Research* 19 (1) (2013): 24–56.

Erik G. Johannesson (PhD, University of North Carolina) has conducted field research on mortuary practices and mobile economies in Mongolia, Russia, Greece, and North America and has taught archaeology at North Carolina State University, Meredith College, and UNC Chapel Hill. He lives in Maine.

P. Nick Kardulias (PhD, Ohio State University) is professor of anthropology and archaeology at the College of Wooster in Ohio. His research interests include Mediterranean and North American prehistory, exchange systems, lithic analysis, and world-systems analysis. Kardulias is associate director of the Athienou Archaeological Project in Cyprus. He has authored one monograph and edited five books, most recently *Crossroads and Boundaries: The Archaeology of Past and Present in the Malloura Valley, Cyprus* (American Schools of Oriental Research, 2011, with Michael K. Toumazou and Derek B. Counts). He has participated in and directed excavations and surveys in Greece, Cyprus, Ohio, Pennsylvania, and Illinois.

Nikolay N. Kradin (PhD, Institute of History, Archaeology, and Ethnology; Dr. Sc., Oriental Institute, Russian Academy of Sciences) is a research fellow at the Institute of History, Archaeology, and Ethnology, Far East Branch of the Russian Academy of Sciences; chief of the Department of World History, Archaeology, and Anthropology at the Far-Eastern Federal University, Vladivostok, Russia; and a member of the Russian Academy of Sciences. His research interests include the anthropology and social history of Eurasian nomads, the political anthropology and theory of state formation, the social archaeology of eastern and Inner Asian cultures, and the evolution of nomadic empires. He has undertaken archaeological and ethnological research in Siberia, the Russian Far East, China, and Mongolia. Kradin is the author of more than 350 publications.

Lawrence A. Kuznar (PhD, Northwestern University) is professor of anthropology at Indiana University–Purdue University Fort Wayne and director of the IPFW Institute for Decision Sciences and Theory. He specializes in the ecological and economic features of traditional pastoral societies and has done extensive research among Aymara herders in southern Peru, as well as with Navajo sheepherders and cattle ranchers. He is the author of several books, including *Awatimarka: The Ethnoarchaeology of an Andean Herding Community* (Harcourt Brace College Publishers, 1995).

Mark Moritz (PhD, University of California, Los Angeles) is associate professor in the Department of Anthropology and the Environmental Science Graduate

Program at Ohio State University. He has been conducting research with pastoralists in the Far North of Cameroon for twenty years. Currently, all his research projects are interdisciplinary and examine pastoral systems within the analytical framework of coupled human and natural systems, using a regional approach that situates the Far North region within the historical context of the greater Chad Basin. His research has been supported by the Wenner-Gren Foundation, the National Geographic Society, and the National Science Foundation.

Mark T. Shutes (PhD, University of Pittsburgh) was associate professor of anthropology at Youngstown State University at the time of his death. His interests included economic anthropology, agricultural production schemes in Europe (with a focus on Ireland and Greece), and comparative ethnographic analysis. He co-edited *Aegean Strategies: Studies of Culture and Environment on the European Fringe* (Rowman and Littlefield, 1997) with P. Nick Kardulias.

Homayun Sidky (PhD, Ohio State University) is professor of anthropology at Miami University, Oxford, Ohio. His research interests include the anthropology of religion, ecological anthropology, anthropological theory/history of anthropological thought, and scientific methods in anthropology. He has conducted field research in Afghanistan, northern Pakistan, western, central, and eastern Nepal, and among the Tibetan exile community in northern India. He is the author of *Perspectives on Culture: A Critical Introduction to Theory in Cultural Anthropology* (Pearson Prentice-Hall, 2004).

Index

Afghanistan, 42, 78, 83; people from, 72, 74. *See also* Wakhi
Africa, xiv, xvi, 3, 6, 11, 12, 27, 56, 66, 104, 171–74, 180, 183, 186, 213, 251; Central, 177,187; colonial states in, 178; East, 3, 6, 9; North, 11, 14, 48, 56, 57; West, 9, 174, 186
Afroeurasia, 269, 272
agriculture, xvi, 2, 8, 20, 38, 83, 152, 268, 270, 272; and emergence of state, 45; Iron Age, 38; irrigation, 29, 71, 119, 125–26, 127, 138, 151, 158; on islands, 252, 254; among Kazakh, 29; Navajo and, 199, 205; origins of, 29; sedentary, xvi, 2, 3, 11, 176, 200; unsuitability of steppe for, 41, 43, 51–52, 60, 82
agropastoralism, 19, 35, 38, 153, 204; in Chorasmia, 117, 123; ethnoarchaeological research on, 20; among FulBe, 174, 176, 181, 187n6; and gender roles, 11; production in, 76; ritualized calendar for, 87, 89. *See also* nomadism, pastoral
Akcha-gelin, 131–32(table), 138, 152
Almaty (Kazakhstan), 17, 18, 36
altitude, 75, 85
Amu Darya delta, 118–21, 123, 124–25, 139, 153, 158
analysis: analytic-scientific, 225–27, 230–32; emotional/ethical, 226; ideological, 232–36; linguistic/symbolic, 236–37
Andes: herding in, 9, 259; highlands of, 7
animal husbandry, xv, xvii, 1–5, 10, 18, 36, 98, 123, 211; archaeological evidence for, 101, 103, 106, 112; in Cameroon, 173; definition of, 4, 82; and flexibility, 213, 214; in Hunza, 71, 74, 83–85, 88–89; in Mongolia, 43, 52, 62. *See also* pastoralism/pastoralists; transhumance
Apachean (people), 198, 199, 204
Arab(s), 1, 57; agropastoralists, 174, 187n6
Arabian peninsula, 11
Argolid, 244(map), 245(photo), 247, 252, 256
Asia, 27, 42, 56, 251; Central, xiv, xvi, 3, 6, 26, 27, 31, 42, 43(map), 73, 82, 99–101, 103–4, 109, 117, 118(map), 120–27, 152, 160n5, 269, 271, 272; East 114, 272; Inner, 41, 42, 46–49, 57, 59, 60; Minor, 261; South, xiv, 73(map); West, 270, 272
Athapaskans, 198

Bactrian, 31; Bactrian-Margiana Complex, 122, 152
Basseri, 19, 21, 87
battue, 51
bioarchaeology, 102
biogeography, 249
Black Sea, 6, 57

bricolage, 273
Bronze Age, 11, 18, 20, 123, 152, 259; Early, xvii, 122
burial, 17, 100, 102, 106, 146, 148; chamber, 108; horse/livestock, 103; khirigsuurs, 105–6, 110, 112; Kuiusai, 143; monuments, 122, 125, 137, 138; mound, 20, 28, 121; Muzdean tradition, 142, 146; ossuary, 127, 143, 157; practices, 102, 106, 109, 111; secondary, 146; slab, 104, 106–8, 109, 110, 111, 112. *See also kurgan(s)*
Buryatia, 60, 61
Byzantium, 57, 245; Early Byzantine period of, 245, 246, 255, 256, 259, 260(photo)

camel(s), 4, 6, 23, 26, 34, 47, 48, 49, 53, 61, 159n3; Bactrian 6, 48, 49; camelids 7; dromedary 6; excrement 49; milk 49, 50
Cameroon, 172, 173, 174, 176, 178, 179, 183, 187nn3–4, 187n6, 188n8, 188n12; Cameroonian 173, 179, 180, 181, 183, 187n5; Cameroon People's Democratic Movement (CPDM), 179
canal(s), xvi, 77, 78, 118, 125, 126, 138, 139, 142, 159n2, 182
CAP. *See* Common Agricultural Policy
cattle, xv, 4, 6, 52, 112, 159n3, 201, 205; breeding of, 42, 44, 50, 51, 54, 56, 58, 60; in burials, 103, 108, 110; in Chad, 173, 174, 180, 183, 187n6; corrals, 122, 139, 143, 154, 156, 157, 158; and Eurasian pastoralists, 19, 23, 26; horns (bucrania), 104, 110(photo); among Hunzakutz, 75, 80, 81, 83, 84, 85, 86, 88, 89, 90n7; husbandry, 71, 74, 81, 84, 85, 89; in Ireland, 211–22, 227–39; loss of, 43–44, 61, 180; in Mongolia, 48, 49, 52; pastoralists, 3; pastoral zone, 6; raids, 174, 176, 188, 198; ranching, xvi, 204; theft of, 184, 188n10; trade, 174; sacredness of, 88, 103; sacrifice of, 11; value of, 53; among Xiongnu, 47, 50; zebu, 187
Chad Basin, xvi, 171–74,177, 179, 181–88
cheese, 19, 49, 81, 82, 253, 256, 260
chiefdom(s), 3, 57, 59, 73
China, 26, 42, 46, 47, 52, 57, 58, 72, 78, 270; luxury goods from, 109; People's Republic of, 60, 62
Chinggis Khan, 44, 46, 51, 54, 59
Chorasmia, xvi, 117–27, 137–40, 143–55, 157, 158; pottery of, 140, 142, 143, 145, 146, 147, 149, 150, 152, 154, 155, 157; sites in, 128–36(tables)
cistern(s), 246, 255, 257, 260(photo)
civilization(s), xvii, 47, 48, 56, 57, 159n1, 272; agricultural 41–42, 45, 46, 56, 58, 269, 272
climate, xv, 2, 5, 26, 42–43, 48, 60, 198, 200; change in, 45–46, 271, 275
colonialism, 12, 41, 195, 226, 270
Common Agricultural Policy (CAP), 226, 227
cooring, 213, 214, 222, 234
core-periphery relations: core-periphery differentiation 250, 274; core-periphery hierarchy 250, 274; interaction between, 244; semiperiphery(ies), 25, 56, 57, 250, 271; theory of, 196, 205. *See also* periphery; world-system(s)
cotters, 235
cow(s). *See* cattle
cultural divergence, 7
culture history, 252
Cyprus, 249

dakhmas, 140, 142
demography, 56
desert, 6, 19, 23, 159n1; Arabian, 6; desert-steppe, 71, 74, 75, 77, 79, 90n3, 98, 117, 124, 125; Gobi (*see* Gobi Desert); islands, 259; Kara Kum, 124; Sahara (*see* Sahara Desert); southwest, 201
development: economic, 1, 12, 52, 171, 189n18, 195, 197; of ethnos, 25; of pastoralism, 78, 111, 172, 180–86; societal, 10, 25, 29, 45, 248
Didyma (Greece), 19
diet, xv; of animals 255; of Hunza, 80–81, 86; of nomads 50–53; of Xiongnu, 102, 113
Diné. *See* Navajo
Dinétah, 198, 200
Dokos Island (Greece), xvii, 243–45, 247–48, 250–61
dzos, 6

ecology, xiii, 24, 34, 78, 214, 271, 273; human behavioral, 2, 196; island, 249; nomadic/pastoral, xiv, 1, 42, 98, 243; political, 56–62
economy(ies), 21, 24, 33, 41, 45, 52, 56, 57, 60, 74, 76, 119, 178, 180, 206n1, 237, 251, 253, 260; adaptive strategy for, 197; agropastoral/pastoral, 2, 34, 36, 42, 46, 48, 58, 61, 97, 123, 125, 138, 154, 174, 201; capitalist, 250; cash, 214, 215, 216, 222; cattle-breeding, 56, 58; economic system, 2, 5, 8, 9, 10, 44, 71, 82, 195, 204, 252, 259, 261, 273; Greek, 243, 254; Hunza, 74, 80;

industrial, 42, 59; Kazakh, 31; market, 62, 180, 214; Mediterranean, 244; mixed, 102; Navajo, 197–206, 273; nomadic, 56; peasant, 196, political, 98, 113, 227, 230; regional, 258, 261; subsistence, xv, xvi, 215; world, 9; Xiongnu, 98, 113, 118
egalitarianism: in relationships (Ireland), 220–36; in social structure, 3, 10, 24, 33, 153
Elassona (Greece), 19
enclosures: animal, 18, 255; fortified, xvi, 51, 118, 122, 124, 126, 128–30(table), 137–39, 146, 148–50
Engels, Friedrich, 24, 25, 33
equilibrium, dynamic, 275
ethnicity, 25, 37n1, 274; symbolic, 273
ethnoarchaeology, 18, 34, 251–52
ethnocide, 25
ethnogenesis, 7, 23, 25, 30–32, 37n3, 72, 273
ethnology, 22, 25, 32, 36, 124
Eurasia, xv, xvi, 3, 4, 11, 42, 46, 48; identity in, 37nn1–2; pastoralism/pastoral nomads of, 4, 17, 18, 20, 22, 23, 32, 34, 36, 51, 53, 57, 200, 271; scholarship on, 117; steppe of, 4, 6, 7, 11, 21, 22, 25, 26
European Union (EU), 180, 226, 228, 238, 258; and milk quotas/production, 228–35
Evans, John D., 246–48
evolution, social, 10, 22, 24

farmer(s), 3, 4, 23, 29, 32, 35, 122, 125, 151; Hunzakutz, 74, 76–90; Irish dairy, 214, 217, 225, 227, 249, 258; relations with pastoralists, 5, 6, 42, 51, 55, 56, 59, 61, 119, 120, 153, 185, 186, 270–71; "strong"/"big" vs. "small" (Ireland), 215, 217–22, 229–39
feudalism: patriarchal, 24–25, 30, 33
food: pastoral, 8, 10, 48, 49–52, 143; production of, 4, 24, 45, 74, 77, 79–80, 83, 89, 231, 232, 233; supply of, 45, 86–87, 249, 252, 260, 261. *See also* diet
food supply index (FSI), 52, 53
foraging, xvi, 7, 21, 24, 33, 84, 198, 250, 268, 269
fortress, xvi, 51, 56; Chorasmian, 117–22, 125, 128–36(tables), 137–38, 140–51, 153–60. *See also kala*
frontier, 25, 42, 73, 97, 269–70; ecology of steppe, 56–59
FSI. *See* food supply index
FulBe, xvi, 171, 172–79, 183–89

Gilgit (Pakistan), 75; agency 73; valley 72
globalization, 9, 11, 250, 251, 254
goat(s), xv, 2, 4, 6, 18, 20, 23; advantages of over sheep, 47, 84–85; in burials, 103, 106, 108, 110, 112, 159n3; in Greece, 20, 243, 252, 254–56, 260; among Hunzakutz, 75, 79, 80, 81, 83–89, 90n3; in Hunzakutz cosmology, 90nn6–7; in Kazakhstan, 19, 26; in Mongolia, 47–50, 53; among Navajo, 199(table), 202, 203; numbers of, 61–62; products from, 82, 89. *See also* livestock; pastoralism/pastoralists
Gobi Desert, 34, 49, 54, 55, 98
Golden Warrior (Kazakhstan), 17
Great Plains (US), 7, 271, 276
Greece, xvii, 244(map), 252; ethnoarchaeological survey in, 19, 20; modern pastoralism in, 9, 18; transhumance in, 21, 35, 243. *See also* island(s); Koutsovlach
Grevena (Greece), 19, 20, 21
Güyüg (Mongolian Khan), 55
Gyaur-kala 1 Chermen-yab, 129(table), 134(table), 152, 154

handicraft, 56, 57
Han Dynasty, 57, 97
Hellenistic states, 57
herding, xiv–xvii, 1, 11, 25, 97, 213–14, 244, 267, 268, 269, 272; in Eurasian steppe, 23; flexibility of system, 2–3, 189n16, 212; in Greece, 18–21, 243, 244, 250, 253, 254, 258–61; households, 18; among Hunzakutz, 81, 85, 89, 274; in Kazakhstan, 26, 31, 33; model of, 24; among Navajo, 197, 200, 203, 204; strategies of, 102, 103, 271; and Xiongnu economy, 98
Herodotus, 21, 22
hierarchy/hierarchical system, 3, 10, 33, 56, 58, 77, 153, 176, 274–75. *See also* core-periphery relations
horse(s), 17, 23, 45, 61, 159, 271, 272, 273; clay sculptures of, 147, 148; among Hunzakutz, 75, 78; in Ireland, 216–18, 220; among Kazakhs, 19, 22, 26; meat of, 50; among Mongols, 44, 47–49, 52, 53; among Navajo, 201; in defining pastoralism, 4, 5, 6, 7; relay messenger service, 58; remains of (in graves), 100, 103, 105, 106, 108; riding pastoralists, 22, 26; trappings, 103, 107, 112. *See also* livestock
horticulture, xvi, 196, 198, 201, 268, 269
Hsiung-nu, 21

humidification, 56. *See also* climate
hunting, 7, 50, 51, 52, 102, 271, 272; and gathering, 11, 198 (*see also* foraging)
Hunza (place), 71–72, 78, 79, 80, 89; history of, 72–74; Valley, 74–76
Hunzakutz (people), 71, 74, 76–80, 90n3, 90nn6–7, 271, 275; cattle husbandry by, 81–82; pastoral rites of, 87–89; transhumance by, 82–86

ibex, 84, 90
Ibn Khaldun, 59, 270
identity: ethnic, 3, 5, 17, 35–36, 37n1, 37nn4–5, 112–13, 173, 186, 273, 274; personal, 204, 243
ideology, 25, 98, 100, 113, 227, 230; change in, 236; egalitarian, 235; pastoral, 90n7
Idhra (Greece), 247, 252, 254, 256–60
Ili River Basin (Kazakhstan), 22
India, 42, 57, 73, 78, 103
Inner Asia. *See* Asia
Inner Mongolia. *See* Mongolia
insecurity, xvi, 172, 174, 182–84, 188n10
Ireland, xiii, xvi, xvii, 212(map), 215, 225, 227, 229, 235; County Cork, 225, 230; County Kerry, 212(map), 214, 227; and European Union, 228, 230–31, 238
irony, linguistic use of, 236–37, 239
irrigation, xv, 8, 182; in Chorasmia, 118, 119, 121, 125–26, 127, 138, 142, 145, 146, 151, 152, 153, 158, 159n2; in Hunza, 71, 74, 76, 77–78, 83; as hydraulic system, 77, 79, 84; among Kazakh, 29; state model based on, 118
island(s), xiii, xvii, 243–45, 251, 252–59; pasture on 260–61; study of 246–50
Issyk (Kazakhstan), 17

kala, 120–21, 124, 128–36(tables), 137–60; definition of, 117, 122–23. *See also* fortress, Chorasmian
Kanga-kala, 130(table), 133–34 (table), 138–39, 141, 146–47, 149(map), 153–56
Karakoram, 55; Highway, 74, 79; Mountains, xiii, xv, 71–72, 84–85
kastro, 245, 256
Kazakh-Kirghiz, 22, 27, 28, 37n2
Kazakhstan, Republic of, 17–18, 20–23, 25–27, 42, 53, 61, 153; study of pastoralists in, 27–34
Kermani, 19
Khakassia, 61
Khazars, 57
khirigsuur, 105–6, 108
Khorezm, 117, 155 (*see also* Chorasmia); Khorezm Expedition, 118, 119, 124, 125, 126, 143
Khubilai (Mongolian Khan), 55, 59
Kirghiz, 26, 29, 31; land of, 61. *See also* Kazakh-Kirghiz
Kitans, 51
Kiuzely-gyr, 127, 130–33(table), 142, 151–52, 154
kochevniki, 23
Koutsovlach, 19, 20, 21, 35
Kuiusai: ceramics, 143–46, 151, 155; culture, 124, 127, 143, 147, 159n2, 159n4; Kuiusai-kala, 138, 154, 156, 158; sites, 131–36(table)
kumiss, 26, 48, 49; mare's milk, 50. *See also* diet
kurgan(s), 31, 100, 117, 119, 121, 124, 137, 139, 142, 146, 159nn3–4; cemeteries, 120, 127, 128–36(tables), 138, 143, 144(map), 145, 148–50, 152, 156–57; fortified sites and, 149–52, 154–56, 160nn6–7; as nomadic monuments, 125–26, 138, 153. *See also* burial
Kutpatshari, 19, 21, 35

labor, 8, 19, 21, 35, 77, 80, 108, 250, 251, 269; compulsory, 73, 77, 198; dependency, 221–23; division of, 90n7, 258; human, 24, 30, 84, 215, 229, 231; and machines, 232; and socio-economic relationships, 213–14, 218, 220–21, 233–35
land, marginal, 244
Lenin, Vladimir I., 22, 30; theories of, 23, 30
livestock, xv, xvi, 18, 71, 36, 43, 80, 101, 103, 123, 52, 56, 78, 197, 227; and burial, 102, 112; cultural reasons for owning, 204; and ecology, 81–89, 90n3, 111, 250; herders/herding, 23–25, 31, 33, 36; kinds of, 47–50, 53, 75, 195 (*see also* cattle; dzos; goats; horse(s); sheep; yaks); knowledge of, 233; markets, 173–74, 181, 183, 214; moving/relocating, 54 (*see also* transhumance); numbers of, 62, 197–203, 205; and ritual, 104; ruminant, 80–81, 84, 85; and social relations, 212–13; tax, 179; and wealth, 77, 196

Macedonia, 243
Manchuria, 42
Mangyr-kala, 124, 131–32(table), 138, 139, 148–51, 153, 155–56, 160n6
manure, 80, 81–82, 86, 89, 90n5, 140, 143

Marx, Karl, 24, 25, 33; theories of, 21, 23, 24, 30, 37n4, 56, 119, 267
Mediterranean, xiv, 6, 18, 104, 246, 248, 249, 254, 260; economy, 244; pastoralism, 20, 21
meithal, 213, 214, 234. *See also* labor, and socio-economic relationships
Middle East, 12, 42, 82, 109, 152. *See also* Near East
migration, 2, 21, 46, 54, 138, 198, 200, 274; emigration, 214, 222, 233; seasonal, 55, 82, 123; mass, 56, 249; out-, 79; theory, 159n2. *See also* movement; transhumance
milk, 2, 5, 6, 20, 48, 49, 50, 60, 80, 81, 218, 237, 256, 260; from mechanical cow, 225, 229, 238; goat, 85, 255; and labor dependency, 235–36; mare's (*see kumiss*); quotas, 232; sale of/price, 215, 216(table), 222, 227–28, 230–32; as secondary product, 111, 250
mobility, xiv, xv, 3, 6, 7, 8, 19, 21–23, 29, 31, 36, 54, 58, 83, 103, 201, 268; definition of, 123; functions of, 33–35, 78, 82, 272, 276; limits on, 30, 212; range of, 20, 48, 112, 171, 203. *See also* transhumance
modernization, 11, 41, 59–60, 182
Möngke (Mongolian Khan), 55
Mongolia, xv, 4, 5, 6, 26, 34, 42, 43, 57, 97; adoption of pastoralism in, 103, 112; animals of, 48–50, 52–53; archaeology in, 100, 104–8; empire of, 41, 46; food of, 50; historic periods in, 98; mortuary data in, 99, 103, 106–8; post-socialist period in, 60–62; seasonal roaming in, 54–55; steppes of, 42, 47, 51–52; Xiongnu culture in, 108–11
Mongols, 1, 21, 43, 44; food of, 50–51 (*see also* diet); herds of, 47–50, 52–54; movements of, 54–55; unification of, 46; yurts of, 54
monument(s): fortresses as, 125, 151; mortuary, 98–99, 104–9, 111–12, 120, 122, 124, 125, 137–38, 151, 153, 157. *See also* burial and *kurgan(s)*
movement: longitudinal, 23; vertical, 5, 19, 20, 23, 79, 83, 88, 89. *See also* mobility; transhumance

Nahondzod, 202
Navajo, xiii, xvi, 195, 213, 270; adaptability of, 196, 205–6; and ethnogenesis, 273; herding, 197, 201–4; history of, 198–200; stock reduction, 30, 203; values of, 196, 204
Near East, 18, 57, 152. *See also* Middle East
nomad(s), 21, 25, 41–44, 120, 125, 146, 154–55, 159n1, 177, 267–68, 269–70, 272, 274; animals of (*see* livestock); and China, 467; confederacies of, 20, 22, 26, 45; defined, 5, 32; equestrian, 6, 7; Eurasian, xv, 6, 18, 22, 29; food of, 50–53 (*see also* diet); Kazakh, 17, 27, 30–31, 37n2; modern culture of, 59–62; monuments of (*see* monuments); movement of, 54–56 (*see also* migration); political ecology of, 56–59; and the state, 171–73, 178–81, 184, 187n1; studies of, xv, 17–18, 26, 33–36; transhumant 82–83
nomadism, 123, 153, 211, 267–68; pastoral, 18, 23–26, 30–31, 33, 35, 43, 45, 47, 59, 82 (defined), 83, 268. *See also* agropastoralism; nomad(s)

Ögödei (Mongolian Khan), 55

Pakistan, xiii, xv, 71, 72–73, 79, 83, 90n7
pastoralism/pastoralists, xiii–xvii, 3, 106, 112, 243; as adaptive, 195; changes in, 61–62; defined, 18, 82, 103, 123, 250, 267–68; ecology of, 2–3, 24, 56; extensive, 180; flexibility of, 51; ideology for, 90n7; identifying in past (archaeology), 98, 102, 117, 119, 125, 138, 154, 159n3; and impact on other institutions and values, 1, 5, 19, 197; on islands, 243, 259; among modern Navajo, 204; nomadic/mobile, xv, 17, 18, 57, 60, 78, 82, 103–4, 111, 113, 200–203; and other subsistence practices, 158, 249; study of, 2–12, 17, 21–24, 35, 267; transhumant, xv, 5, 6, 9, 18, 21, 35 (*see also* transhumance); and use of landscape, 20. *See also* agropastoralism; nomadism
pasture(s), xiii, viv, 4, 5–7, 8, 9, 18–21, 23; alpine, 19, 75, 79, 84, 88, 89
Peloponnesos, xvii, 243–44, 259
People's Republic of China: *See* China
periphery, 59, 119, 124, 174, 182. *See also* core-periphery relations
Pindos Mountains, 19
population density, 45, 56, 77
privatization, 61, 62, 180
Puebloan(s), 198–99, 204, 205

raiding, 2, 33, 44, 78, 127, 200, 201–2, 205, 267, 269, 271–74; to obtain slaves, 176
regime(s) (political), 138; authoritarian, 12, 32, 178, 179; Soviet, 4; Tsarist, 27, 29

relationships: economic, 5, 83, 261; social, xviii, 8, 212–14, 222
Rockies, New Mexican, 198
Roman Empire, 57
ritual(s), xv, xvi, 11, 21, 71, 86–88, 89, 104, 106, 111, 148, 155, 247, 252, 275; centers, 119; feasting, 101; funerary 104, 108–9, 112–13, 138, 142, 159n4
roaming, seasonal, 54
Rourans, 58
Rubruck, William, 41, 55
Russia, 22, 28, 30, 42, 61, 72
Russification, 27

Sahara Desert, 6
Sahel, 6
Saka, 17
San Juan Basin, 198
Sarakatsani, 243
Sarykamysh/Prisarykamysh delta, 121, 124, 125, 127, 128(table), 132(table), 137, 141, 143, 147, 151, 152–54, 156–58
Savanna (East Africa), 6, 116
scheduling, 87, 89
scholarship, 32, 34; Eurasian, 117, 120, 149; Russian, 4; Soviet, xv, 18, 23
Scythian(s), 17, 21, 22, 26, 32, 34, 57, 144, 147
sedenterization, 42, 51, 60, 62
semiperiphery. *See* core-periphery relations
Semirech'ye, 22, 26, 34, 35, 38
settlement(s), 20, 180; Chorasmian, 128–36(table); 138, 141–42, 152, 153–54; fortified, 143–46, 245, 259 (*see also* sites, fortified); Kazakh, 27, 30; Khangai, 54; Mormon, 201; Navajo, 190; pastoral, 157; pattern/regime/strategy/system, xvi, 121, 137, 139, 151–52, 158, 257; sites, 101, 103; Xiongnu, 113
sheep, xv, 2, 4, 6, 13, 19, 20, 23, 26; in burials, 103, 106, 110, 112, 159n3; in Greece, 19, 20, 243, 252, 254–56, 260; among Hunzakutz, 75, 80, 81, 83–89, 90n6; in Ireland, 214–16, 229; in Kazakhstan, 19, 26; in Mongolia, 47–50, 52; among Navajo, 7, 195, 198–200, 202–205, 272–74; numbers of, 52–53, 61–62. *See also* livestock; pastoralism/pastoralists
sheepherders, xvi, 3, 282
Siberia, 27, 41, 42, 66, 68, 98, 99, 101, 102, 106, 107, 108, 165, 282
Sima Qian (Chinese historian), 97
Sioux, 7
site(s), 214, 252, 261; archaeological, 99–100, 121, 124, 126–27, 128–36(tables), 137; fortified, xvi, 118–22, 127, 137–41, 143, 146–57; Iron Age, 34, 125; Kuiusai, 127, 143; mortuary, 120, 137–38, 140 (*see also* burial; *kurgan[s]*); nomadic 120; pastoral, 5, 10, 21, 117, 120, 124, 140, 243; ritual use of, 148; Sarmatian, 144; seasonal, 60–61; settlement, 103, 139, 142, 146; Xiongnu, 101
social organization, xvii, 2, 7, 8, 33, 35, 103, 153, 252, 275; FulBe, 187n6; Kazakh, 23, 24, 26–27, 31; nomadic, 58
society(ies): agrarian, 23, 46, 58, 72, 78, 118, 122, 268–69; maritime, 269; transhumant, 82–83
Spanish, 7, 198–201, 204–5, 272
Ssu-ma Ch'ien (Chinese historian), 47
state(s), xv, 3, 12, 17, 20, 22, 24–25, 28, 29, 31, 32, 35, 37nn1–2, 55, 56, 57, 60, 87, 171, 251, 269; agricultural, 42, 45, 55, 268 (*see also* societies, agrarian); buffer, 46; Chorasmian, 118–19, 122, 147, 148, 153–56, 158; Hunza, 72–74, 79, 89; neo-patrimonial, xvi, 172–73, 176–80, 182–86, 188n8, 270; rise of, 77–78, 274; rituals of, 71, 88; taxes paid to, 77, 80
steppe(s), xvi, 7, 17, 41, 55, 271; in Arizona, 201; confederacies, 35, 270; contemporary societies of, 59–60; desert-steppe, (*see* desert); elements in art, 124; environment of, 42, 43, 44, 46, 58; Eurasian, 4, 6, 7, 11, 21, 22, 25, 26, 34, 55; fortified sites in, 122, 153; in Kazakhstan, 19–20, 26, 27; Mongolian, 47, 51–52, 54; nomads/pastoralists of, xv, 11, 17, 21, 22, 23, 29, 32, 34, 42, 47, 97, 271; political ecology in, 56–59; semiperiphery, 57; Steppe Statute, 29; steppe vs. sown paradigm, 117, 120, 121, 268; study of, 18, 37n2
stratification, 45, 77, 234, 273
subsistence, xii, xv–xvii, 1, 3, 4, 7, 8, 10, 43, 44, 51, 62, 71, 111, 127, 212, 217; as category for classifying societies, 103; divergence in, 102; on islands, 249, 261; limits on, 89; local strategies of, 113, 251, 258; production strategies for, 122, 123, 158, 214–15
Syr Daria (Kazakhstan), 26

Talgar (Kazakhstan), 19, 20, 21, 36, 153
Tartar(s), 37n1, 47, 50–51, 54
Tarym-kala, 145, 159n4
taxonomy: folk, 103; Linnaean, 102–3

Temujin. *See* Chinggis Khan
Thessaly (Greece), 19
Thum, 72, 74, 77–78, 86–87, 88, 90n6
Tian Shan Mountains (Kazakhstan), 18, 19, 21, 22
Tibetan Plateau, 6, 7
tillage, 214–16, 218, 227, 231, 233
Tolstov, Sergey P., 31, 118, 119, 127, 146, 154, 155, 157
trading, 82, 122, 156, 157, 199, 216, 267, 269; among Comanche, 272; posts, 157; and raiding, 33, 273, 274
transhumance, 8, 11, 18, 19, 20, 21, 23, 123, 268; in Chad, 173, 174, 176, 179, 182, 187n6, 188n10, 188n12, 188n17; cycle of, 86, 87; defined, 82–83, 103, 267; in Hunza, 71, 74, 89; in Ireland, 214; mobility goals of, 212, 243, 269
tribe(s), 25, 26, 35, 44, 123, 154; Navajo, 197; nomadic/steppe, 57, 58, 97, 154; pastoralist, 59
Turkic: language, 17, 26, 27, 28, 31, 33, 37n2; khaganat, 46
Turkmen, 19, 31, 153
Turkmenistan, 61
Turks, 31, 43, 47, 57, 245
Tuva, 43, 60, 61
Tuz-gyr, 125, 130–31(table), 135(table), 137, 146, 152, 154

Uighurs, 43, 58
Ural: River, 27; Mountains, 42

Vaynberg (Vainberg), Bella I., 127, 144, 157
verticality. *See* movement

Wakhi, 74, 78
water, xv, 5–6, 8, 18, 21, 36, 47–50, 53, 55; in Chad, 174, 181, 182; in Chorasmia, 125, 139, 151, 152, 157; fowl, 50; fresh, xvii; glacial melt, 74, 75, 87; holding capacity, 79; in Hunza, 76, 77, 80, 82–83, 85, 89, 90n3; irrigation, 79; on islands, 244, 246, 252, 255, 258; management of, xvii; sources, 42, 55, 75; supply canal, xvi; troughs for, 20; valley, 34
war, 45, 122, 176; in Chinese astrology, 42; civil, 58–59, 173, 183, 188; FulBe, 176; in Inner Asia, 51–52, 54, 58–59, 97; Navajo, 201; war-like, 48, 56; War Zones, 276; World War I, 178; World War II, 31, 215, 254
warming (climatic), 46–47
wool, 49, 53, 60, 61, 82, 85, 111, 215, 250, 256
world economic system, 202, 204, 205
world-system(s), 254, 271, 274, 276; Chinese, 46; edges of, 25; exchange networks and, 247; incorporation into, 205, 250–51; status of pastoralists in, 11, 56, 59, 261. *See also* core-periphery relations; world-systems analysis
world-systems analysis/theory (WSA), xvii, 9, 12, 244, 250, 275; examining Navajo using, 196, 205
WSA. *See* world-systems analysis/theory
Wusun, 18, 20, 22, 35

Xianbei, 47, 51
Xiongnu, xv–xvi, 97–98, 101–102, 274; and climate, 43, 47; empire of, 57–59; fortified sites of, 122, 152; lifestyle of, 50–53; tombs of, 104, 107–13, 138

yaks, xv, 4, 6, 23, 75
Yasy-gyr, 129–30, 133, 137, 139–42, 155–57
Yuezhi, 20, 35
yurt, 6, 20, 22, 53–55, 82

Zailiisky Alatau, 21
zooarchaeology, 21, 100, 102, 103